Anonymous

American Plumbing Practice

From the Engineering Record (Prior to 1887 the Sanitary Engineer.)

Anonymous

American Plumbing Practice
From the Engineering Record (Prior to 1887 the Sanitary Engineer.)

ISBN/EAN: 9783337186951

Printed in Europe, USA, Canada, Australia, Japan

Cover: Foto ©berggeist007 / pixelio.de

More available books at **www.hansebooks.com**

AMERICAN

PLUMBING PRACTICE.

FROM

THE ENGINEERING RECORD.

(Prior to 1887 THE SANITARY ENGINEER.)

A SELECTED REPRINT

OF ARTICLES DESCRIBING NOTABLE PLUMBING INSTALLATIONS IN THE UNITED
STATES, AND QUESTIONS AND ANSWERS ON PROBLEMS ARISING
IN PLUMBING AND HOUSE DRAINAGE.

WITH FIVE HUNDRED AND THIRTY-SIX ILLUSTRATIONS.

NEW YORK:
THE ENGINEERING RECORD,
1896.

PREFACE.

THE ENGINEERING RECORD, prior to 1887 THE SANITARY ENGINEER, has for 17 years given much attention to domestic water supply, house drainage, ventilation, and plumbing. Beside the frequent illustrated descriptions of notable and interesting current work, a great variety of questions in this field have been answered. In 1885 "Plumbing and House Drainage Problems" was published.

The present volume. "American Plumbing Practice," is a compilation of illustrated descriptions of plumbing installations in modern buildings of every character, together with Notes and Queries touching interesting points developed in practice, from articles which have appeared in THE ENGINEERING RECORD since the publication of "Plumbing and House Drainage Problems." Within this period the towering office building has been developed, involving special problems of drainage and plumbing. The equipment of hotels, hospitals, amusement halls, swimming baths, and other public buildings has been upon the most thorough and elaborate scale, and in the description of the plumbing of residences examples may be found of nearly every class of dwelling. Its division of Notes and Queries is intended to supplement "Plumbing Problems," bringing these queries well up to date. The greater part of the book consists of descriptive matter nowhere else available in permanent form.

TABLE OF CONTENTS.

PLUMBING IN OFFICE BUILDINGS.

PLUMBING IN AMUSEMENT HALLS AND PUBLIC BUILDINGS.

PLUMBING IN THEATERS.

PLUMBING OF SWIMMING AND RAIN BATHS.

MISCELLANEOUS:

NOTES AND QUERIES ON PLUMBING OF SWIMMING BATHS.

NOTES AND QUERIES ON CIRCULATION FROM KITCHEN BOILERS.

PLUMBING OF RESIDENCES.

DRAINAGE AND WATER SUPPLY OF A HARTFORD HOUSE.

(PUBLISHED IN 1891.)

In the recently finished house of C. K. Forrest, Esq., Hartford, Conn., of which Andrews, Jacques & Rantoul, of Boston, were the architects, and Albert L. Webster, Assoc. M. Am. Soc. C. E., of New York, was the sanitary engineer, the general plans for the drainage and water-supply systems were carefully drawn to a scale of one forty-eighth, and showed the correct position and length of all pipe lines, their sizes and principal fittings. Accurate detail drawings were also made of special or complicated arrangements, and diagrams of each room bearing marginal specifications, notes for all fixtures, workmanship, and details required there,

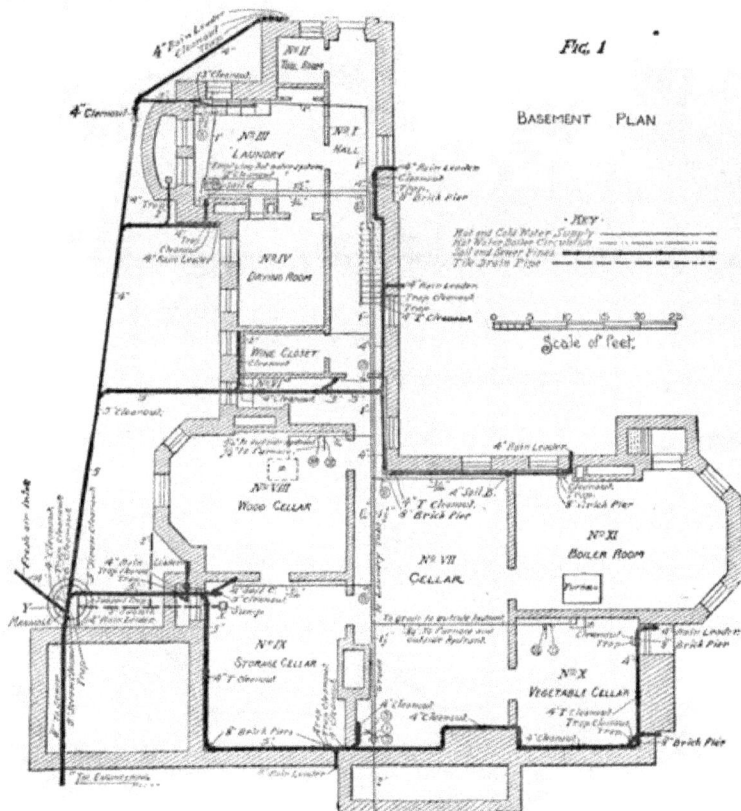

DRAINAGE AND WATER SUPPLY OF A HARTFORD, CONN., HOUSE.

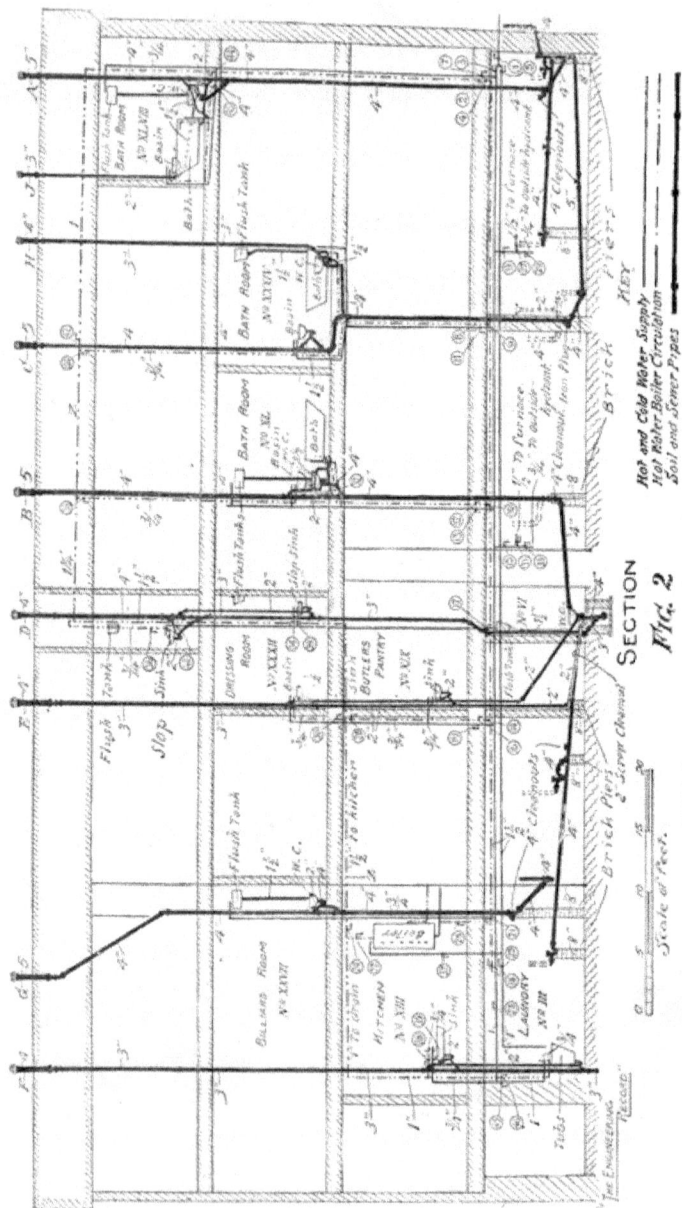

SECTION

FIG. 2

KEY

Hot and Cold Water Supply
Hot Water-Boiler Circulation
Soil and Sewer Pipes

DRAINAGE AND WATER SUPPLY OF A HARTFORD, CONN. HOUSE.

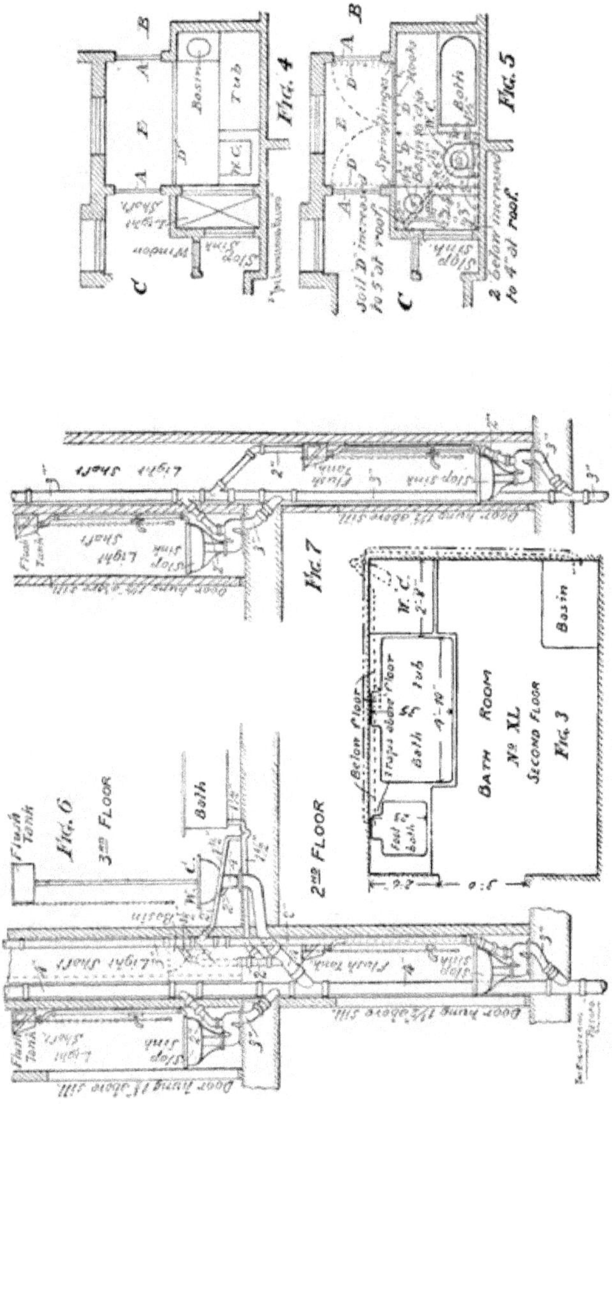

DRAINAGE AND WATER SUPPLY OF A HARTFORD, CONN., HOUSE.

were also furnished by the engineer to the contractors. The requirements and specifications were thus made unmistakable, and were neatly and conveniently incorporated with the plans. All main drains were run on side walls, and supported on brick piers, and had direct vertical columns to fixtures above. Main cold-water pipes were run on the basement ceiling, with risers following the soil columns. The hot-water circulation pipe Z, Fig. 2, was carried in a covered safe trough along the attic floor. All pipes are accessible, generally exposed, and have numbered tags, corresponding to a diagram, attached to all their main cocks and valves. All parts of all soil and drain pipes are commanded by suitably located cleanout holes.

Figure 1 is a general cellar plan and Fig. 2 a vertical section and elevation of the house. As will be seen by the key on Figs. 1 and 2, hot and cold water pipes are here indicated by light full lines, circulation hot-water pipes by a line broken by two dots, soil pipe and sewer pipe by heavy full black, and tile drain by heavy broken lines. In the original drawings, from which we have prepared these illustrations, these different lines were respectively

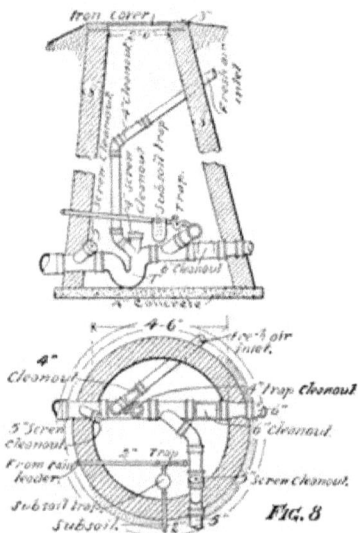

FIG. 8

shown by single full blue lines, single full red lines, double full blue lines washed with red, and full heavy green lines. On the original drawings tags 6, 10, and 14 designate special valves regulating the hot-water circulation flow; all others designate ordinary stop cocks, or a few globe valves. Numbers 18 and 25 empty the hot-water supply system and the cold supply to the cold kitchen boiler, and are so noted on the drawings.

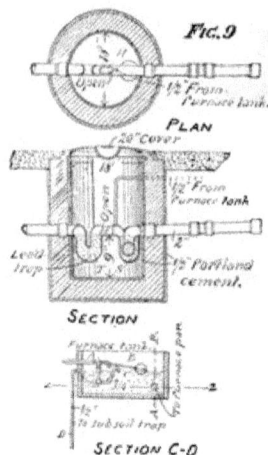

FIG. 9

PLAN

SECTION

SECTION C-D

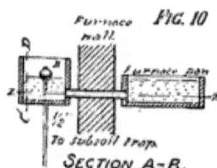

FIG. 10

SECTION A-B.

In the detail drawings for this work special pains were taken to furnish detailed information of the work on the general plans and diagrams (similar to Fig. 3) of each room, were accompanied by written notes, in which each fitting or fixture was named, the method of fitting prescribed, and the character of all the details and work concerned was stated. Obviously this is of great advantage where the engineer or architect has to rely upon a distant contractor with whom he cannot be in daily communication. It leaves nothing to be assumed and saves the very frequent embarrassment of permitting violations of the specifications to stand, or of having work pulled out after it is once erected.

Figure 4 shows the plan of the third-floor bathroom as originally arranged, with one door D into ante-room E, which communicated by the open doorways A A with the principal chambers B and the servants' quarters C. This plan made the room so crowded that it was modified, as shown in Fig. 5, where the washbowl was set on brackets and double doors D₁ D₁ were set, so that when they were opened to the positions D¹ D¹ they closed the portals A A, and practically added the ante-room E to the bathroom. Figure 6 shows the connections to soil column O (Fig. 1), and Fig. 7 shows the slopsinks at the second and third floors on soil column D (Fig. 2). Figure 8 shows vertical and horizontal sections of the outside manhole Y, Fig. 1, which gives access to the running trap T of the main sewer pipe and to the several cleanout holes at this point. Figure 9 is a horizontal and a vertical section of the brick sump X, Fig. 1, which contains the traps for subsoil drains laid under the cellar floor. The sewer air is cut off by an ordinary running trap T, which is vented by an open handhole H, and an additional hole trap S is put in to prevent back-flow of sewage into the subsoil drain in case of stoppage in the house drain beyond the point where the subsoil drain joins it. The opening between the two traps and the bolt in the iron cover would permit the back-flow to come out on the cellar floor and make it known that there is a stoppage in the house drain.

Figure 10 shows details of the automatic supply to the furnace pan at Z, Fig. 1, in connection with a supply to the subsoil trap just described. Water is admitted to the receiving tank B, through ball cock A, and is always replenished when its level falls below Z Z. Every time the ball cock opens a little excess of water is received, which discharges through the overflow D and seals the subsoil traps S and T, Fig. 9. Overflow D has a bent inlet C, so as to preserve a water seal of about 2 inches against the cellar air from the trap. The water pan in the furnace being connected with the supply tank below the water surface, prevents dust and air from the cellar entering the furnace. A small loss of water by evaporation opens the ball cock and fills the tank

and furnace pan, at the same time giving a small supply to the subsoil traps through the pipe D.

Robert Garvie, of Hartford, was the plumber, and the work was done by day's labor.

PLUMBING IN A NEW YORK RESIDENCE.

(PUBLISHED IN 1891.)

THE plumbing in the residence of R. A. C. Smith, Esq., in West Seventy-second Street, New York City, of which Mr. Smith himself was the architect, was executed by John Renehan, of New York City, and conforms to the current practice in careful metropolitan work. It has, however, special provision for local vents, for the disconnection of the sewer system for an interchangeable service of the kitchen and laundry boilers, and some other details and arrangements. Some of the these may be held to illustrate one of many practical solutions of the general problem which is more or less modified in the familiar requirements and conditions of all large cities, and is similarly handled by different designers. Each designer may have a particular style, excel in some individual detail, or more carefully provide for emergencies or conditions generally neglected Thus, while no important part of the work under consideration may be said to be novel, it has an individuality, and the frequent description of different instances of the same standard class of work empha-

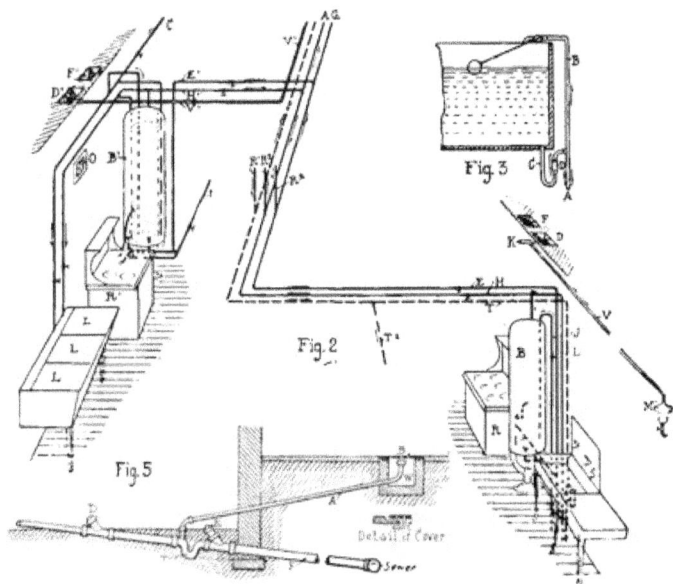

PLUMBING IN A NEW YORK RESIDENCE.

sizes its important features and may be held to be suggestive of combinations, variations, and possible merits or faults which, although they may seem evident in the finished work, are not accomplished without study and the aid of comparison.

The house is a brownstone building about 25 feet front and four stories high above the basement. The plumbing comprises a sink in the cellar, laundry boiler and three tubs, kitchen boiler and sink, a servants' water-closet and bath in the basement, a toilet-room with urinal and hot and cold water washbowl, a butlers' pantry sink and plate warmer on the first floor, a bathroom with water-closet and washbasin and two washstands on both the second and third floors, and a 1,000-gallon storage tank, a maids' slopsink and a washbasin on the fourth floor. The soil pipes are of calked cast iron and the vents and water pipes are of screwed galvanized iron, except where they are exposed in the upper stories, where all the metal-work is silver-plated. All the soil and vent pipes were subjected to the water and smoke test. The trap vents are carried above the overflows, and all the soil pipes and traps are accessible through screw caps, Y's, etc. for cleaning.

A diagram plan of the hot and cold-water pipes on the basement ceiling is shown in Fig. 1. Here B is the kitchen boiler and B¹ the laundry boiler, C is the main cold-water supply, R R¹ ranges, L L L laundry tubs, V and V¹ local vent pipes, T T¹ and T² are trap vents, H H¹ hot water from the boilers, E E¹ hot-water circulation pipes, A and G hot water and circulation connections between boilers, R R¹ and

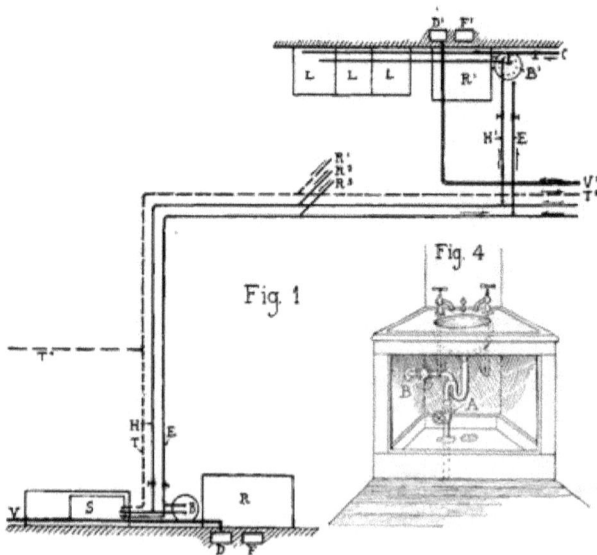

Fig 6

Fig 7

Fig. 4

Fig. 1

PLUMBING IN A NEW YORK RESIDENCE.

R¹ are risers, S the kitchen sink, F and F¹ smoke flues, and D and D¹ ventilation ducts.

In the perspective of the hot-water circulation system, Fig. 2, I and I¹ are sediment pipes, J is a cold-water supply, L is a check valve opening upward, K is a cleanout, M is a receiver, and O is a ventilating register to the duct D¹. The other reference letters have the same significance as in Fig. 1, and in both figures the water pipes are shown by single heavy lines, the trap vents by broken single lines, and the local vents by double lines.

By the location of the smoke flues F F¹ adjacent to the ventilation ducts D D¹ it is intended to promote circulation in the ducts by radiation of heat from the flues. Pipe V is commanded by cleanout K and terminates in an open receiver-bell M, beneath which a gas jet heats the air and causes it to be drawn out of the servants' bathroom just above the water-closet seat. Pipe V¹ terminates in a tin box set in a wall flue in the first-floor toilet-room. This box, accessible through a glass door, contains a gas flame and has an open pipe at the bottom through which air is drawn up through the urinal bowl.

Although the house is provided with a storage tank for future need, if required, it is at present unnecessary, as the city pressure is sufficient, and the double laundry boiler is connected up for a single pressure, though it is so connected with the kitchen boiler that either of them could be used under street and the other under tank pressure, and they could work separately or alone to supply the basement and upper floors.

The arrangement by which one pipe A is made to serve for supply and delivery for the tank is shown in Fig. 3. The filling is through branch B and a ball cock, and the drawing is through the same city pressure or pump pipe, and the U branch C, which is made to have an upward flowing current so that the check valve D closing downward will prevent any discharge of water from A, and the consequent overflowing of the tank through its delivery.

The chamber washbasins are situated in pairs between the front and rear rooms, and their arrangement is shown in Fig. 4. Valves A and B are set in the waste and vent pipes respectively, and are intended to be closed when the rooms are left long disused, so that by this means all danger of sewer connections through broken trap seals is obviated. If any obstruction occurs in the waste, it is likely to be at or above the trap, and if, as is frequently the case, it is due to gummy accretions on the internal surface, valves A and B can be closed and the pipes filled up to the top of the overflow with a strong potash solution, which will usually cut them out clean.

The fresh-air inlet for the main house trap is on the walk near the curb in the front area and is shown in Fig. 5. A being the inlet and T the trap. The pipe terminates in an open vertical leg about 12 inches long, which is set in a brick well W and protected by an iron grating B, removable by turning hook C with a key. The well W may collect considerable dirt and rubbish without obstructing the

inlet. The main trap T is commanded by cleanout D, and the house sewer from the trap to the street sewer is commanded by another cleanout E. A little way above trap T, Fig. 5, the house sewer, in following the cold-air chamber Z, of the ventilation system, and as this room is exposed to a current of the coldest external air, a special masonry wall W, Fig. 6, was there built up around it to protect it and an extra cleanout R was provided to command the built-in pipe.

The details of support for the main house sewer pipe F, Figs. 5 and 6, are shown in Fig. 7. The pipe is carried in wrought-iron yokes G, which are screwed into sections of old pipe H, which, with the cast flanges I, make solid pillars easily cut to any desired length. Where pipe F is near the wall, brickbats J are filled in alongside and plastered over to form a smooth sloping top K, which presents a neat finish and prevents the lodgment of dirt and rubbish.

SOME PLUMBING DETAILS IN A WASH-INGTON RESIDENCE.

(PUBLISHED IN 1891.)

WE illustrate herewith some of the details of recent plumbing-work in a new house at Massachusetts Avenue and Twentieth Street, Washington, D. C., that, without being novel or remarkable, indicates the style and character of plumbing executed in some of the new work of construction which is now rapidly increasing in Washington.

Figure 1 is a view of the kitchen and laundry boilers. The supply from street mains is through pipe A, with branch B to pipe F for direct house supply, and C to the filter D, through which all the water usually passes and is delivered, through branches G and E, to the kitchen sink S and the general supply F. I is the supply to the kitchen boiler J, which delivers hot water through pipe H, with branches K to the kitchen sink and L to the laundry boiler N. M is the hot-water circulation pipe, connected by branch O to the laundry boiler, and with circulation pipe P to the kitchen range Q. R is the sediment pipe for emptying boiler J.

The laundry boiler N is supplied through pipe T and can be emptied by pipe X. It is heated from a laundry stove W by circulation pipes V V, and delivers hot water through pipe U to the laundry tub and a servants' bathtub. It supplies also a distribution pipe for the rear part of the second floor. It is connected with kitchen boiler J by pipe L. Y Y are pipes supplying hot and cold water to a first-floor basin. L L are supplies to an upper bathroom, and O is a branch connecting the circulation pipe M to the laundry-stove water-back.

Ordinarily boiler J alone is heated, and branches L L and Y Y are supplied from it. When, however, boiler N is in use, its supply may be supplemented by boiler J. Z Z are drips for emptying rising lines. Generally, valve *a* is closed and *b* and *c* are open, so that the whole supply passes through the filter D. Reversing these valves cuts out the filter, which

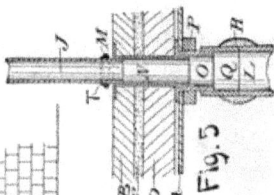

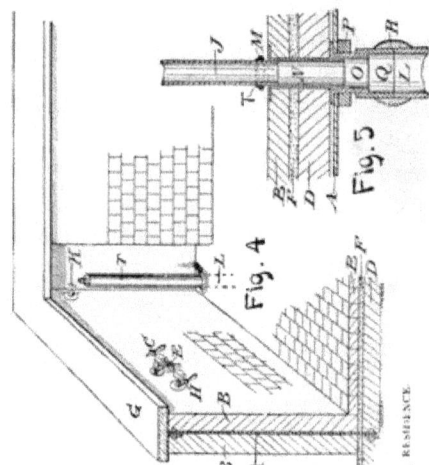

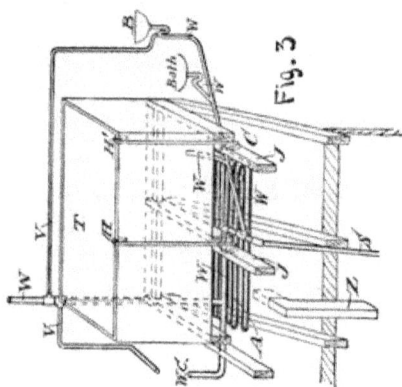

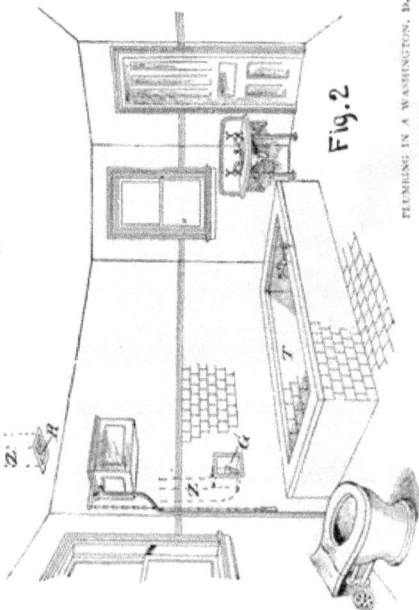

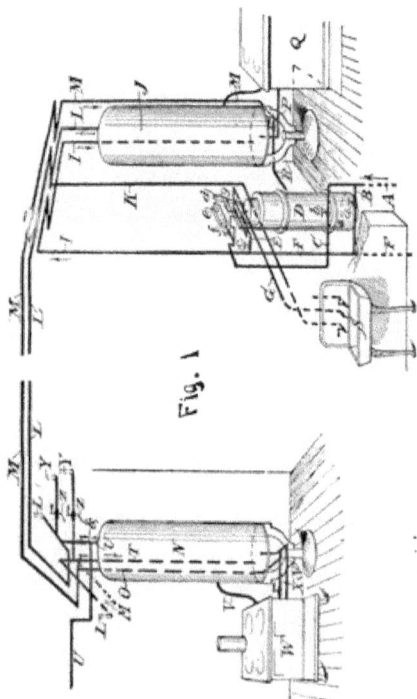

Fig. 3

Fig. 5

Fig. 4

Fig. 1

Fig. 2

PLUMBING IN A WASHINGTON, D. C. RESIDENCE.

can be washed out by reversing lever *d* and opening valves *f* and *g*.

Figure 2 is a perspective view of the sunken bathtub in the private toilet-room, which is in a second-story bay window. Figure 3 shows the setting and support of the tub and arrangement of supply and waste pipe, etc. T is a tank of ¼-inch iron, about 7x4x3 feet deep, supported on projecting joist, and secured to the main floor joist J J by wrought-iron hooks H H. S is the soil pipe receiving waste W W, etc. from bath, basin B, and water-closet. V V V are 2-inch vent pipes, branched from 4-inch stack W, which extends above the roof. Z is a 4x8-inch galvanized-iron local vent duct, which is shown in dotted lines in Fig. 2 and ventilates the bathroom through register R. G is a gas jet to promote a constant circulation. A is a steam coil to prevent freezing in the chamber C, which is lined with galvanized iron.

Figure 4 shows one corner of the bathtub T, Figs. 2 and 3. A is the riveted iron tank, B B B the inside and outside tiled surface, F the cement setting; D a layer of brick; E is the inlet cock with hot and cold handles H and C respectively, G is a marble top, and I is the overflow pipe, which can be raised and hung on hook K, to empty the tub through waste pipe L.

Figure 6 shows the detail of connection of waste, etc. A, B, D, F, J, and L are the same as in Fig. 4. The brass overflow pipe fits closely into the brass ferrule N, and is supported from its upper flange by the rubber wing N, which is confined by the shoulder M. Ferrule N screws into brass sleeve O, which is finally secured by its top flange and the jam nut

P. Another ferrule Q is secured to O, and connected to the 2-inch lead waste pipe L by a wiped joint R.

The plumbing was executed by Reynolds & Murphy, of Washington, and is comprised chiefly in one laundry-room, one servants' water-closet and bathroom, one butler's pantry, two tiled bathrooms, one chambermaid's sink-room, one dressing-room, and the kitchen. A private 6-inch iron sewer runs outside three sides of the house and receives all leaders and soil pipes. Harvey Page, of Washington, was the architect of the house, and John S. Larcombe the builder.

MARBLE PLUNGE BATH IN A PRIVATE RESIDENCE.

(PUBLISHED IN 1895.)

THE following description illustrates an arrangement of a sunken bath lately executed by Mr. Paul S. Bolger in a residence in Fifty-eighth Street, New York City. Figure 1 is a perspective view of the owner's private bathroom, which is floored and wainscoted with white marble and contains a 420-gallon sunken bathtub, or pool, which is lined with polished white marble slabs. It is entered by two steps at S, descending from the floor level. The principal clear dimensions of the bath are 7'x3'x3' deep, with an overflow set to permit a depth of about 32 inches of water. Hot and cold water supplies are brought up through the floor at A A, and passing, as indicated by the arrows, through pipes H and C respectively, are carried behind the wall panels to the dolphin's head D. From the dolphin's

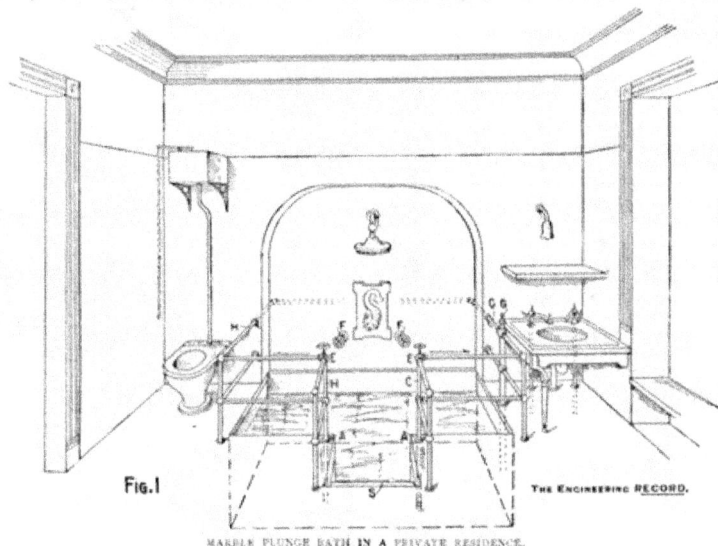

FIG. 1

THE ENGINEERING RECORD.

MARBLE PLUNGE BATH IN A PRIVATE RESIDENCE.

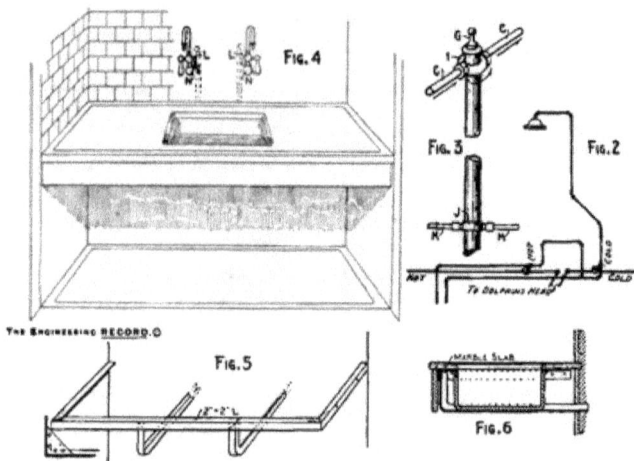

MARBLE PLUNGE BATH IN A PRIVATE RESIDENCE.

mouth the stream is delivered which fills the tub. Pipes C and H form part of the silver plated brass posts and rails which inclose and protect the pool. The dolphin's head delivery is controlled by valves E E, which are within easy reach of an occupant of the bath. Valves F F command the shower bath, the connections of which are so arranged that the hot water must pass through the cold-water valve, as shown in diagram, Fig. 2, thus insuring the certainty of cold water being mixed with any hot that may be drawn through the shower. Cold water may be delivered by the shower, but not hot water alone. If the shower hot valve only be opened, no delivery follows, nor can it be secured until the cold valve also is opened, thus preventing any possibility of scalding the bather. G, Fig, 1, is the handle of the waste and overflow for the bath, and Fig. 3 is a detail showing how it passes through a hollow ring I, which carries the cold water from one length to another of the supply pipe rail C. The lower rail sections K K do not serve as water pipes and are connected by ordinary nipples to a solid guide ring J, through which the vertical pipe slips freely. Figure 4 shows the arrangement at a pantry sink between two chambers. Here a square porcelain sink is set in a handsome marble slab 6 feet long, which extends across the full width of the alcove. The curtain panel in front is only about 6 inches deep, and the slab and bowl are supported by a special wrought-iron frame, shown in Figs. 5 and 6, which clearly show in perspective and cross-section the manner of bolting the 2x2-inch angle bars to the wall above and building them into it below,

The hot and cold water supplies were brought up behind the tiling to a point above the slab, where they were brought through and connected to angle

pieces N N, to which sink faucets L L are attached, so as to leave an unobstructed surface on the table slab, and incidentally to permit the placing of pitchers, etc. beneath the faucets. The long marble slab of the kitchen sink is supported by an iron frame similar to that shown in Figs. 5 and 6.

A SPECIAL BATHROOM.

(PUBLISHED IN 1891.)

In making recent improvements in a house at New York City the owner wished to utilize a very small, narrow room for a bath and water-closet, and therefore designed the arrangement shown here.

Figure 1 is a plan of the room, showing the new position of the water-closet and of special bath B, with shower ring S and needle posts N N N N The old fixed washstand was discarded and a marble side shelf substituted for a movable bowl, if required. Sufficient room was thus obtained for toilet and dressing purposes, and for shaving-cabinet, gas stove, etc.

Figure 2 is a view from Z, Fig. 1. The bath B is about 3'x4'x15" deep, the bottom A A being raised 9 inches above the floor, so as to be at a convenient height for a child to stand upon while being bathed by a person seated in a chair. The bathtub may be filled with water up to level P P of the overflow O, and is intended as a tub for children and a shower and spray for adults. S is a 14-inch ring shower, and N N N N are water tubes perforated for needle sprays that pitch toward the center of the tub. Around the bath the room is paneled 8 feet high with Italian marble wainscoting; elsewhere it is 6 feet high all around. The floor is also marble-tiled and the tub is lined with marble. All the pipes and fixtures are exposed and are of brass, silver-plated.

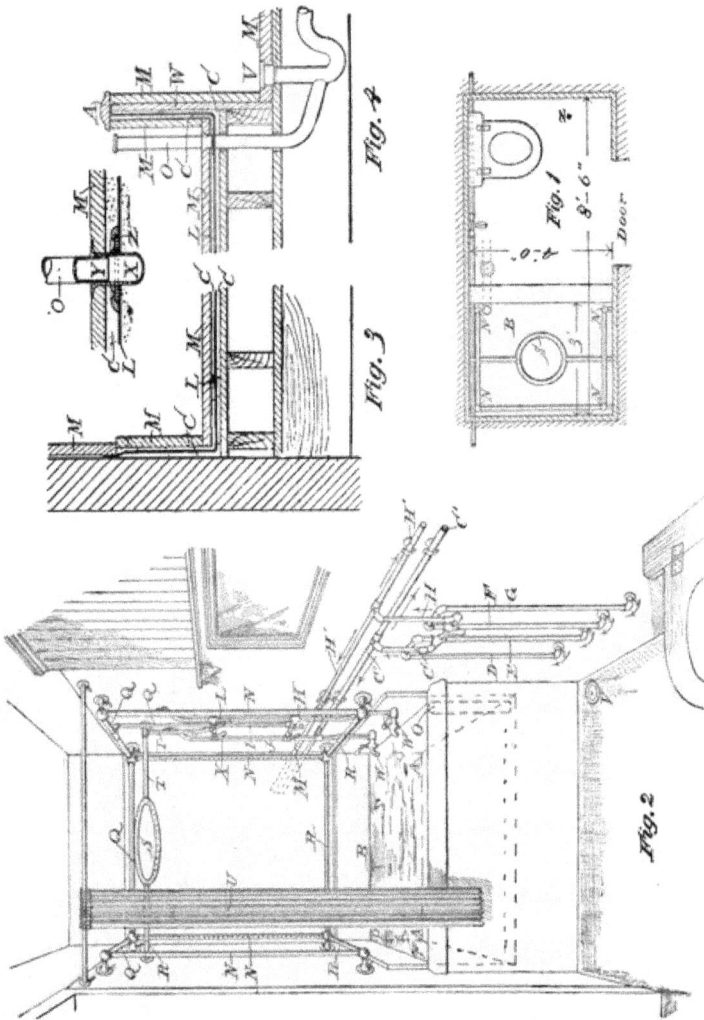

Fig. 1

Fig. 2

Fig. 3

Fig. 4

Water under both street and tank pressure is supplied to the room, and its distribution there and for
the upper part of the house is controlled by a Mor-
rison cut-off M, that receives cold water under street
and tank pressures through pipes D and E, and delivers it through C and its horizontal branches C and
C', and receives hot water under street and tank
pressures through F and G, and delivers it through
H and its horizontal branches H' and H'. The bath
fixtures are supplied with hot water through branch
I, the delivery to the needle spray N N N N being
controlled by cock K, and the delivery to the shower
S by cock L. Branch J supplies cold water, and
cock M controls its delivery to the needle spray; cock
N controls its delivery to the shower. The needle
sprays are supplied through pipes Q Q, etc., and the
shower ring through pipes T T. Pipes R R, etc. are
blanks, merely for stiffening braces; U is a rubber
curtain; W W are drawcocks, and V is a trap screw
for cleaning the bath waste. The bathtub really
consists of a deep, soldered, lead tray, set in cement,
and with a cement coating all over inside, in which
the marble surface is set. The marble lining itself
would probably be nearly or quite water-tight, but
the lead pan is put outside to give absolute assur-
ance.

Figure 3 is a section through the side of the tub,
and Fig. 4 is a section through the front of it. L is
the eight-pound lead tray; C is Portland cement; M
M, etc. are marble slabs; W is a wooden faceboard,
and A is a 4-inch marble cap. The overflow stand-
pipe O fits into the waste pipe X, with a ground
joint at Y, and a flange Z is soldered to both L and
X to prevent leakage.

This work was done by Messrs. Rossman & Bracken.
New York.

PLUMBING DETAILS IN R. P. FLOWER'S
RESIDENCE, NEW YORK.

(PUBLISHED IN 1891.)

In a recent remodeling of Mr. R. P. Flower's residence, at New York, all the piping and many of the
old fixtures were replaced, and a new supply, waste
and ventilation system were adapted to the existing
construction and its requirements. The following
details show how some of the work was made to conform to the necessities encountered.

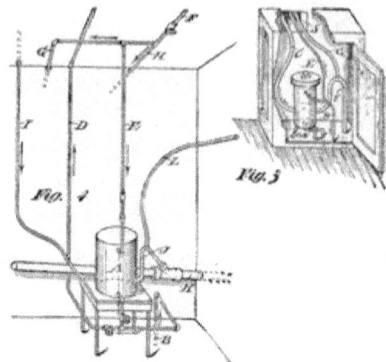

Figure 1 shows the 1,000-gallon wooden storage
tank T in the attic. It is filled through the pump
pipe A, or, when city pressure suffices, through the
ball cock B. C and D are circulation and relief pipes
from the high and low-pressure boilers. They are
both connected to the check valve E that opens with
a downward current and lets steam escape, but
closes with an upward current, thus preventing
escape of tank water through it by syphonage. G is
the house supply, with a check valve H that closes
with a downward current to prevent the filling and
overflowing of the tank through it when city pressure exceeds tank pressure. I is a lead safe, and J is
a soil pipe carried, with the other risers, in the pipe
shaft K.

Figure 2 shows the connections of the gas-engine
pump A and hand pump B in the subbasement. C

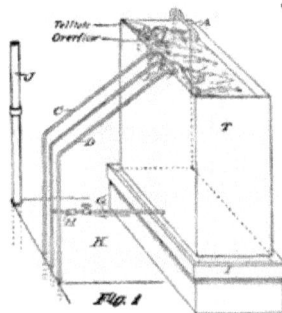

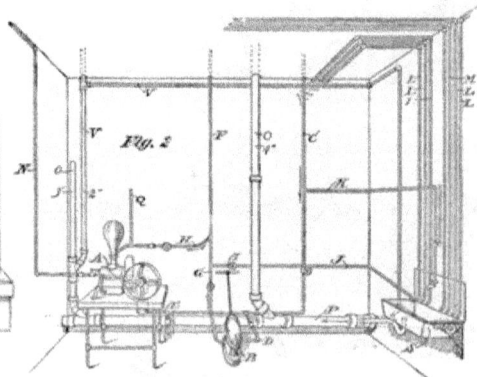

PLUMBING DETAILS IN MR. R. P. FLOWER'S RESIDENCE, NEW YORK CITY.

is the cold-water supply from city mains, with branches D and E for the suctions to hand and gas pumps which deliver through branches G and H respectively to the 1¼-inch tank pipe F.

Valve I is usually closed, but by opening it riser F may be drained through waste pipe J into the drip sink S. K is a cold-water supply to sink, M is a 2-inch refrigerator waste, and L L, etc. are drip and safe wastes, all provided with flap valves at sink. N is a gas pipe to engine, Q is an air chamber, V is a vent pipe, O O are soil pipes, and P is main 6-inch house sewer, supported on iron hooks from cellar wall at each joint.

Figure 3 shows the arrangement, in a basement cupboard, of a Tucker grease trap T, S is a duct up to butler's pantry, C is a 2-inch waste from butler's pantry sink, A is the 1-inch supply from city mains, B is the 1-inch outlet with ½-inch branch E to butler's pantry and ¾-inch branch B to kitchen sink, F is the 2-inch waste, and G the 1½-inch vent.

Figure 4 shows the large grease trap in the sub-basement for kitchen waste, B is the 1-inch cold supply under tank pressure with branches C to trap and D to kitchen. E is the delivery with branches F to laundry, H to servants' water-closet, and G to first-floor basins, I is the 1½-inch waste from kitchen sink, and J the 1½-inch outlet to sewer K; L is a vent pipe.

The work was executed by John Toumey & Son, of New York.

SOME PLUMBING IN BOSTON.

(PUBLISHED IN 1889.)

SOME interesting plumbing has recently been done by Henry Hussey & Co., of Boston, in a large building in that city, and we illustrate some of its details from sketches recently made by a member of our staff.

Figure 1 is a view of the main attic tank. It is filled by pumping or from the street pressure, through

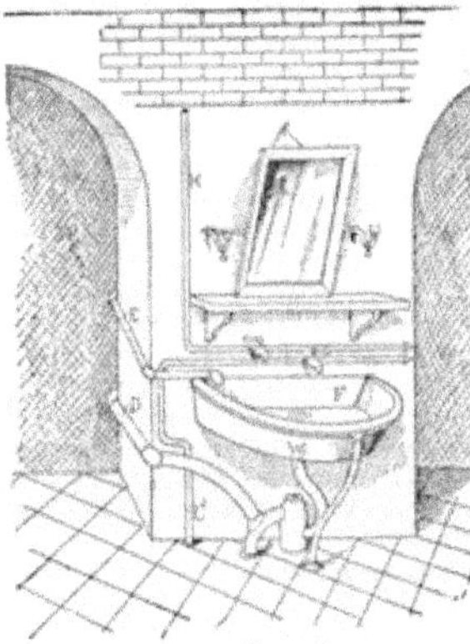

FIG. 4

the two ball cocks on the 1¼-inch brass pipe A, and overflows through the 1¼-inch pipe B.

The pipe A, where it comes down inside, is attached to the side of the tank by the foot E, which is soldered to the copper lining of the tank, as will be seen by the sketch. The two ball cocks are attached to branches from the tee D. It is intended ordinarily to keep the tank about two-thirds full of water, as shown, but in summer it is used temporarily to supply water for the infrequent use of the freight elevator, and it is arranged to be then filled up to the top of the overflow pipe B. The ball cocks are then removed from tee D to tee F, and the plugs now in the branches of the latter are transferred to tee D and the overflow outlet at G is closed.

H H, etc., are ½-inch brass vent pipes from the heating boiler, main lines, etc. I is the waste pipe for emptying the tank. J is its valve, which can be raised and opened by the chain K. Valves similar to

Fig. 2

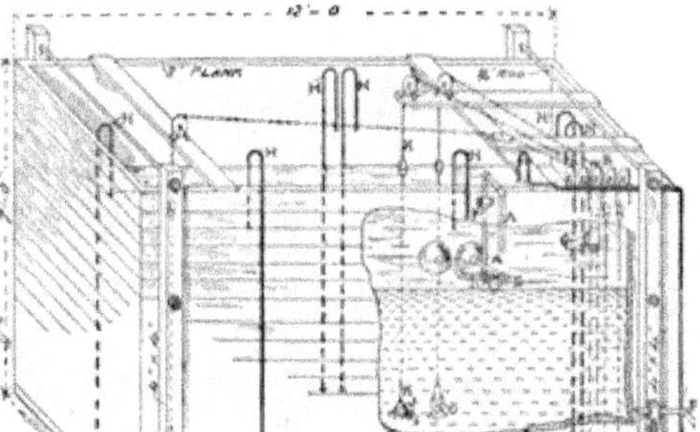

Fig. 1

SOME PLUMBING IN BOSTON, MASS

J are at O O, etc., controlling the supply to the various pipes; lead weights are attached to the chains to insure the closing of the valves.

Figure 2 shows the face of the tank, which is broken away in Fig. 1. The tank supply through pipe A will, if the key valve N be opened, supply the elevator through the 1½-inch brass pipe P.

R is a locked glass case containing the handles 1, 2, 3, 4, and 5, of the chains from supply valves J O O O O, Fig. 1. When the handles are pulled down, as 1, 3, and 5 are, their valves are opened. When raised, as are 2 and 4, their valves are closed.

Figure 3 shows the hot-water boiler A in the basement that supplies hot water to the kitchens, bathrooms, etc., throughout the house. It is jacketed with asbestos, and contains a coil of brass pipe which receives live or exhaust steam through pipe B, and returns it through the 2-inch brass pipe C to the trap D and thence through the 1-inch iron pipe E to the boiler.

T is a thermometer.

F is the cold-water supply pipe.

H and H are hot-water supply pipes and G is the hot-water return circulation pipe.

I is the 1½-inch brass sediment pipe to empty the boiler.

Figure 4 shows the sink in the boiler-room. It is neatly supported by a frame made of polished brass pipes nicely curved.

H is the hot, and C is the cold-water pipe; W is the waste, and D the trap vent pipe.

E is a waste pipe from the receiver into which all the safes and refrigerators in the house are drained.

DANGEROUS BLUNDERS IN PLUMBING.

ANOTHER WAY TO MAKE A BY-PASS.

(PUBLISHED IN 1889.)

E. H. Kendall, architect, of New York City, has called our attention to the ingenious arrangement

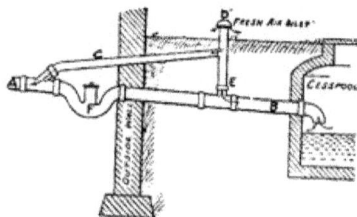

A BY-PASS FROM CESSPOOL.

which he recently discovered, whereby the plumber of a country residence contrived to admit cesspool gas where it was most necessary to exclude it.

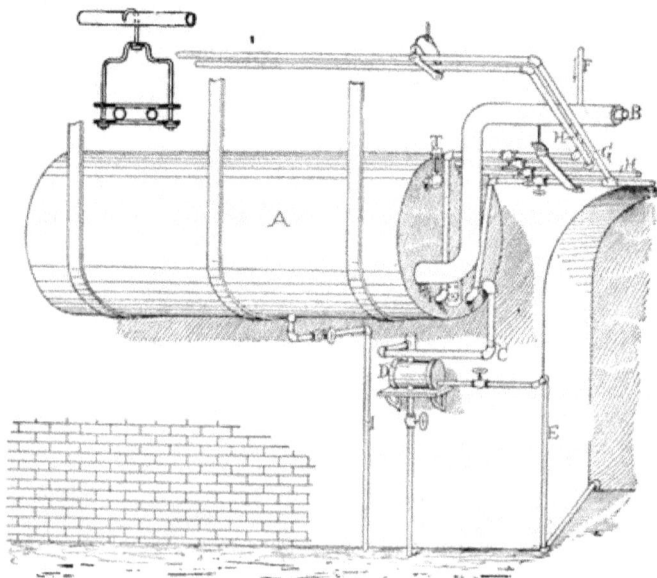

Fig. 3

SOME PLUMBING IN BOSTON, MASS.

A is the house drain, F the main trap cut-off drain from cesspool. C is a fresh-air pipe from inlet D. To ventilate the cesspool the plumber conceived the idea of extending the inlet pipe to E, where he connected it with the drain pipe and thus provided for the free escape of cesspool gas under the windows, and past the main trap F, as indicated by the arrows, thus making his main trap useless.

KITCHEN BOILERS IN RESIDENCE OF GEORGE VANDERBILT, ESQ.

(PUBLISHED IN 1887.)

THE accompanying illustrations are from the residence of Mr. George Vanderbilt, 9 West Fifty-third Street, New York. Figure 1 shows the boilers in the kitchen, one on either side of the range. That on the right is supplied from the street pressure, while that on the left receives its water from the tank on the top floor. The water-back was made specially for this work, and has a partition through the center allowing each boiler to be heated independently of the other. The water pipes throughout are tinned brass, giving the work a very bright and pleasing appearance. Gate valves are used in place of the ordinary stop-cocks, and reverse cocks are placed directly over the kitchen sink where they can be readily reached. Each boiler contains 60 gallons, and is supported on specially made brass rims with four brass legs. The arrangement of the supply pipes to the sink is very simple and neat, with air chambers evenly spaced and a plug in the pipe to prevent the mingling of hot and cold water. The arrangement of the pipes will be evident on examination of the illustration, making further description unnecessary.

The master plumber was Mr. Alexander Orr, of New York.

AN OVERHEAD ARRANGEMENT OF BATH-ROOM PIPES.

(PUBLISHED IN 1884.)

A MEMBER of our staff some time ago saw, in a Boston building, the work that is here illustrated from sketches he made. Each suite of apartments has laundry tubs, kitchen sink, butlers' sink, two washstands, a large bathroom and a servants' bathroom.

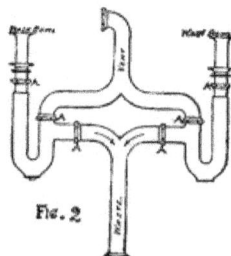

Fig. 2

All the pipes are of brass, either polished or silver-plated, and all either entirely exposed or readily accessible by doors commanding all inclosed portions.

Figure 1 is a view of the arrangement in a large bathroom. There are marble safes underneath all the fixtures, and the washbowl table and panels are of marble. The room is wainscoted with white ceramic tiles, and the cabinet-work is in polished cherry, except that on the ceiling, which is painted white.

The bathtubs and bowls are of porcelain, and all the exposed metal-work is of brass silver-plated.

C and C are cold-water supply pipes for the washbowls, and H and H are hot-water pipes. D is a

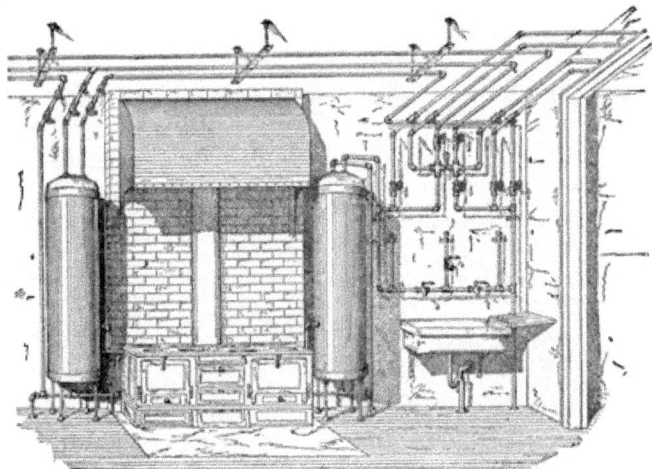

KITCHEN BOILERS IN RESIDENCE OF GEORGE VANDERBILT, ESQ.

Fig. 1

OVERHEAD ARRANGEMENT OF BATHROOM PIPES

special brass pipe, with wastes B B and trap vent A brazed on.

E and E are trap vents; they can easily be removed, and access had through the cap plates G G to their traps by unscrewing the unions F F.

The supply and waste pipes and traps are all arranged underneath the floor in the same manner as those of the room above, which are shown at the ceiling of this room. P is a cupboard containing the flush tank for the water-closet in the same room and the trap for the water-closet in the room above. K and K are cupboards containing the other pipes and traps for the upper room.

O is the soil pipe, M is the hot and N is the cold-water distribution pipe, and L is the trap for bath-tub waste.

Figure 2 shows a special double waste and trap vent connection that is somewhat similar to the one shown at D, Fig. 1, and was used by the same plumbers for their work in a club-house. It is made of 1½-inch brass pipe, nickle-plated, and has ground unions A A, etc.

PLUMBING IN A NEW YORK BATHROOM.

(PUBLISHED IN 1894.)

In the recent reconstruction of the plumbing in a residence on West Eighty-first Street, New York City, some special work was done by John Tucker in arrangement and floor construction of a handsome bathroom adjacent to the owner's bedroom. There

were half a dozen other bathrooms in the house that presented no especial features, but this one was more elaborately equipped and elegantly finished, and as its location is directly over some costly ceiling decorations, unusual pains were taken to prevent any possibility of leakage or moisture from percolating below the floor.

The room was paneled 6 feet high with Italian marble wainscoting, and had a floor of large white marble slabs. The fixtures were of marble, porcelain, and plated metal, and are designed to be both handsome and rich. Figure 1 is a floor plan showing the general arrangement and the pitch and jointing of the different sections of the floor, which sloped about one-half inch to the line A B, besides which point B was about 1 inch lower than A, so that it receives all the drainage and conducts it to the strainer-plate B. The floor surface is in three planes, A D D and A D E B. A D E B intersecting in the lines A D. A D, and A B. On the regular rough floor a sloping platform F was built on joists J J at 1-inch pitch and entirely covered by a large safe pan of heavy sheet lead, in which a layer of Portland cement from 1 to 2 inches thick bedded the marble floor slab at a pitch of one-half inch. A deeper pan was made in the safe to receive the 4-inch slab of the needle bath, and to its sides on the outside corner was soldered a vertical sheet of 12-ounce copper about 8 feet high and weighing 60 pounds. This was intended to prevent any possibility of water from the bath spattering and running down behind the wainscot.

Figure 2 is a drawing of the lead safe, dog-eared at the corners. Figure 3 shows the arrangement of soil, waste, and water pipes. The soil pipe is 4-inch cast iron with calked joints and lies on the rough floor. The waste pipes are 1½-inch and 2-inch screwed wrought iron, also laid on the rough floor. The water pipes are laid in channels grooved in the sloping platform, into which the lead safe has been smoothly beaten down. Figure 5 is a diagram of the arrangement of the trap vent pipes behind the wainscoting and over the ceiling of the bathroom. Figure 6 shows the method of making traps on the screwed waste

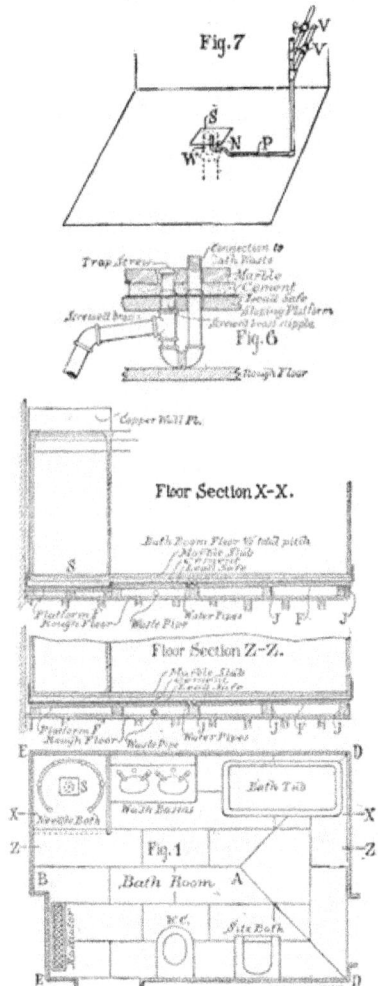

Fig. 7

Fig. 4

Fig. 6

Fig. 2

Floor Section X-X.

Floor Section Z-Z.

Fig. 1

Fig. 3   Bath Room

Fig. 5

PLUMBING IN A NEW YORK BATHROOM.

pipes and carrying them through the lead safe. In Fig. 7, S is the strainer-plate S, Fig. 1, and W is the waste pipe from the needle bath. P is a ½-inch pipe supplying hot or cold water at a considerable pressure to the nozzle N, which is concealed inside the waste pipe and furnishes a vertical jet commanded by valves V V.

PLUMBING IN JOHN D. ROCKEFELLER'S HOUSE AT TARRYTOWN, N. Y.

(PUBLISHED IN 1891.)

THE systems of hot and cold-water supply, drainage and trap ventilation in this house and its stables, etc., illustrate the design and execution of extensive work for a large and costly establishment, having nearly all the requirements of a city house in addition to some others belonging to a country place, and under the conditions imposed by its isolated location.

Water is brought from the Pocantico water supply of Tarrytown, in a special 4-inch main, about 4 miles long, which supplies it at a pressure of about 140 pounds per square inch to a reducing valve. This maintains a constant delivery at about 40 pounds, which is sufficient to raise it to the storage tanks in the attic.

All water pipes throughout are of galvanized iron. The branches to all fixtures are ¾-inch AAA lead pipe; all boiler connections are brass. Two toilet-rooms on the first floor have exposed silver-plated pipes; elsewhere the washstands, etc. have the space underneath occupied with drawers. Lead safes are everywhere placed under the pipes and fixtures. All soil pipes are perfectly vertical from sewer to above roof; most of them are built in recesses left in the walls and subsequently bricked and plastered up. All are 4 inches except one line 5 inches in diameter. All water supplies rise in elevator shafts to attic or fourth floor, where they are distributed in lead-lined troughs, see Fig. 1, between floor and joist, and have vertical branches with stop cocks to all fixtures on second and third floors.

For the basement and first floor all branches for water supply rise vertically from distribution pipes, suspended below the basement ceiling.

All the drainage pipes are suspended below the basement floor, overhead in cellar.

In the main part of the house there are, on the third floor, six toilet-rooms, each containing a water-closet, a bathtub and a washbasin; another toilet-room with water-closet, bathtub and two basins, and a chambermaid's slopsink. On the fourth floor is a photographers's sink. On the second floor is a toilet-room containing a water-closet, two washbasins, a long bathtub and a hip bathtub; another toilet-room with the same fixtures except the hip bath; three toilet-rooms with one basin, a bathtub and a water-closet in each, and one with a basin and water-closet only. On the first floor is one toilet-room with water-closet, washbasin, and urinal, another with a wash-basin and urinal only. There are also two butlers' pantry sinks, one servants' hall sink, one kitchen sink, one scullery sink, and one slopsink. In the basement there are two servants' water-closets, one engine-room sink, seven laundry tubs (in two sets), and one laundry sink.

Besides the above, there is, in a portion of the house designed for the servants' hall, a water-closet and bathroom on the third floor; water-closet, bath, and slopsink on the second floor; and in the kitchen a house boiler, laundry boiler, and a water heater.

The two stables have accommodations for the superintendent and coachman and their families, and bachelors' apartments for the stable hands. The superintendent has a bathroom, water-closet, kitchen boiler, sink, and two laundry tubs. The coachman has a water-closet and ordinary kitchen fixtures. The bachelors' apartments are furnished with a bathtub, water-closet, and a sink.

In the stables there are 26 box stalls and 16 open stalls. The latter drain into gutters at the rear with bell-trap waste pipes for every two stalls. Each box stall has separate waste and trap. There are two horse troughs, one harness-room, sink, one wash-basin, and two water-closets in the stable. Outside, under a roof, is a concrete platform, 30x60 feet, pitched to two wastes, with bell-trapped strainers and hose cocks for washing carriages. In the adjacent engine and dynamo house there is one water-closet and one sink. There are about 25 fire hydrants distributed through the grounds, and 15 street washers for sprinkling lawn.

All the sewage is discharged through an 8-inch main iron pipe into the river between high and low water. All the soil pipe was tested by water pressure up to the roof after the connections were made ready for the fixtures.

Figure 2 shows the location of the water tanks A and B in the attic, and Figs. 3 and 4 are separate views of A and B respectively. Each tank is filled through a ball cock from the 2-inch pump pipe G, and overflows through funnel heads H on the branches of a 4-inch pipe D open above the roof and discharging into the engine-room sink. F is a 2-inch equalizing pipe which connects the two tanks, and through which they may be emptied directly into the sewer. J is the 2½-inch house-supply pipe, and E is a 4-inch supply to the elevator tank. K, Fig. 3, is a special 2-inch branch to the kitchen boiler, and L is the 1½-inch supply for the distributing pipes throughout the house. M is a 1-inch supply for the photographer's room on the attic floor. N N are ¾-inch safety pipes from the kitchen and laundry boilers. O is an iron safe, and P is its waste pipe.

Figure 5 shows the 250-gallon boiler A in the kitchen. It is about 40x70 inches and has a domed top, flat-bottomed, and is supported by a frame of brass pipes B, screwed into a floorplate C. The cold supply is through D, and the hot delivery through E, both 1½-inch pipes. Circulation is through the 1-inch pipe F, which runs from the third-floor pipes to the 1½-inch circulation pipe G. The branch I connects with the water-back in the kitchen range R, which returns the heated water through the 1½-inch branch J of the upper circulation pipe K. All these pipes are of polished brass.

Figure 6 shows the heater M (Bramhall, Deane & Co.'s No. 2), which is placed in the basement under-

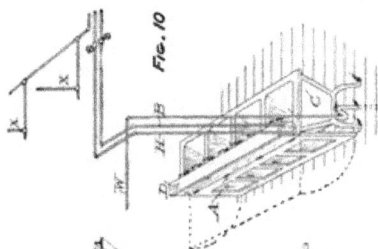

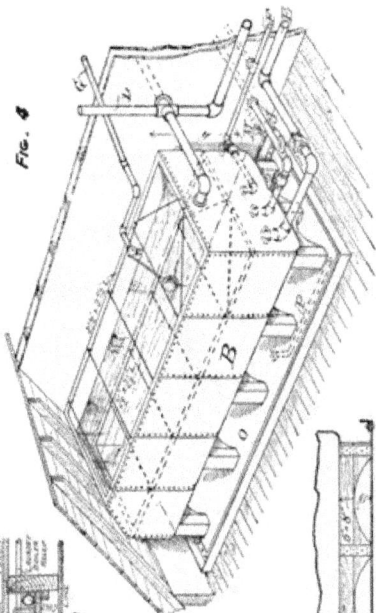

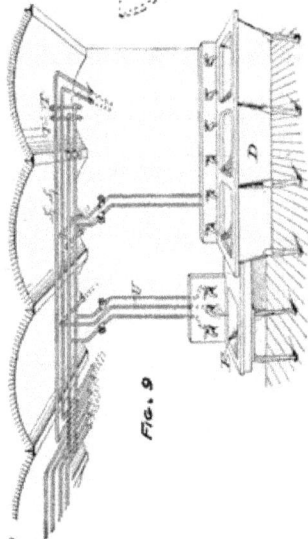

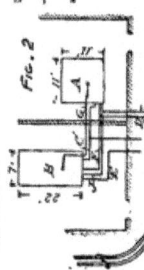

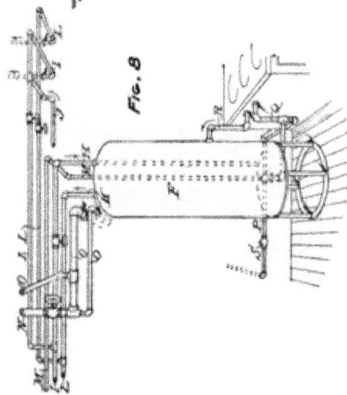

PLUMBING IN MR. JOHN D. ROCKEFELLER'S HOUSE AND STABLE, TARRYTOWN, N. Y.

neath boiler A, to help heat its water when the demand overtaxes the range water-back. The 1½-inch brass pipes L and H are connected to the water jacket and above, as shown in Fig. 5, with the circulation pipes G and K, in which part of the water flows through branch I to kitchen range and returns through branch J. The remainder flows through branch H to the heater and returns through branch L. N is a key valve through which, and the pipe O, the hot-water system may be emptied into the sewer.

Figure 8 is a diagram of the overhead arrangement of the hot and cold water pipes in the basement.

A A are the laundry rooms, B is a closet, C and D are sets of enameled tubs, E is a sink, F is the 100-gallon laundry boiler, G is a steam pipe.

Figure 7 shows the connections to laundry boiler F, Fig. 7. Cold water is supplied to it through the 1¼-inch branch H from the tank pipe I, and as the main part of the house may be closed and the tanks emptied in the winter, it has also a special branch J, through which it may be supplied direct from the 4-inch main. Hot water is delivered to the tubs through 1¼-inch pipe K, which has also a branch L, connecting with the bathtub and washbasin supply, so that boiler F may assist the kitchen boiler to operate the house system if necessary, or the kitchen boiler can be connected with the laundry tubs. M is the return circulation pipe, and P is the sediment pipe to the sewer, with a key valve S; Q and R are the circulation pipes to the water-back.

Figure 9 is a perspective from Z, Fig. 7, and shows the exposed arrangement, under the fireproof floor, of the distribution pipes to the tubs D, sink E, and risers T T, which are like the numerous other lines, taken from the basement ceiling vertically to the first floor

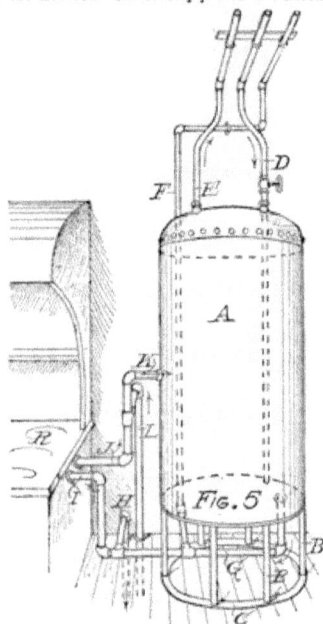

FIG. 5.

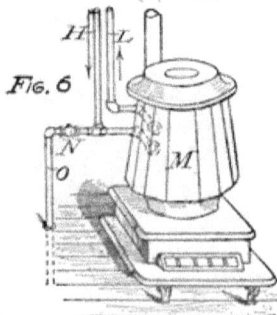

FIG. 6

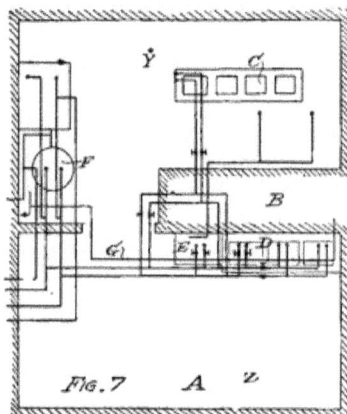

FIG. 7

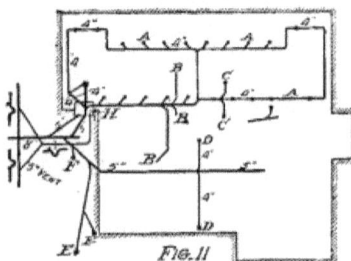

FIG. 11

PLUMBING IN MR. JOHN D. ROCKEFELLER'S HOUSE AND STABLE, TARRYTOWN N. Y.

and servants' fixtures, or lines of fixtures they supply. S S are steam heating pipes, and U is the refrigerator waste which terminates above the sink in a flap valve V.

Figure 10 is a view from Y, Fig. 7, of the detached tubs C, which were especially designed by Mrs. Rockefeller to stand in the middle of the room and be provided with an adjustable table A, that, when not in use, should hang down as shown, to form a side panel, and may be raised to the position shown by broken lines. The hot and cold pipes H and B are protected by a case D, and are carried on the ceiling as shown in Fig. 9. X X are refrigerator drip pipe, and W is a trap vent.

Figure 11 is a diagram showing the sizes and arrangement of the soil and waste pipes and leaders under the stable floor. The stable is in the second story of the building, and all the pipes are suspended from its floor joists in the upper part of the basement. A A, etc. are the branches to the open and box stall; B B, etc. are branches to rainwater leaders; C C are wastes from hose troughs; D D are drains from the carriage-washing platform; E E are soil pipes from the employees' rooms; F is a drain pipe from the exercising room; and H is the elevator shaft through which the trap ventilation pipe passes to above the roof.

E. M. Roberts was the architect. The plumbing and gasfitting was executed by John Toumey & Son, of New York.

AN ATTEMPT TO DECEIVE A PLUMBING INSPECTOR.

(PUBLISHED IN 1889.)

SOME time ago we had occasion to note the attempt of some plumber, we think in Cincinnati, to deceive the inspector in regard to the tightness of his drain pipes, by putting a plug in the upper part of the soil pipe and filling the part above with water, thus giving the false impression that all the pipes were filled and subjected to pressure.

A similar trick is reported in the *Asbury Park Journal* as having been recently tried in that place.

One of the requirements of the health laws governing Asbury Park is that all plumbing, upon completion, shall be examined and tested by the Inspector of Plumbing. The principal test is to see that the whole system in a house is perfectly air-tight, so that sewer gas may not escape, and for this purpose the ends of the pipe are sealed up and an air pump with pressure gauge is attached. Air is forced in the pipes until a pressure of five pounds to the square inch is attained. As soon as the pump is stopped the gauge will quickly indicate the presence of a leak; if it stands at the required notch for a certain length of time the test is deemed satisfactory.

In this case, a plumber doing business in Asbury Park informed the Board of Health that he had completed the work on a house and was ready to have it inspected. Inspector Lippincott went to the building and was met there by the foreman plumber, who remarked to him that he had a hard time to get the job tight. The air pump was attached, and the first

two or three strokes indicated pressure on the gauge. This was unusual, and excited the suspicion of the inspector, for in a line of large pipe it requires more time to compress the air sufficiently for it to show on the indicator. The pressure went up to five pounds and stood there firmly, and the inspector then went upstairs to look over the line of pipe. He requested the foreman to open one of the capped inlets, but before doing so the foreman made an excuse to go downstairs after his tools. When the cap was removed no air escaped, and the inspector then knew that some trickery was being practiced. Returning to the pump he found there was no pressure on the gauge. The foreman declared that he had not touched it when he came down after his tools. The pump was again worked and the pressure immediately went up to five pounds and stood there, although the pipe was open upstairs.

The pump was then removed, and in the pipe, about 2 feet from the end, was found a plaster of Paris plug, calked in with oakum, that effectually shut off the passage of air, so that not a single joint of the house drain pipe was subjected to pressure. When the foreman saw that the trickery was exposed he confessed that he and his helper had plugged the pipe. The boss plumber claimed to know nothing about it, and charged it all on the two men.

Inspector Lippincott deserved praise for his quick detection of the fraud, and the owner of the house can thank the Board of Health for defending him against having a dangerous construction erected in his house, when he had contracted for and paid for a first-class job.

The houseowner doubtless appreciates the efficiency of the inspector and the value of the plumbing law.

PLUMBING IN THE RESIDENCE OF MR. ELBRIDGE T. GERRY.

(PUBLISHED IN 1894.)

THE new house of Mr. Elbridge T. Gerry at Fifth Avenue and Sixty-first Street, New York City, is a large and costly building designed by Richard M. Hunt, architect, of New York, and equipped with an ample provision of hot and cold water supply and abundant fixtures for the comforts and necessities of the domestic and toilet requirements. The pipe lines are all screw-connected and pressure-tested. The water supply is received through two 2-inch connections with different street mains and is all filtered through a battery of four pressure filters, whence it is pumped by electricity to an attic tank which is intended to provide a day's storage and furnish uniform pressure. There are complete systems of trap and local ventilation and an abundant supply of hot water from several separate interchangeable sources. The installation, which was executed by James Muir, Sons & Co., of New York City, comprises nine bathrooms, a butler's pantry, kitchen, scullery, and two laundries, together with metering, filtering, pumping and storing and heating apparatus, with its necessary machinery. There are in the house four waterclosets, 10 washbowls, three slopsinks, one urinal,

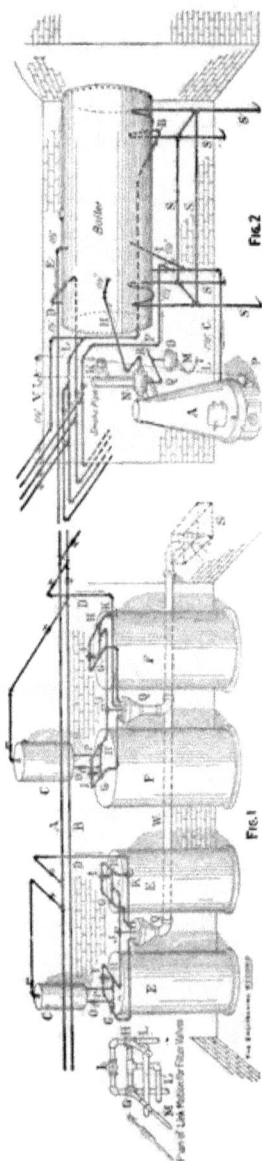

one sitz bath, two butler's sinks, a kitchen sink with special marble table, two sets of five earthenware washtrays, and two iron sinks for washing, drips, etc. Most of the bathrooms have white tiled floor and walls and oak cabinet-work, but the boudoir bathroom is finished with rose-colored tiles, mosaic floor, and rosewood cabinet-work, the metal-work being of ornamental design, silver-plated, and the porcelain decorated in white. The bathtubs are of porcelain and enameled iron.

Figure 1 shows the arrangement and connection of the Continental filters in the basement. A and B are 2-inch supplies from different street mains which deliver to the filters through the independent branches D D. Filtered water is delivered through valves I and pipes, here omitted to avoid confusion, to the pump suction tank and basement distribution lines. The valves G and H of each filter are connected and operated by a link arrangement commanded by handle M (omitted in the general view, but shown in a detached detail) which operates both together, so that when turned into one position the filter is in service, delivering into the house system, and when reversed washes out the interior under tank pressure and delivers the water into funnels Q Q of a waste pipe which empties freely into an adjacent trapped sink S.. Filters E E are intended to be used for the supply to the kitchen and butler's pantry, and filters F F for the general house supply.

Figure 2 shows the hot-water boiler, about 3x8 feet, for supplying the kitchen and bathrooms. Cold-water supply is received through pipe B and check valve V, which closes away from the boiler to prevent back flow. The coldest water in the boiler circulates through pipe C to the Hitchings greenhouse heater No. 3, A, and returns to the boiler at a higher temperature through pipe H. The boiler delivers hot water through pipe E to the basement and first-floor distribution lines and through pipe D to the upper floors. F is the return circulation and I the emptying pipe. S S, etc. is a pipe frame supporting the boiler. The cold-water supply to the hot-water boiler passes through five Tucker grease traps from sinks in the basement before supplying the boiler.

If steam forms in the heater A it is immediately carried over into the drum N, and separating from the hot water that is delivered through pipe H, its pressure is transmitted through pipe Q to a diaphragm at O, which actuates a lever M and operates the chain L commanding the dampers. The illustration shows the draft on, but if the water becomes too hot the lever M will move in the direction T, closing draft damper P and simultaneously opening check damper K. When the temperature of the water falls sufficiently and the steam pressure is relieved, the lever M will return to its former position and the dampers will be reversed. R is a support for diaphragm case O.

Figure 2 shows the connections to the attic tank, which is made of riveted steel angles and plates, and is about 5'x8'x6' deep. It rests on three rolled 12-inch beams A A A that distribute its weight upon the transverse beams B B B, beneath which are sup-

FIG. 9

FIG. 8

FIG. 6

FIG. 5

FIG. 4

FIG. 3

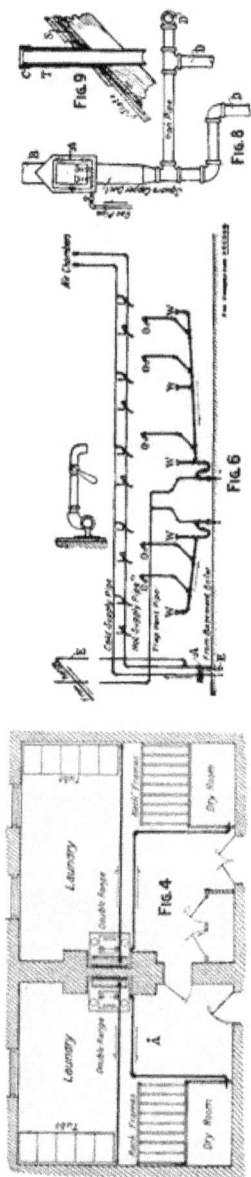

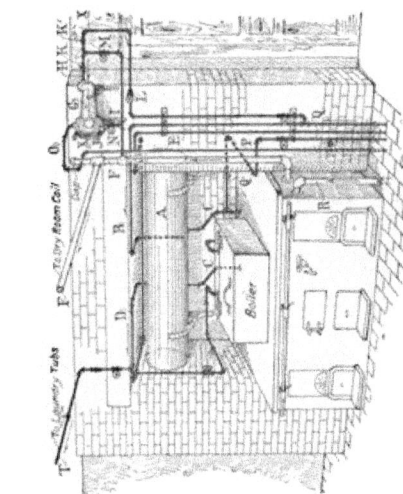

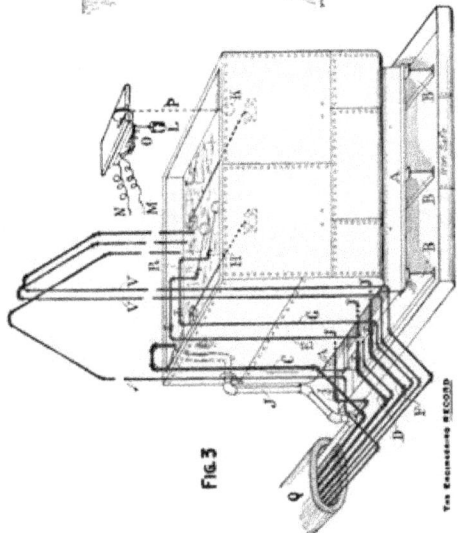

PLUMBING IN THE RESIDENCE OF MR. ELBRIDGE T. GERRY, NEW YORK CITY.

ported on the iron ceiling girders of the upper floor.
C is the 1½-inch delivery from the pump. D is the
1½-inch general house supply, down-vented by the
¾-inch pipe E, which promotes its emptying, and is
fitted with a ball cock to close when the tank is full
and prevent discharge through it. F is a 1½-inch
riser from the city mains under street pressure. It
has a check valve opening with a delivery from the
tank, so as to serve as a supply from the tank to the
city pressure pipes when the adjacent gate valve is
open, and it delivers into the tank through pipe G
and ball cock H. I is the emptying and J is the overflow pipes discharging into the open sink in the cellar,
and V V V are hot-water expansion pipes carried 10
feet above the tank to the highest point under the peak
of the roof. R is a copper float which, with its accompanying counterweight L, is attached to the chain
P which operates the spring lever O that makes and
breaks the electric circuit between wires M and N,
and sounds a high and low water alarm bell near the
pump in the basement. Q is a canvas covered asbestos
conduit through which the main pipe lines are run
underneath the roof above the chamber ceilings.

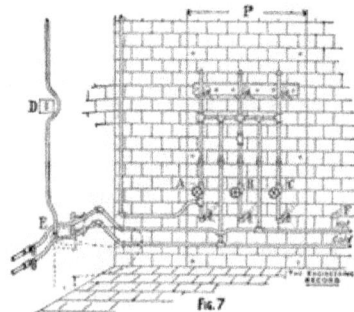

Fig. 7

Figure 4 is a sketch plan showing the arrangement
of the connected adjacent laundry-rooms. Figure 5
is a view from A, Fig. 4, showing the pipe connections to one of the double ranges. Its 60-gallon
boiler A is supplied through the 1-inch street pressure pipe B, and its circulation from the rear water-
back is through the two pipes C C. It delivers hot
water through pipe D, and E is its vent pipe, carried
down and under the floor across to the riser shaft.
The circulation pipes of the front water-back
are connected directly to the flow and return pipes F
and R that serve the radiator coil in the laundry dry-
room. The water consumed in this system is replaced by an automatic feed-water regulator G, furnished by the Duparquet, Moneuse & Huot Company, that is supplied under street pressure through
pipe H. When the inclosed float falls below a certain
level X X it opens an interior valve and delivers
water through I to pipe F. J is a gauge glass, and
K K' are key valves that are ordinarily open. If it
is desired to feed the coil by hand valves K and N
are closed and valves M and L are opened. O is a

vent pipe, P is an overflow pipe, and Q is an emptying pipe.

Figure 6 is a diagram of the piping for one set of
five laundry tubs. Ordinarily valve A is closed and
the hot water is supplied from the range boiler, but
by opening valve A supply may also be taken from
pipe E from the basement general boiler, Fig. 2.
W W, etc. are waste pipes, and O O O, etc. are overflow connections.

Figure 7 shows one of several switchboards located
in different bathrooms to control the supply of hot
and cold water to the different fixtures in the room.
Valve A commands the flush tanks, B the hot and C
the cold supply. Offsets, intersections, and changes
of direction are made by curving the plated brass
pipe, which is so skillfully done as to present a not-
ably handsome and elegant appearance. Special
examples are at D, where the flush pipe is carried
around a register handle, and at E. where the bathtub pipes are carried around the water-closet bowl.
P is a panel extending to the ceiling, on which the
tile are secured, and which is removable throughout
its entire length, together with the pipe on its face,
to give access to the riser lines in the recess behind it.

Figure 8 shows the arrangement for inserting circulation for local vent pipes. The box A is set in the
attic wall, and inside it are burned two gas flames,
visible through the glass door. These tend to exhaust the air from the vent-pipe branches and discharge the lighter warmed air through conduit B
above the roof.

Figure 9 shows the method of flashing the soil and
trap vent pipes. A copper plate S is laid under the
slates and the copper thimble T soldered to it, inclosing the wrought-iron pipe up to its first joint, or
to the top, where in either case a coupling C, cham-
fered out and beveled on the lower edge, is screwed
down over the sleeve and holds it firmly and caps it.

PLUMBING IN SEVENTY-SECOND STREET HOUSES, NEW YORK.

(PUBLISHED IN 1891.)

THE plumbing of two new houses on West Seventy-
second Street, New York, provides for an unusually
complete control of the hot water, which may be supplied from either or both of two boilers to any or all
of eight subdivisions of the house at will. There are
also some special details designed for this work by
Paul S. Bolger, who executed the plumbing and gasfitting. The largest house is four stories high with a
basement and cellar, and its three upper floors are
divided by central halls, to each side of which the
water supply is independent.

Figure 1 shows the two 90-gallon boilers B and C,
which are heated from water-backs in the kitchen
range R. Both of them are at present supplied with
water under street pressure, but it is intended that
either of them shall be supplied from a roof tank if
the city pressure becomes insufficient. Cold water is
received from the street through pipe C and from
the tank through pipe A. To connect either boiler,
as B, with tank pressure, its supply pipe D must have
the street pressure cut off and tank pressure admitted

by closing its valve at Z and opening its valve at X. Delivery pipe E and circulation pipe G must be cut off from the street pressure by closing their valves at X and Z, and the supply, delivery, and circulation pipes d, f, and h, of street pressure boiler b, must be cut off from tank pressure by closing their valves at x and z. Then boiler B will deliver through pipe F and boiler b through pipe e, pipes E and f and circulation pipes G and h being cut out. By reversing the valves, boilers B and b would work under street and tank pressures respectively, and pipes e, F, g, H would be cut out.

I is street hot, J is tank hot, L is street circulation, and K is tank circulation for the upper stories. M is the street hot and N is the tank hot supply to the first floor and basement fixtures. By closing the valves at X and x on pipes D and d, and opening all others, both boilers are put under street pressure, or by closing the valves on C, at Z and z, and opening all others, both are put under tank pressure.

The circulation pipes G g and H h are united below the kitchen floor to pipes O for the tank and P for the street pressure systems, and they and the boilers may be emptied through valves at Z and z, and in the cellar. Q and q are check valves to prevent the escape of tank water into the street mains. U, u, etc. are eight unions. All the pipes, boilers, fixtures, and other metal-work are nickel-plated, and the kitchen walls are finished with white ceramic tiles.

Figure 2 shows the continuation into an adjacent hall of the tank, hot-water, and circulation pipes A, I, J, and K L, and their division into two groups of risers, one supplying each side of the house. Pipe A_1 is connected with the tank, and both A_1 and A_2 serve for general tank supply. Each group of risers

has from the foot of the vertical shaft an additional pipe C_1 C_2 (not shown here), supplying cold water under street pressure. Y is a drip pipe and S S are 10 special valves through which the riser lines may be emptied. B B are five unions and D D are 10 special angle valves controlling the riser lines.

Figure 3 shows a glass cabinet G, on the wall of the back stairway, which contains a cut-off M and three-way cock O for connecting the fixtures on one side of the second floor, with either the tank or street pressure system. The risers are carried in a wall shaft X, accessible throughout through panels W W, but are deflected to one side at V, just below the cut off, to afford more room for connections, as shown in d agram, Fig. 4. L_1 is the street and K_1 the tank pressure circulation; J is the tank and I the street pressure hot-water supply, and A_1 the tank and C_1 the street pressure cold-water supply; H is the hot-water and Q the cold-water distribution to the second-floor fixtures, and P is the return circulation pipe from them.

When the cut-off handles are turned down, the corresponding pipes H and Q are connected with tank pressure, when turned up, as shown, they are connected with the street pressure. When the handle of the three-way cock O is turned to the right, as shown, the circulation is under the tank-pressure system; it is under the street-pressure system, however, when it is turned to the left. There are six switch arrangements like the above, three in the back stairway halls for the second, third, and fourth stories of one side of the house, and three under the washbasins for the same stories on the other side of the house.

The system is essentially the same in the adjacent smaller house, except for the arrangement and con-

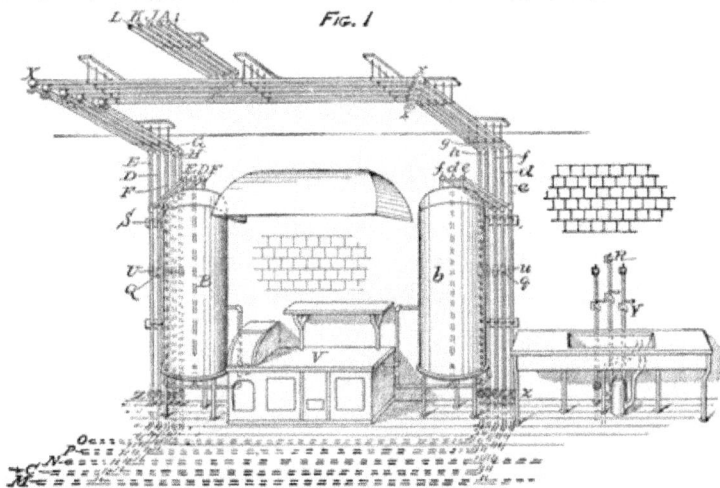

FIG. I

PLUMBING IN SEVENTY-SECOND STREET HOUSES, NEW YORK.

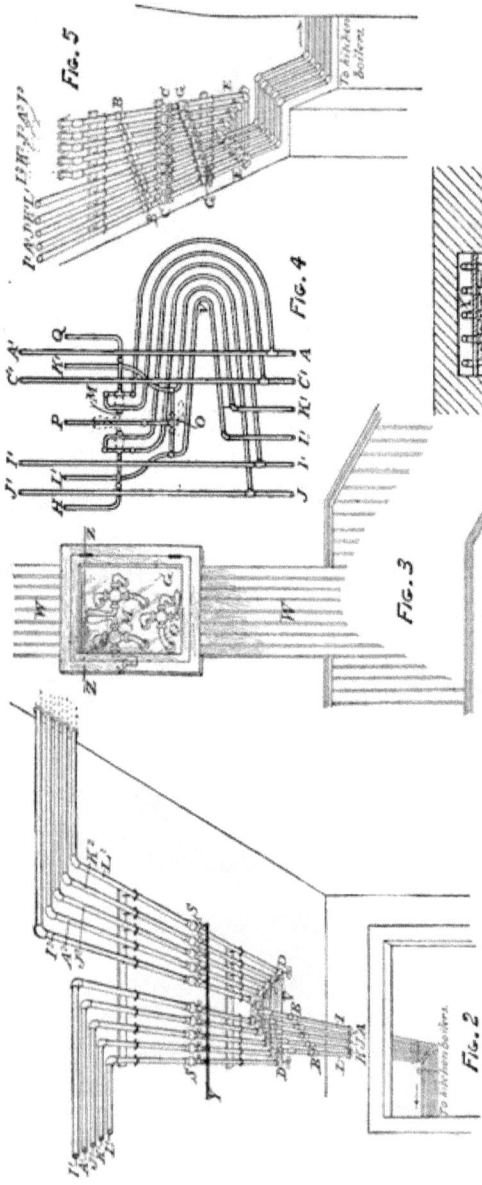

PLUMBING IN SEVENTY-SECOND STREET HOUSES. NEW YORK CITY.

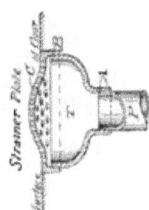

Fig. 9

Steamer Plate

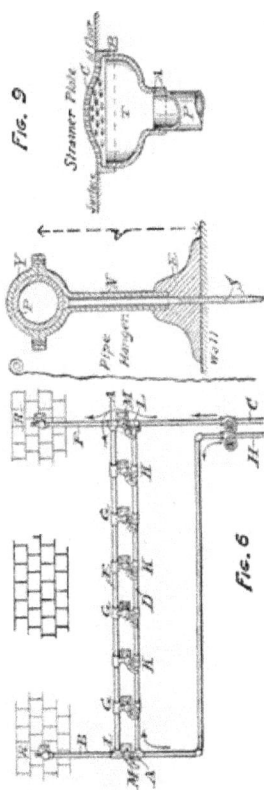

Fig. 6

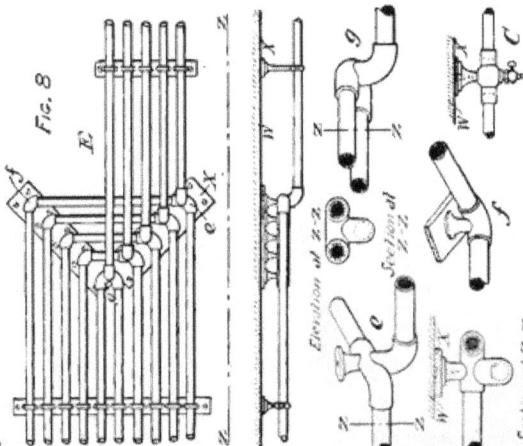

Fig. 8

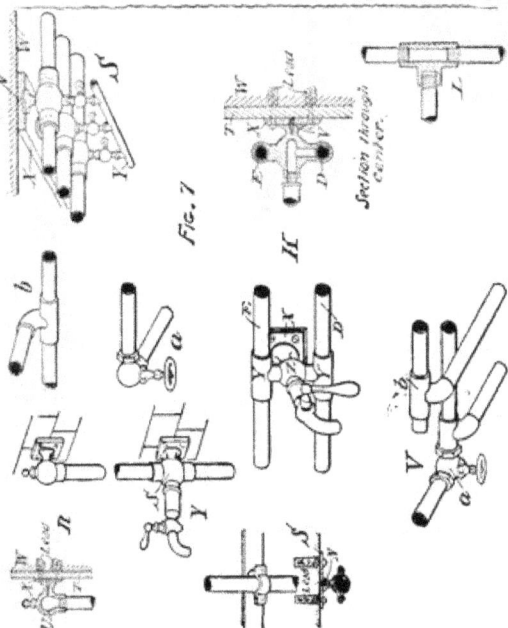

Fig. 7

nection of the supply lines in the basement hall, corresponding to Fig. 2. This is shown in Fig. 5, where the same reference letters have the same significance as in Fig. 2, and G G are ordinary valves controlling the rising lines, C C are 10 special drip cocks for emptying the rising lines into a pail, since it was not desirable to run a drip pipe from this point. The work throughout the house is well executed and is beautifully finished, corresponding with the costly materials and fixtures, but the most noticeable features are the symmetry and neatness in the pipe and valve work, as shown in Figs. 1, 2, and 5, where the unions, valves, connections, etc. are set in exact regular lines and the pipes compactly and attractively arranged.

Figure 6 shows the hot and cold supplies, H and C, for a set of four laundry tubs, adjacent to the kitchen. Hot water flows into distributing pipe D and air chamber B, and cold water flows into distributing pipe E and air chamber F. A A are ordinary tees, but L L are tees with one branch left solid, as at L, Fig. 7. Faucets G and K, etc. are specially made for this work, as shown at K, Fig. 7, the sleeves Y Y being cast on ball Z, which has a waterway from the faucet to only *one* of the sleeves, this being set *downwards* at G G G and *upwards* at K K K. Ball Z has a flange bearing on escutcheon plate X, to which it is fastened by screw V, and the whole is secured to the wall W, which is faced with white ceramic tiles T.

At faucets M M the balls Z have the water through from one sleeve to the other. The tops of the air chambers R R are finished like those at the kitchen sink, Fig. 1. The ball U being finished and fastened to the wall similarly to Z, above described, as shown at R, Fig. 7. T, in Fig. 7, shows the connection of the kitchen sink faucets which have extra long and heavy barrels screwed into ball S, which is similar to that at M, Fig. 6. S, Fig. 7, is a detail of the wall fastenings at S, Fig. 1. The ceiling hangers in that figure are the same, except that shank N is made longer.

V, Fig. 7, shows the arrangement of any two of the pipes at V, Fig. 2. The special corner valve *a* and tee-bend *b* are clearly shown at *a* and *b*, Fig. 7. S, Fig. 7, shows the connection of drip pipe Y, Fig. 2. E, Fig. 8, shows the arrangement at E, Fig. 5; *e*, *f*, and *g* show the details of special fittings at E; C, Fig. 8, shows the drip cock at C, Fig. 5. In Figs. 7 and 8 the letters W, T, X, and V have the same meaning, and it will be seen that care was taken to secure uniformity and simplicity in their design, and to avoid unnecessary separate parts. The ball connections are all essentially similar, except that the internal waterways and the special bends, tees, and unions admit a very simple and close arrangement of intersecting pipes, while they dispense with separate additional hanger pieces, and secure strength and stiffness.

Figure 9 shows an adjustable pipe hanger and an adjustable floor strainer. The former was used where a single pipe was to be supported from a wooden joist or board. Its stem N was made long, and fitted the socket in the top of the escutcheon

base E. The stem N was cut to any required length and fastened in position by a screw S, of any desired length to penetrate to the proper firmness. After S is driven pipe P is put in place and yoke Y screwed on. For lengths A, of less than 3½ inches, stem N and base E were cast in one piece, and for very long lengths of A stem N was screwed into base E, which was tapped to receive a screw in the bottom, as shown at S and in other details, Figs. 7 and 8.

In the strainer-plate, for bathroom flow wastes, etc., the brass waste pipe P has a flaring tap T and separate perforated cap C, and the two screw joints A and B afford sufficient play to adjust the cap exactly for a difference of an inch or more in the position of the floor slab, without altering the length of the pipe.

PLUMBING IN A COUNTRY RESIDENCE AT SEABRIGHT, N. J.

(PUBLISHED IN 1889.)

" ROHALLION," the residence of Edward D. Adams, Esq., at Rumson, near Seabright, N. J., comprises a large, isolated house, with conservatory, stables, etc., remote from public water, gas, or sewer lines. Modern conveniences and improvements are supplied throughout, and as illustrative of an entirely independent and self-contained system of plumbing, a description of the general features and some details are given that are not otherwise remarkable.

The buildings were designed, and their construction supervised by the architects, McKim, Mead & White, of New York, and the plumbing and drainage was done by S. & A. Clarke, also of New York

Abundance of hard water is obtained from a private well, 6 feet in diameter and about 50 feet deep, through red clay.

Rainwater from the house roofs is stored in a cistern.

The house-drains, soil pipes, stable drains, etc., empty into the drain pipes, 6 inches in diameter, that empty into sealed brick cesspools, whose contents are automatically disposed of by subsurface irrigation through 2-inch unglazed tile pipes.

The house and stable has each an independent sewerage system with about 1,000 feet of irrigation pipe, occupying a total area of about one acre of meadow land.

The cistern water is raised to a 1,000-gallon roof tank by a Ryder gas engine pump, and is distributed thence throughout the house for all purposes, except flushing cisterns and slopsinks, which are supplied with well water at a head of about 15 feet at the highest fixture.

Well water is also supplied to the butler's pantry and kitchen sink.

A wooden water-tower furnishes a high and a low-pressure service and has a windmill to operate the pump. The tower is built in a conspicuous location, and is designed to present an attractive appearance and afford a commanding prospect from its observatory gallery.

Figure 2 is a vertical section of the tower at Z Z, Fig. 3, and Fig. 3 is a horizontal section at Z Z, Fig. 2. A is the Hercules windmill, built by the George

L. Squires Manufacturing Company, Buffalo, N. Y., for an estimated daily service of raising 10,000 gallons of water 90 feet. The mill is about 14 feet high and 12 feet in diameter, with vertical axis. Its sails are vertical wooden slats whose angle with the circumference is easily adjustable; they may be closed so as to receive no impulse from the wind, or set open to work at different speeds.

At the right of the figure they are shown closed, so as to stop the mill, and at the left open as for working. B is the observatory gallery, reached by stairs not here shown. C is a 3,500-gallon tank for house and fire service. D is a 17,500-gallon tank for all other needs—viz.: stables, conservatory, irrigation, etc. The lower story of the tower is used for storing garden implements, etc.

C and D are both coopered wooden tanks that receive their supply from the adjacent well through the 2-inch pump delivery pipe T. In winter only tank D is used, and tank C is cut off by closing valve G. A branch from pump pipe T has a ball cock H and a valve at U (accidentally omitted in Fig. 2).

In summer both tanks are used, and their operation is as follows: Valve G being opened and U closed, pipe T delivers into tank C. When C is nearly full it overflows into D through the stand-pipe Y until D is full and the ball cock I closes; the water then rises above Y in C and closes ball cock E. The pump, still working, exerts a pressure in the small cylinder N, and raises its piston M, which operates a connecting-rod (shown broken off at V) that stops the windmill A that drives the pump. As soon as any water is used from tank, ball cock E is opened, the pressure is relieved in chamber N, and the weight of piston M pulls it down and set the windmill sails

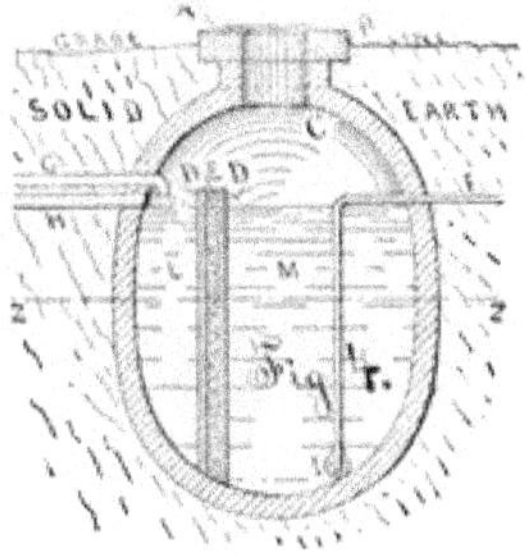

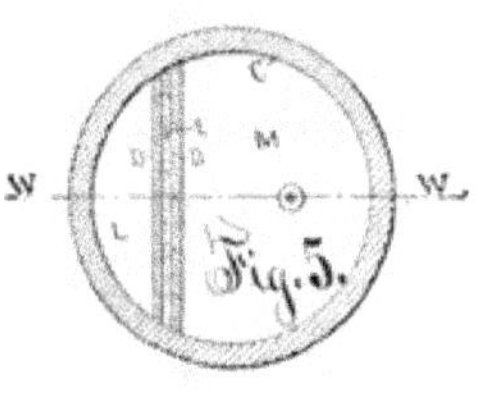

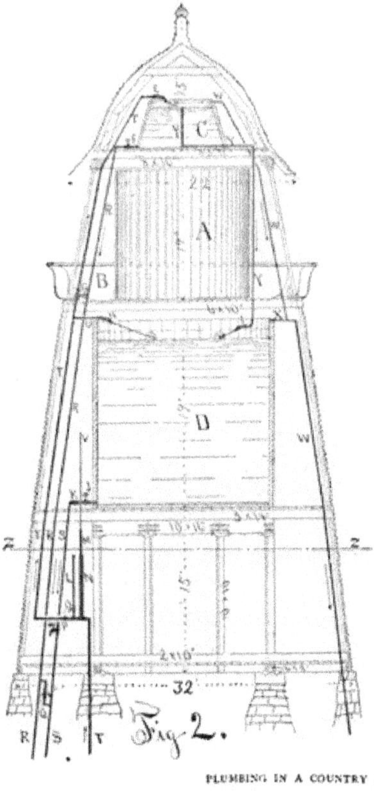

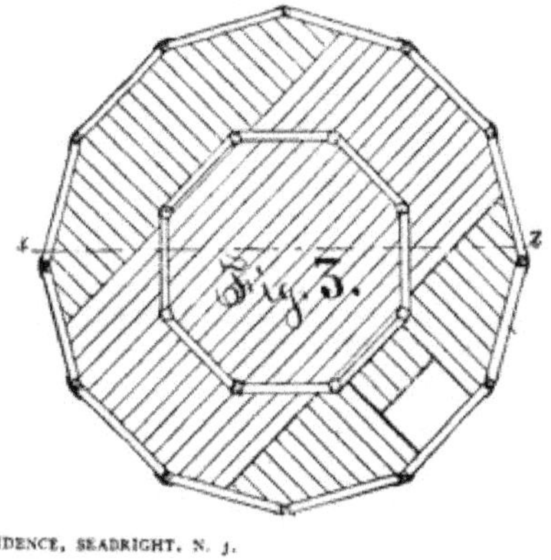

Fig. 4.

Fig. 5.

Fig. 3.

Fig. 2.

PLUMBING IN A COUNTRY RESIDENCE, SEABRIGHT, N. J.

ready to run again; by this arrangement tank C is always filled first. In winter valve G is closed and U opened and tank D only is used. W is a 1½-inch overflow, with branch X from tank D. R is a 2-inch supply pipe from tank C to the house, and is controlled by valve F. S is a 2-inch supply pipe from tank D to the stables, and it is controlled by valve J. Q is a cut-off valve ordinarily closed, but it may be opened and give high pressure to the stable system, or an additional low-pressure supply to the house

There is a flagstone cap B with a manhole closed by the iron cover A.

Figure 4 is a vertical section at W W, Fig. 5, and Fig. 5 is a horizontal section at Z Z, Fig. 4. Up to a point above the level of the 6-inch overflow pipe H, the cistern is divided by a filter partition composed of two 4-inch Portland cement-laid brick walls, D and D, with the 4-inch space E between them filled with charcoal. The water is received in chamber L, and percolates rapidly through walls D and D to chamber

Fig. 6

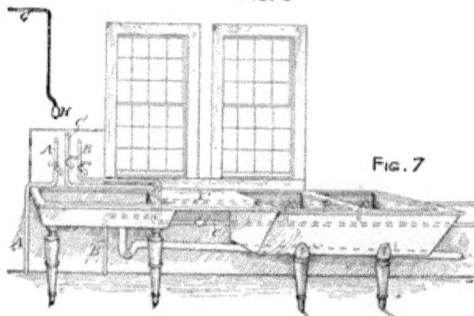

Fig. 7

PLUMBING IN A COUNTRY RESIDENCE, SEABRIGHT, N. J.

system. K is a check-valve to prevent tank C from emptying into D when Q is opened. O is a key valve, and P is a draw cock.

Figures 4 and 5 are sections of the house cistern for rainwater, which is received through the 4-inch leader G. The cistern is 15x20 feet, with 12-inch cement-lined brick walls C, laid directly on the smoothed clay surface of the excavation near the house.

M, whence it is pumped to roof tank through the 1½-inch suction pipe F with basket strainer I.

The house is supplied with well water by the 2-inch galvanized-iron pipe R, Fig. 2, and with cistern water by a 1½-inch pipe from roof tank. Pipe R serves as a direct fire line to three hose cocks in the house, one on each floor, and to one inside and one outside of the stable.

The 2-inch pipe S, Fig. 2, is direct to the stable, and has ¾ and 1-inch branches supplying water for sprinkling the lawn and garden and for the conservatories, etc.

In the kitchen, and elsewhere that they are exposed, the water pipes are of brass, tin-lined and nickel-plated; where not exposed they are of galvanized iron.

There is a kitchen boiler and sink, independent laundry boiler and tubs, butler's sink, one chambermaid's slopsink, one toilet-room with washbowl only, one with water-closet only, and one private and one guests' bathroom with washbowl, bathtub, and water-closet, and one servant's bathroom with tub, washbowl, and water-closet.

The stable, built for 10 horses and four cows, has one watering-trough, a number of convenient draw cocks and hose cocks, a washroom sink and a water-closet. In the upper parts of the building are apartments for the coachman's and gardener's families. These each contain a set of laundry tubs, kitchen boiler and sink, one washbowl, one bathtub, and one water-closet.

Figure 6 shows the arrangement of kitchen boiler A, whose cold-water supply is through pipe C.

D is hot-water upstairs, F is hot-water to lower floor (and may be connected to supply or receive from laundry boiler), and E is the hot-water return-circulation pipe.

H H are circulation pipes to the water-back, and I is the sediment pipe emptying boiler and water-back, G is a relief pipe from boiler and terminates in a vacuum drip valve K over the kitchen sink, Fig. 7. B B are heavy wrought-iron brackets bolted through the wall.

Figure 7 shows the kitchen sink and two wash-tubs provided for use when it is not convenient to use the laundry.

A is pipe supplying well water, B is hot-water pipe, and C is cold-water pipe.

FIG. 8.

Figure 8 shows the washbowl in private bathroom. The table and wall and floor slabs are of red Knoxville marble, and the metal-work is nickel-plated.

Figure 9 shows the hostlers' water-closet on the first floor of the stables. A is a slate slab and B is asphalt tiling. The room is ceiled with yellow pine, oiled, and there is an automatic cistern flushing the urinal.

Figure 10 is a view of the washroom on the first floor of the stables. The floor is tiled with asphalt and the walls ceiled with yellow pine, oiled. The large iron sink B is furnished with hot and cold water, and is intended chiefly for washing the harness.

The cold-water supply is controlled by cock A, and has branch C to sink, and I to the boiler. H is the hot-water from the boiler. The water circulates from the boiler to the water-back and return through pipes F and E. G is a sediment cock for emptying the boiler.

FIG. 9.

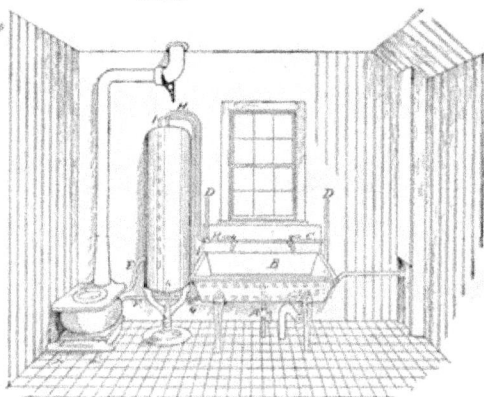

FIG. 10.

PLUMBING IN A COUNTRY RESIDENCE, SEABRIGHT, N. J.

PLUMBING IN A NEW YORK CITY RESIDENCE.

(PUBLISHED IN 1892.)

THE residence of Kalmar Hass, Esq., on East Sixty-ninth Street, New York City, is a four-story brownstone-front house, which has just been completed according to plans and specifications of John H. Duncan, architect. Its elaborate plumbing-work, which presents interesting arrangements and detail, was executed by Paul S. Bolger, of New York. The entire second floor of the house is shown in plan in Fig. 1 of the accompanying illustrations, and in vertical section at Z Z, Fig. 1, in Fig. 2. Figures 3 to 6 inclusive show plans at U U, V V, X X, and Y Y,

Fig. 2, respectively. From these it will be seen that, excepting the boilers and sink in the basement kitchen, there are no plumbing pipes or fixtures whatever in the main building, and, excepting the basement laundry tubs and the butler's sink, all the plumbing is in the special brick-walled tower, which absolutely isolates all the baths and water-closets and the soil and vent pipe risers to which all fixtures are connected.

In the second and third floor toilet-rooms the washbowls W B, W B, Fig. 1, are entirely unconnected with any plumbing. Stationary oval bowls are set in marble tables, but are fitted only with overflow and waste terminating in a long, vertical ferrule

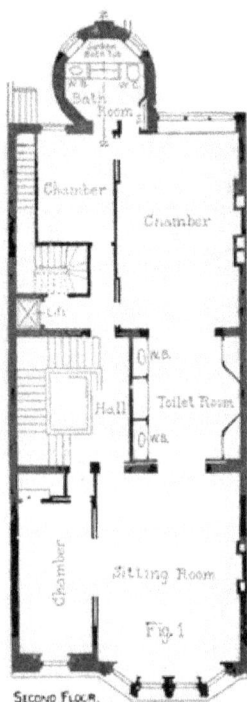

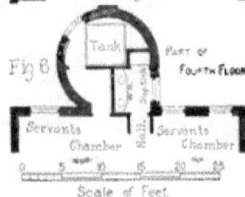

PLUMBING IN A NEW YORK CITY RESIDENCE.

which discharges into a special movable brass nickel-plated sloppail, the bottom of which is cushioned by a rubber-bearing ring sprung over a groove made for it in the flange. The water is supplied by pitchers. All the rising lines, 12 to the second floor and nine to the tank floor, are carried together in the corner S of the tower, where they are run behind a movable panel, which incloses them and leaves them entirely accessible.

A double water-back in the kitchen range is connected with two 100-gallon boilers. One of them is double, providing hot water under street and tank pressure for all purposes except for the second-floor bathroom, which only is supplied by the other boiler under street pressure. Connecting pipes and valves are so arranged that this boiler may easily be put under tank pressure, or its hot water may be turned into the delivery pipe from either of the other boilers, as for an excessive demand by the laundry tubs, or that the double boiler can at pleasure be made to deliver to the second-floor bathroom, the principal

consideration being, however, to afford a discharge for the bath boiler if its water is not required for the second-floor bathroom and becomes too hot.

The kitchen boilers are shown in diagram in Fig. 14. A is the double boiler for general house supply, the inside part being under tank and the outside part under street pressure. B is the special bath boiler. Both boilers are heated from the kitchen range C by the circulation pipes D D D D. The cold-water supply comes from the city mains by the pipe E, with branches F F F to supply the boilers and G to supply the basement fixtures and the butler's pantry. The cold supply from the attic tank is by pipe H, filling the boilers through pipes K K. The special hot delivery L connects with the second-floor bathroom and may be put into communication with the street-pressure house boiler through pipe M by opening valve N, which is usually closed. The street-pressure hot-water delivery O connects with all basement and cellar fixtures. Tank pressure hot water is delivered to the upper floors through P, and Q Q are hot-water return-circulation pipes. The sediment pipes R R discharge into a cellar sink. S are the soil and vent risers, etc. shown in shaft S, Figs. 1, 4, 5. Ordinarily, valves T T T N and U, which admit tank pressure to the special boiler, are closed and all others are open. Check valves I I are provided, opening up with street pressure against the tank

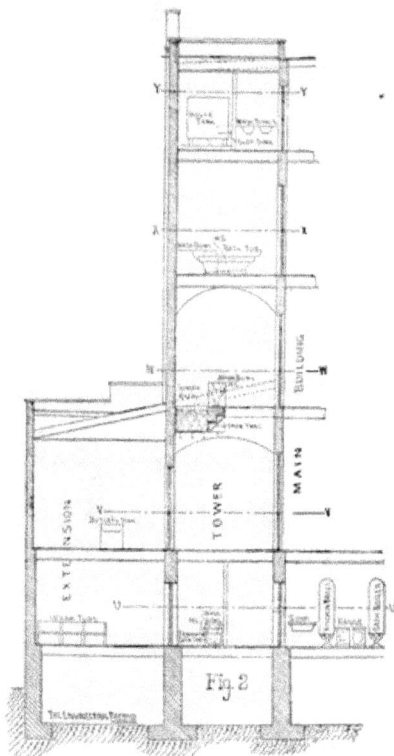

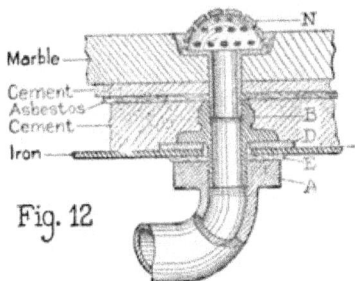

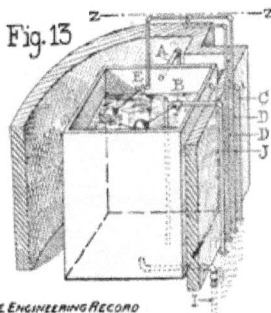

Fig. 2

THE ENGINEERING RECORD

Marble
Cement
Asbestos
Cement
Iron

Fig. 12

Fig. 13

THE ENGINEERING RECORD

PLUMBING IN A NEW YORK CITY RESIDENCE.

pressure, to prevent the possibility of the escape of tank water into the street mains. The check valve J, which is soldered shut, acts as a stop between the tank and street pressure hot water, and is placed there to preserve a symmetrical appearance. V V are dummy valves with the wings removed so that it is impossible to close them. They are placed there only for the sake of uniformity of appearance.

The second-floor bathroom, shown in enlarged plan in Fig. 7, has a white marble floor, high wall panels and an ornamental domed ceiling. It is attractive for its elegance, spaciousness, and for its unique sunken bath, which is lined with white marble and forms a large pool below the floor level, and is reached by descending marble steps. It is supplied with hot and cold water issuing from a dolphin's head at R, the supply being controlled by valves P P.

Fig. 9

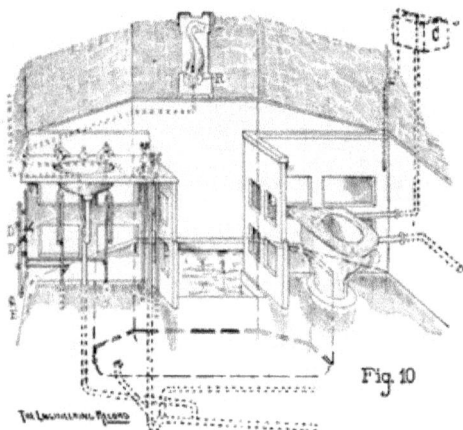

Fig. 10

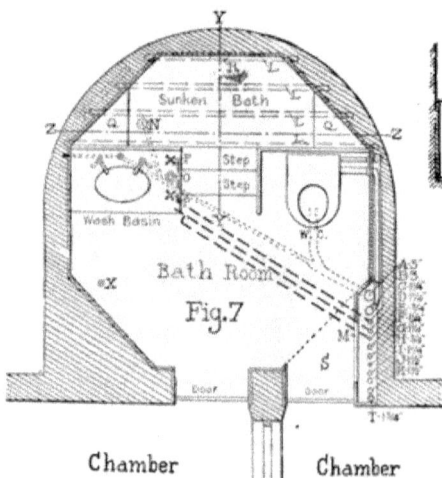

Fig. 7

Chamber   Chamber

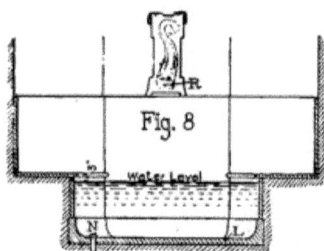

Fig. 8

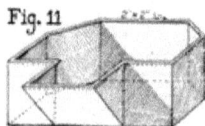

Fig. 11

PLUMBING IN A NEW YORK CITY RESIDENCE.

An overflow and waste valve O allows of emptying the bath through strainer N. The movable panel M allows of access to all the risers. Of these A is the soil pipe, B back-air pipe, C D and E respectively special cold and hot supply and return-circulation pipes from the separate boiler serving this room only. The hot and cold-water supplies and the hot-water return-circulation pipes serving the upper floors are indicated by F G and H. Other reference letters signify: I, the cold-water supply pipe from the tank; J, the pump delivery pipe to the tank; K the safe waste pipe; and T a telltale pipe.

An outline section and elevation at Z Z, Fig. 7, of the sunken bath is given in Fig. 8, showing the deeply countersunk 8-inch marble slab L, lining the smoking-room, and about 3 feet below the wooden joist of the bathroom floor. This tank was carefully calked water-tight and painted; then a thick bed of cement mortar was spread on the bottom, a thick layer of loose fibrous asbestos was pressed down on it before it set, and another thin layer of cement mortar immediately placed on top of the asbestos to receive the marble bottom slab of the bathtub lining. The sides were done in exactly the same manner, and finally the floor marble was laid, covering the edges. The asbestos was designed to prevent radiation of heat, and it was intended to have its fibers cohere thoroughly to the cement.

Figure 12 shows the connection of the waste pipe and strainer N, Fig. 6. The brass waste A, connected to a unique overflow valve O, Figs. 7 and 9, has an extra heavy flange threaded to receive the

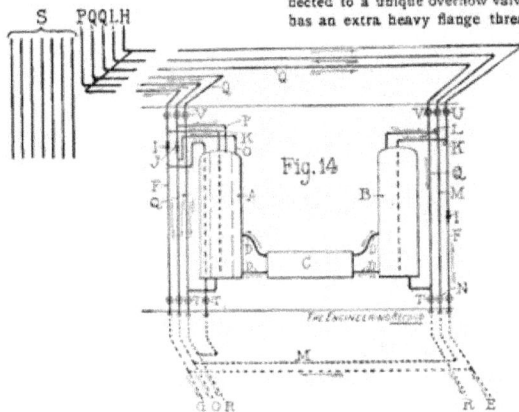

PLUMBING IN A NEW YORK CITY RESIDENCE.

bottom of the tub. A perspective from Q, Fig. 7, is given in Fig. 9, showing one end of the tub and its valves P P O, which may be readily operated without leaving the tub. The perspective from point X, Fig. 7, given in Fig. 10 shows the exposed silver-plated piping for the hot and cold-water supplies and indicates the arrangement of the waste and vent pipes in the chambers formed by the double flows between the marble tiling and the bottom of the bathtub. The connections to the risers of the branches serving this room are indicated by dotted lines on Fig. 7. E is a connection for a gas stove, and C is the water-closet flush tank entirely concealed behind the marble wainscoting. As it is impracticable to draw water here for household use in pitchers or other portable vessels from the bath or basin cocks, the two self-closing hot and cold-water cocks D D have been set at the side of the washstand for the housemaids' use, and a similar arrangement is provided in the third-floor bathroom.

The sunken bath is surrounded by a special boiler-iron tank, Fig. 11, calked water-tight and supported on four special rolled iron I beams L L L L, Fig. 7, which are built into the tower walls, just above the dome of

special sleeve B, which screws down and tightly grips the bottom plate of the tank, making a tight joint by compressing the gasket D and packing E. The sleeve B receives the extra heavy neck of strainer N, which may be removed at will. Figure 13 is a view of the 1,000-gallon house tank on the fourth floor of the tower. It is built of 2-inch boards, and stands in a lead safe. It is filled through the 1¼-inch pipe J and ball cock B, which admits water from the street mains during the night when the pressure is sufficient to rise to this height, but provision has been made to convert J into the delivery pipe for a force pump, if necessary, when the cock B would be removed. A is a 1¼-inch overflow discharging into the fourth-floor slopsink. C is a ¾-inch telltale discharging, together with the safe waste, into the basement sink. D D are relief pipes from the highest points of the hot-water circulation systems, and are provided with a ball cock F, which is set 3 inches higher than B so as always to remain open unless the street pressure in the boiler should cause their water to rise above Z Z, and escaping through D D to fill the tank to nearly the height of telltale C, when it would close until some water was drawn.

PLUMBING IN MR. C. P. HUNTINGTON'S RESIDENCE.

(PUBLISHED IN 1894.)

THE house just being completed for Collis P. Huntington, Esq., at Fifty-seventh Street and Fifth Avenue, New York City, by Architect George B. Post, is a large and costly edifice designed to provide every modern requirement for the convenience of the family, guests, and numerous servants, and is equipped with a complete and elaborate plant for heating, ventilating, lighting, drainage, elevator service, and gas and water distribution. The sanitary arrangements are extensive and complete, conforming to standard advanced metropolitan practice, and embrace kitchen, laundry, and bath and toilet-room service, besides an extensive swimming pool and Turkish bath installation, and the necessary heating

meters, pumps, etc., and the approximate arrangement of the horizontal distribution pipes in the cellar. The cold-water distributing pipe A, Fig. 1. supplies 16 separate lines or groups of fixtures through branches three-fourths, 1. 1½, and 2 inches in diameter, which in general are carried along the cellar ceiling and are connected with vertical risers direct to the line of fixtures. Alongside each street pressure pipe is an equal sized tank water cold supply pipe and a hot water supply and a ¾-inch circulation pipe under both street and tank pressure, making a group of six parallel adjacent pipes on all lines, which are generally carried up in boxes in wall recesses which are afterwards covered with wire lath and inaccessibly plastered in. At the foot of each riser is a controlling valve and just above it an emptying valve and waste pipe.

Cellar Plan
Showing horizontal lines of pipes.

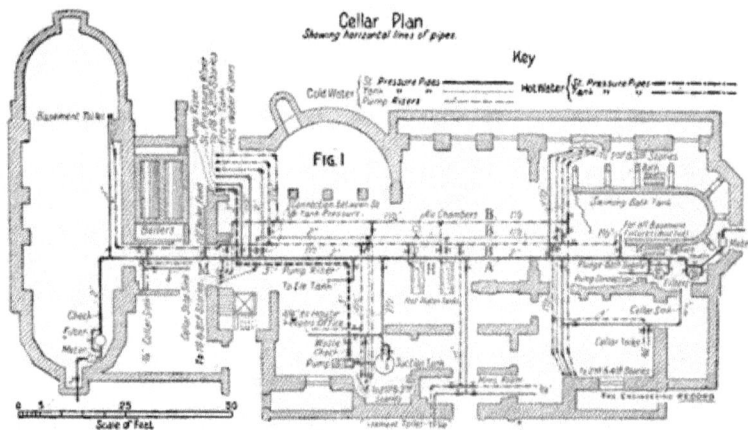

PLUMBING IN MR. C. P. HUNTINGTON'S RESIDENCE, NEW YORK CITY.

boilers, tanks, filters, and pump connections, which comprehend a more complicated and extensive system than is provided in some important public edifices. The contract for plumbing and gasfitting was awarded to Messrs. Rossman & Bracken, and partly from their contract drawings and largely from our notes and sketches made on the premises, we illustrate some of the principal features and details of the work.

Three-inch water supplies are taken from the street mains on Fifth Avenue and Fifty-seventh Street, and each is connected with a 2-inch Worthington meter with inlet and outlet valves. These meters are directly connected by a 2 inch pipe with check valves at each side so that water cannot escape from one meter through the other. The entire supply can thus be drawn from either main, or the supply for the swimming bath can be drawn from either main and the rest of the supply from the other main. Figure 1 is a plan showing the location of the boilers, filters,

Figure 2 is a conventional elevation of riser lines on one of the vertical sections of the house. Each vertical pipe terminates in an air chamber on top, and just below it, on the hot-water pipes, is branched off the circulation return pipe. Then at each floor served, branches from both hot pipes are connected to a cut-off valve as V, from which the distribution branch H is taken Similarly the cold-water tank and street pipes are connected to a cut-off valve W, from which the distribution branch C is taken. All the cut-off valves are set so that when their handles are turned to the right, as shown, tank pressure is on and street pressure is cut off. When they are turned to the left street pressure is on and tank pressure is cut off, and when they are turned half-way, so as to be at right angles to the wall, both street and tank pressures are cut off.

The main 1½-inch tank supply leads directly to the cellar, and there distributes through 12 ceiling 1 and 1½-inch branches. The main hot street and tank supplies are 1¼ inches diameter, and have direct connections to eight and to five lines respectively. Water supply is arranged for direct street pressure to the fourth-story bathrooms and all fixtures below and for tank pressure in the fifth and fourth stories. It is also so that the entire building, or any part of it, can be supplied with either street or tank pressure. The roof tank is filled by an Ericsson hot-air pumping engine with 12-inch cylinder, which draws from an open suction tank supplied from the meters. The air chamber is of galvanized-iron pipe, 6x36 inches, with a 1¼-inch stop and waste cock. Beside this the street and tank main supply pipes each have a vertical 16-inch cylindrical air chamber 5 feet high. The suction and discharge pipes are also connected with an auxiliary electric pump and a 2-inch cylinder double-action Douglas hand-power pump with 1¼-

third, fifth, and cellar stories, each with ¾-inch hot and cold-water supply. There are two iron drip and draw sinks in the cellar and three porcelain sinks, one each in the kitchen, scullery, and butler's pantry. There is one urinal in the first floor toilet-room. There are throughout the house 20 washbasins, all with ground top flange clamped to a marble slab with silver or nickel-plated cast-brass legs and frames and patent wastes.

In the second and third stories are six porcelain roll-rim decorated bathtubs on marble legs, and in other stories are three porcelain-lined roll-rim cast-iron bathtubs. In the boudoir bathroom is a decorated porcelain roll-top sitz bathtub with silver-plated fittings. The other tubs are plain with nickel-plated fittings and the bathrooms are finished with white marble and ceramic tiles. On the first floor is a

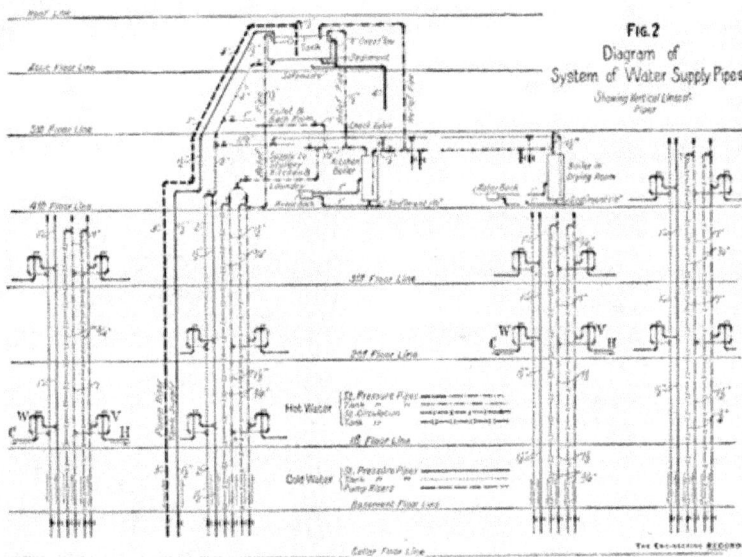

Fig. 2

Diagram of
System of Water Supply Pipes

Showing Vertical Lines of Pipes

PLUMBING IN MR. C. P. HUNTINGTON'S RESIDENCE, NEW YORK CITY.

inch suction, and discharge is provided for the purpose of emptying the contents of the boiler pit cesspool, which is below the sewer level.

The suction tank is 5 feet in diameter and 5 feet high with a ¼ inch iron top and is automatically supplied through a 1-inch pipe and ball cock and open valve, or at will through a 1½-inch pipe and valve. There is also an attic house tank, two hot-water boilers, a laundry and a kitchen boiler, and two swimming-bath boilers. There are 15 water-closets, situated one in the cellar, four in the basement, one in the first story, four in the second story, three in the third story, one in the fourth, and two in the fifth story. There is one slopsink in each of the second,

swimming bath, needle bath, shower bath, and two shampoos. In the laundry are four white porcelain washtrays.

All waste and soil pipe connections to the main lines are 2 inches diameter for washbasins and sinks, 3 inches for slopsinks, and 4 inches for water-closets. Vent pipes are 1½ inches for 1½-inch and 2-inch traps and grease traps, and 2 inches for 3-inch traps and water-closet traps. All 2-inch vent pipes are galvanized or rustless wrought iron; other sizes are cast iron. All brass pipes are extra heavy seamless tubing, tinned inside and outside, except where exposed, where they are polished and nickel-plated, and are subject to a one-year written guarantee. All

horizontal pipes are run in four and six-pound lead
safes which, where in contact with cement or concrete, are painted with two coats. All safe waste
pipes in the cellar discharge through horizontal ends
with hinged flap valves and engraved brass labels.

Waste and soil pipes have 3-foot lead pipe connections of the following weights: 1½-inch pipe, 3½
pounds per foot; 2 inch, four pounds; 3-inch, six
pounds; 4-inch, eight pounds. All lead waste and
soil pipes in contact with cement or concrete, also
have two coats of paint. The concealed traps to
bathtubs are of extra heavy lead with brass screws.
All others are of cast solid brass, nickel-plated, full
S, 1½-inch for washbasins, 2 and 3 inches for sinks
and washtrays, and 3 inches for slopsinks. Traps in
the cellar are cast-iron with brass screws. Tucker
No. 2 grease traps are provided in the kitchen and
scullery sinks. All exposed pipes connected with
brass and porcelain traps are of nickel-plated brass.
All vertical vent pipes are connected to soil or wastes
below the bottom fixtures so as to discharge into
them freely any water, rust, or other accumulation.
All cast-iron pipes and fittings have two coats of oil
paint after testing. All valves are of Ludlow make,
of steam metal, silver-plated where they are exposed to view above the cellar., All cocks are extra
heavy with steam metal ground keys. All cast and
galvanized pipes were tested with water pressure of
a maximum of 40 pounds.

The nine main vertical rainwater conductors
have a handhole and brass cap in the foot trap, and
are connected at the top to an extra heavy 3-foot
length of lead pipe flanged and soldered to the
copper gutter. All drains in area, yard court, and
driveway have trapped and grated cast-iron cesspools. In the refrigerator room are three and in the
butler's pantry is one polished nickel-plated cast-
brass floor pan with strainer discharging into a 2-
inch waste pipe from the refrigerator to an open sink.
In the vicinity of the swimming bath are 11 cast-
brass cesspools with bell traps discharging into two
2-inch waste pipes which empty into the cellar sink.
Under fixtures marble safes are provided with 1½-
inch galvanized-iron wastes to the cellar sink.

Figure 3 shows the construction and connections of
the attic tank, which is made of boiler-iron, stayed,
9'x5'x6' high, and sets in a four-pound lead safe.
The tank is furnished with floats that operate a
Bracken's patent electric high and low water alarm
(not here shown), which rings a bell near the pump
in the basement, where there is also a gauge indicating the height of water in the tank. The tank
can be emptied through a 1½-inch pipe and valve
A, and the house supply is drawn through a fine
copper strainer in the bottom of the tank and commanded by valve B. The top of the 4-inch overflow
terminates in a 4'x4'x5' T. opening upwards so as to
form a kind of funnel to receive the ends of the vent
and expansion pipes. They are so arranged in order
that if any but hot water is discharged through them
it may be wasted instead of mixing with the tank
cold water. The overflow pipe empties directly into
the roof gutter, and its waste is carried off to the
sewer through the rainwater leaders.

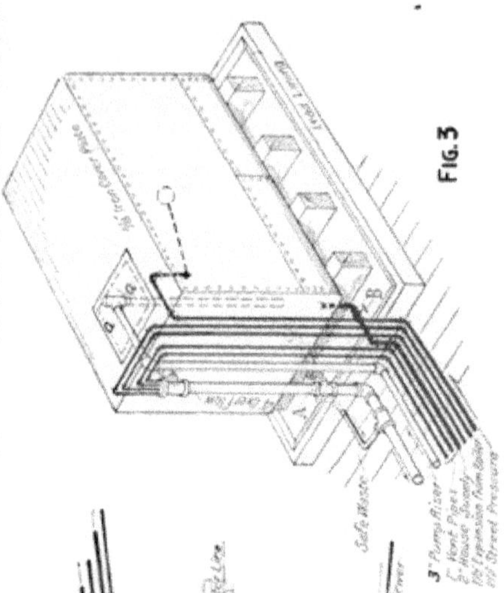

FIG. 3

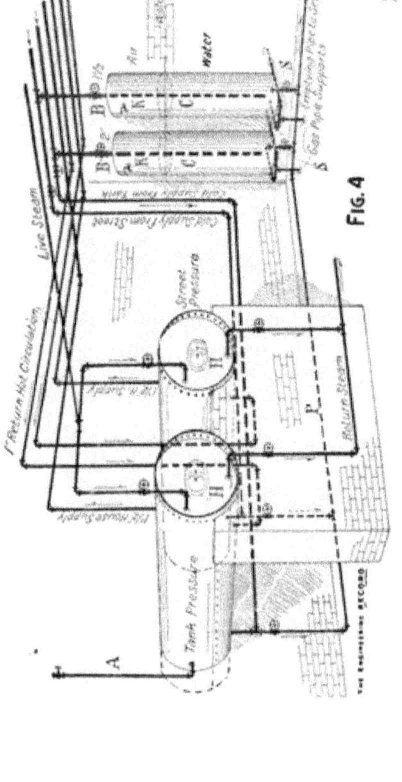

FIG. 4

Figure 4 shows the arrangement and connections of the hot-water heating boilers and air chambers in the cellar, the location of which is indicated at G, Fig. 1. The function of the pipes and the operation of the valves is in general clearly shown in the drawing. The boilers are about 2'6'x6' and hold about 150 gallons each. They are entirely separate and independent, one being connected to the street pressure and the other to the tank pressure system.

Each brass boiler contains a 50-foot coil of 1-inch brass steam pipe which is accessible through handhole H. The boilers are supported at the rear end by being built into the wall a few inches, and in front rest on a brick pier P. A is a special street pressure supply to the tank boiler.

As may be seen from Fig. 1, the main distribution pipes A and B connect all branches, and secure free communication to all fixtures, so that if several faucets should be quickly closed simultaneously, they might produce a cumulative water hammer. To absorb this impact Mr. R. Maynicke provided two large air chambers C C to furnish adequate elastic cushions which should automatically receive all shocks and prevent injurious hammer. These tanks are about 5 feet long and 16 inches in diameter, made of ¼-inch riveted steel plates, and communicate with the mains by vertical pipes, open at the foot and extending to 9 inches from the bottom of the chambers. When the water supply is admitted to the system it enters the bottom of these cylinders, and soon stealing up the ends of the pipes, compresses the contained air until its pressure is uniform with the water, and it occupies a proportionately reduced volume in the top of the chamber and expands and contracts in conformity with the variations of pressure, so as to prevent violent strains in the pipes. The petcocks K K are set a few inches below the tops of the chambers so that if when they are tried they show any air, the system will be efficient. If, however, the air becomes absorbed so that the water rises and escapes from the petcocks, the valves B B should be closed and S S and K K opened so that all the water will escape and be replaced by air at atmospheric pressure, which, when the valves are reversed to original positions, will again fill the upper parts of the chamber as shown.

The fixtures in the house are in sets in approximately vertical lines in the different stories, each group being served by a line of vertical risers from the main horizontal distribution pipes in the cellar, from which branches are taken as required at the different floors. In general each stack of risers consists of six water pipes, besides which the soil, vent, and safe waste pipes may be carried alongside or in a separate place. Figure 5 is a sketch of the foot of one of the typical groups of water pipes at M, Fig. 1, and shows the arrangement of valves for controlling and emptying them.

The kitchen work, cooking, laundry work, etc., are done on the upper floors, and tubs, ranges, refrigerators, etc., are accordingly set there. In the steward's room are two refrigerators, and for these and the tiled floor special trapped floor strainers are provided. They are set on waste pipes which have a combined

length of nearly 100 feet from their junction with the soil pipe, and although every angle is commanded by a handhole through the floor and a cleaning-out screw, a special arrangement was provided for flushing them under pressure and forcing out any obstructions that may find lodgment there. This is accomplished as shown in Fig. 6. The flow water, etc. is received in an 8-inch brass bowl B and flows through perforations in a hollow plug P into the elbow A, which is screw-connected to the waste pipe. If it is desired to flush out the waste pipes, the plug P is removed, and in its place a hose nozzle is screwed into the nipple N of one of the floor bowls B. Then at each of the other bowls B the plug P is screwed down till the shoulder S seats tightly and seals the nipple N so that no water can back up through it, and the required pressure may be safely applied through the hose connection.

Figure 7 shows the connection of laundry tubs to the laundry and kitchen boilers, which, though really remotely separated in different rooms, are here indicated close together for convenience. C is the cold-water supply from tank main, K is an 80-gallon boiler connected to the kitchen range water-back and supplying hot water for the kitchen and scullery sinks, etc., and L is the 80-gallon laundry boiler set about 40 feet from the laundry range and intended to supply the laundry tubs only on ordinary occasions. To avoid unnecessary radiation it is located inside the laundry drying-room (not shown here), the temperature of which is raised by a steam radiator considerably above that of the laundry room. Ordinarily valve D is closed and valve E is open, and the two boilers K and L operate entirely independently, but if it is desired to re-enforce boiler K, valves D and E are reversed and boiler K must be fed by water that has passed through boiler L and been warmed there. The washtrays are served by combination double supply cocks. The supply pipes are run on the opposite side of the partition so that they are accessible, and, as the tubs stand open in front of the tiled wainscoting, these cocks were made specially long in order to reach over the tub. The tail pieces were also made specially long in order to reach through the partition and connect with the supply pipes behind.

After the plumbing had been designed throughout the house it was thought desirable to provide a special convenient toilet-room for the engineman and fireman, and as no room was available, and as the soil and waste pipes could not be depressed below the floor level, a special platform was built 2 or 3 feet above the cellar floor, a bathtub, washbasin, and water-closet set upon it and side walls of light matched boards built inclosing it and reaching to the ceiling. Windows and doors were provided and the cabinet-work was nicely finished to present a neat and attractive appearance. All work was completely accessible and exposed and the appearance was clean and attractive. Figure 8 is an outside view showing the pipes, all carried outside, and Fig. 9 is a sketch of the interior.

In the basement, slightly below the ground level, there is a large suite of rooms, especially comprising

a complete Turkish bath establishment equipped with all the steam, bath, and attendant's appurtenances, and elaborately appointed in every detail. Figure 10 shows the plan and principal sections of these apartments, which are mainly finished in white marble and nickel-plated metal-work. The plunge bath or swimming bath is nearly 10x34 feet in size, and consists of an iron tank lined with marble and set in a cement mortar bed on a brick and iron floor and surrounded by buttressed brick walls. Figure

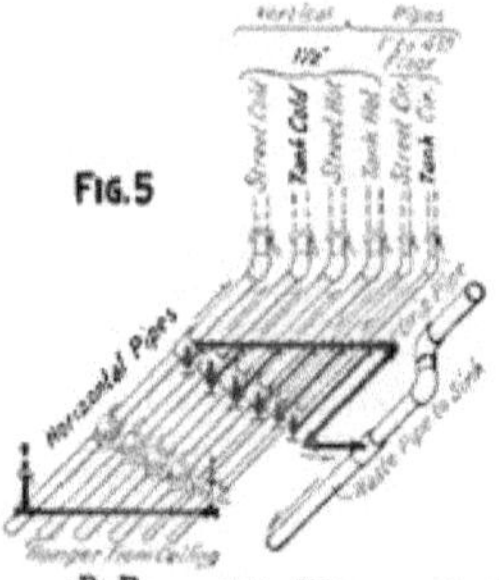

FIG. 5

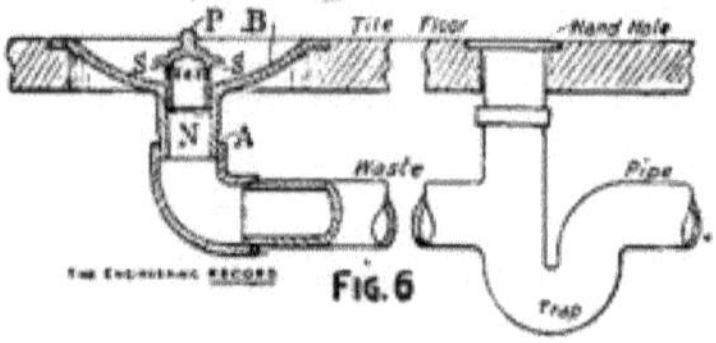

FIG. 6

Fig. 9

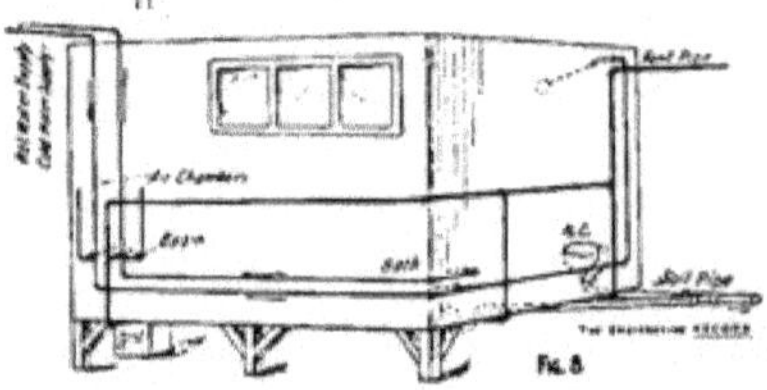

Fig. 8

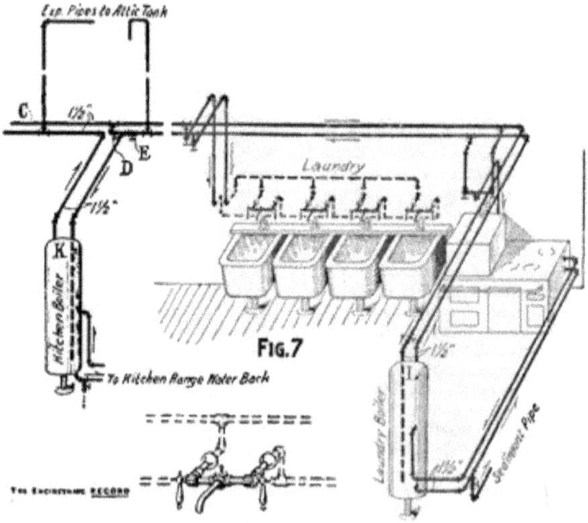

FIG. 7

PLUMBING IN MR. C. P. HUNTINGTON'S RESIDENCE, NEW YORK CITY.

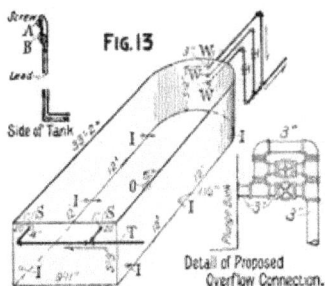

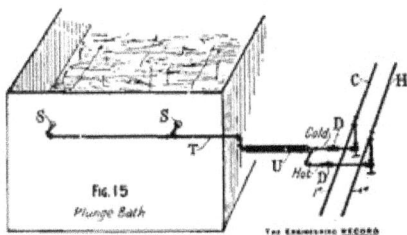

FIG. 13

Detail of Proposed
Overflow Connection.

Fig. 15
Plunge Bath

Section C-C.

Plunge Bath

FIG. 10
Basement Floor

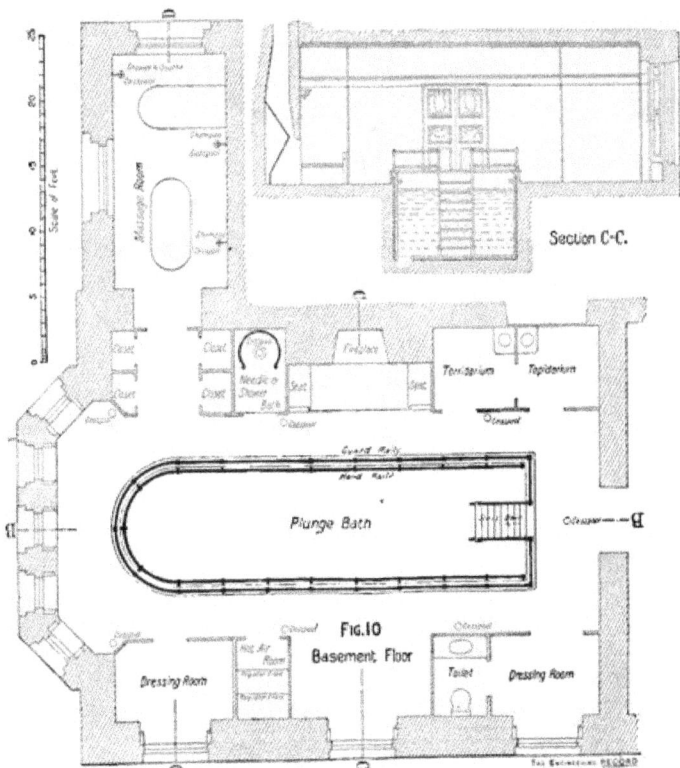

PLUMBING IN MR. C. P. HUNTINGTON'S RESIDENCE, NEW YORK CITY.

11 shows the masonry setting. The whole weight is
carried by the foundation walls W W, and the lateral
pressure, transmitted from the tank through occasional bearing bricks to the side walls, is received by
the buttresses and arches A A, etc.

Figure 12 is a partial section through the wall
showing the construction and water-proofing of the
sides and the attachment of the plated brass stairs
and railing. The horizontal bed having been prepared for it, the iron tank was set in cement mortar

upon it, and the outer walls, 12 inches thick, were
built around it, leaving a 1-inch air space everywhere
except where contact bricks projected to touch the
iron plates. The tank was painted with asphalt and
then five layers of asphalt paper were successively
applied over its entire inner surface and each well
drenched with hot asphalt. Then a lead lining was
secured to the sides and bottom and turned over the
upper edges, and the marble lining was set inside
this The bottom slabs were laid in a 1-inch bed of

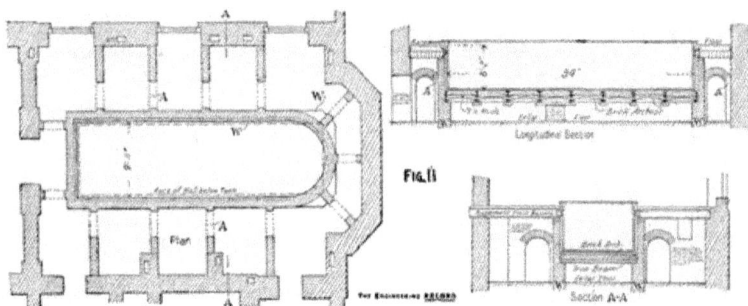

FIG. 11

Longitudinal Section

Plan

Section A-A

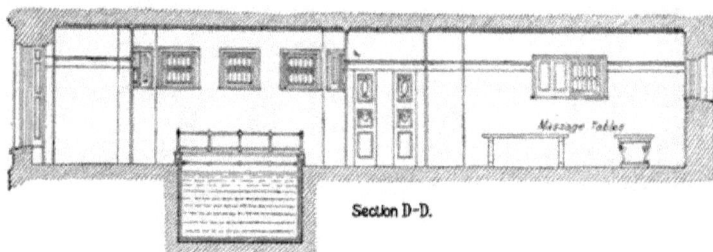

Section D-D.

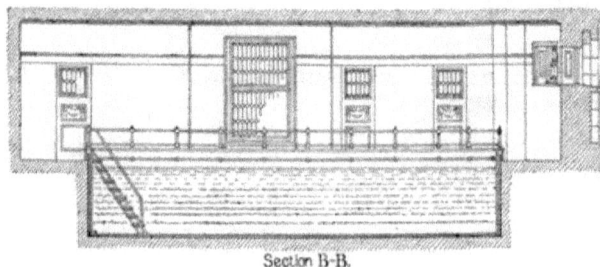

Section B-B.

PLUMBING IN MR. C. P. HUNTINGTON'S RESIDENCE, NEW YORK CITY.

FIG.16

5-Layers of Canvas

4 lb. Sheet Lead Union Tank

Flange

Galvanized Iron Pipe

Air Space

12" Brick Wall

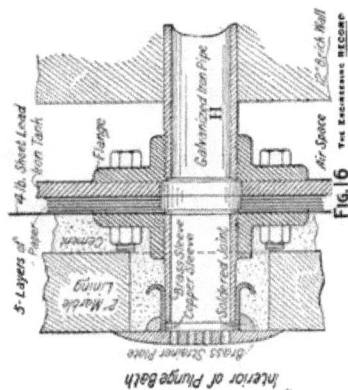

THE ENGINEERING RECORD

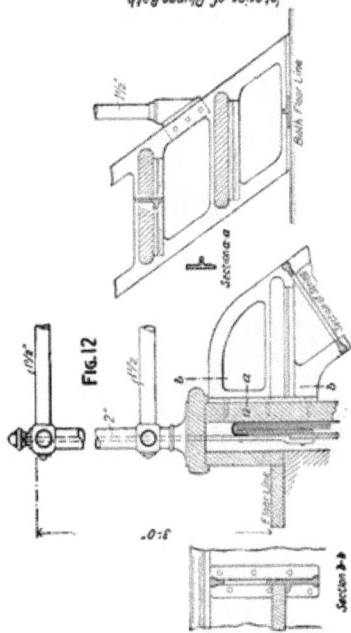

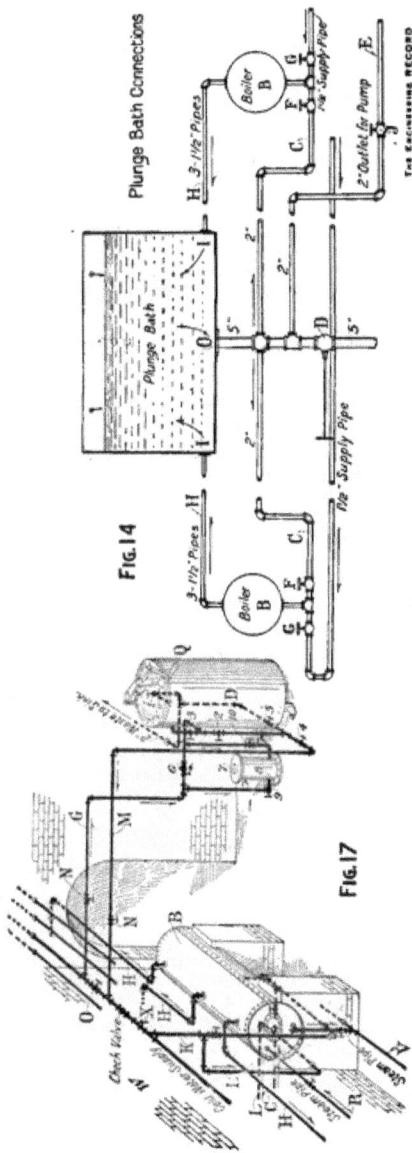

Brass Sleeve Copper Sleeve

Soldered Joint

2" Marble Lining

Cement

Brass Strainer Plate

Interior of Plunge Bath

Bath Floor Line

Section a a

FIG.12

1½"

2"

a

b

b

3'-0"

Section b b

Plunge Bath Connections

H. 3-1½" Pipes

Boiler B

G
G
F
C

1½" Supply Pipe

2" Outlet for Pump E

THE ENGINEERING RECORD

Plunge Bath

5"

O

D

5"

2"

2"

2"

2"

1½" Supply Pipe

3-1½" Pipes H

Boiler B

G
F
C

FIG.14

FIG.17

PLUMBING IN MR. C. P. HUNTINGTON'S RESIDENCE, NEW YORK CITY.

Check Valve

Steam Pipe

Steam Pipe

cement mortar. The vertical side slabs were then placed, and the 1-inch space between them and the sides of the tank were filled with thin liquid cement mortar which was poured in, in about three or four courses, each being allowed to set before the next one was made, thus avoiding excessive hydraulic head, which was further provided for by abundant temporary crossbraces.

Figure 13 is a diagram of the iron tank and shows the different pipe connections, comprising six 1½-inch inlet holes I I, etc., for the water from the heater to enter 6 inches from the bottom, two 1-inch holes S S for the surface spray to enter 4 inches below the top, one 5-inch bottom outlet O for emptying and circulation, and three 3-inch overflow holes W W W for the water to waste through, the upper hole 5 feet 2 inches above the bottom, and the other 6 and 12 inches below it. The overflow was originally intended to be controlled as shown in the detail, but as there was not sufficient space without interfering with the brickwork the valves were arranged as shown on the diagram. The valves are arranged so as to close either of the lower outlets, but never to close the upper one, which under all circumstances can carry off the water to a waste sink that is trapped into the sewer. In order to raise the sides of the tank above the basement floor, as shown in Fig. 10, a plate B was riveted to the side at the upper edge, and the lead lining being turned over on it, was clamped there tightly by the continuous iron bar A, screwed to the plate B.

A special steam boiler is provided on each side of the bath to keep the water heated, and acting virtually like a common range water-back, receives at the bottom cold water from the bottom of the tank, and warming it delivers it back from its upper part to a slightly higher level in the tank, thus keeping up a continual circulation.

Figure 14 is a cross-sectional diagram showing the arrangement and connections. Each pipe H is one of three connected to the inlet holes I I, etc., Fig. 13, and the two pipes C C receive the cold water circulating from the outlet O to the boilers. Ordinarily valves D, J, G G are closed and valves F F open, the circulation being continuous from boilers to tank, as indicated by the arrows, with a small amount of fresh water received and overflowing at the surface. By closing valves G G and opening D the water may be emptied into the sewer. It was originally intended to provide a special filter (E, Fig. 1) for this bath, which would receive its supply from pipe E and then discharge it through a pump into the boilers B B, thus using the same water over and over indefinitely with repeated filterings and a small addition of fresh water to compensate for waste, but this arrangement has been abandoned for the present and the entire contents of the tank are to be renewed from time to time instead.

Figure 15 shows a special device provided for skimming or flushing the top of the tank and removing automatically and continuously any scum or floating objects from the surface of the water where a large part of the impurities and foreign particles collect. At the upper end of the bath two nozzles S S deliver a fine spray of hot, cold, or tempered water above the high-water line, and so directed as to cover the surface of the water and wash all the top part gently towards the opposite end, where it slowly overflows through the waste pipe. This is intended to prevent the accumulation of scum and constantly removes the dirtiest portion of the water. The nozzles S S are supplied by a 1-inch pipe T, which is carried along the outside of the tank wall just below the cellar ceiling and delivers from a 1½-inch mixing chamber U, which is filled from the hot and cold pipes of the regular house system with check valves D D placed so as to prevent the possibility of water from main C backing up into main H, or *vice versa.*

Figure 16 is a partial vertical cross-section through the side of the tank at one of the inlets I, and shows the details of lining and the method of connecting all the attached pipes. Inside and outside flanges are tightly bolted upon the iron shell to receive the screw ends of the outside pipe and a heavy brass sleeve, over which latter is slipped a tight fitting copper sleeve which is soldered firmly to the strainer-plate and slightly flanged at the other end so as to hold securely in the cement mortar which was poured behind the marble after the pipes were set.

Figure 17 shows the connections of one of the two duplicate bath boilers B B, Fig. 1 and Fig. 14. It is placed close to the tank wall W, and contains a 100-foot coil of 2-inch brass pipe which is connected with the supply and return steam pipes R and V, and receiving water from the bottom of the tank through pipe C returns it, warmer, through the three distributing pipes H H H. Fresh cold water is delivered by pipe K and by branch L to the companion boiler on the opposite side of the tank. X is a connection left for the swimming bath special filter pump. D is the Jewell filter (see Fig. 1), which is set adjacent to the boiler and which filters all the house supply. It is connected by pipes M and G to the principal cold-water main A, Fig. 1, and when in operation has valves N N open and valve O closed, but by reversing these valves the filter is cut out; valve O serving as a by-pass. Q is the crank for operating the rotating mechanism when the filter is washed, and the designation of the different valves is as follows: 1. Washout valve. 2. Wash valve. 3. Inlet valve. 4. Pure water valve. 5. Rewash valve. 6. Back-pressure valve. 7. Filling cap. 8 Overflow valve. 9. Tank inlet valve, 10. Regulating valve. To wash the filter, open valves 1 and 2, close all other valves and turn the crank on top of the filter. To filter, close all valves except 3 and 4; valve 5 should be opened only about two minutes after washing to allow a little water to filter in the sewer.

The operation of coagulating attachment is as follows: To fill the tank close valves 9 and 10, open valve 8, remove cap 7, and fill full of crystal alum, allowing the displaced water to overflow at 8, then replace cap 7, close valve 8, open valve 9 full, and adjust valve 10 to deliver the required amount of solution. The back-pressure valve 6 should be weighted sufficient to cause a shunt current as indicated by the arrows.

SOME PLUMBING DETAILS IN THE RESIDENCE OF JOHN J. ASTOR.

(PUBLISHED IN 1895.)

AMONG the costly and magnificent residences which are being extended along the east side of Fifth Avenue opposite Central Park, New York City, to form practically a line of superb private hotels facing the lovely park landscape and comprising the most elegant modern city architecture and sumptuous and elaborate equipment that has perhaps ever been concentrated in a group of so many different houses, there have been several whose construction or installations have been carefully described in THE ENGINEERING RECORD, among them the residence of Cornelius Vanderbilt, C. P. Huntington, J. J. Astor, Mr. Brokaw, and Elbridge T. Gerry. As Mr. Astor's new residence on the corner of Sixty-fifth Street progressed towards completion a member of the RECORD'S staff made the notes and sketches from which the following description of some features of the plumbing has been prepared.

The house really consists of two separate and distinct establishments adjacent under one roof, with complete and independent equipments that, though generally of corresponding appearance and arrangement, are designed to be entirely separate in their operation and maintenance. The house, 125x150 feet, occupies the northeast corner of Sixty-fifth Street and Fifth Avenue, and has five floors. The lowest one or basement is devoted to servants' rooms, storerooms, kitchen, etc., the next to social purposes, and the upper ones to chambers, etc. All of these are shown, together with the location of bath and toilet rooms, closets, and washbowls, in the plan diagrams of Fig 1. The plumbing system comprises the supply and filtration of the water, its elevation to roof tanks and distribution throughout the house, the system of hot-water heater and supply, the waste and drain pipes, and the local and trap-vent system, together with the installation in the stable and carriage house adjoining. The cast-iron pipes are extra heavy and not coated with tar. At each joint in cast-iron pipe 12 ounces of lead is used to each inch of diameter of the pipe. All wrought-iron pipe used for the drainage system is thoroughly coated with asphaltum and all its joints are made with red lead. All branch lead soil, waste, and vent pipes, including bends, have the following weights per lineal foot: 1½ inches, three pounds eight ounces; 2 inches, four pounds; 3 inches, six pounds; 4 inches, eight pounds. All connections of lead with iron

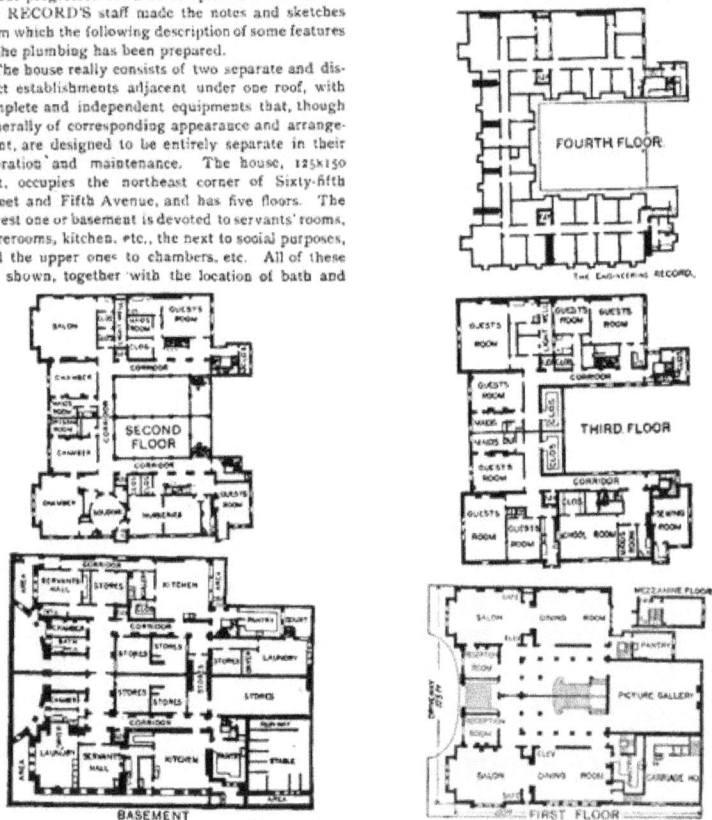

FIG. 1.—PLUMBING DETAILS IN THE RESIDENCE OF JOHN JACOB ASTOR, NEW YORK CITY.

pipes are made by heavy brass ferrules of the same size as the lead pipe, set in the hub of the branch of the iron pipe and calked in with lead. All the soil, waste, drain, and vent pipes and supply lines were subjected as soon as set to a hydraulic test of about 30 pounds pressure maximum, and after the fixtures were all set and the drainage system completed it was tested by the plumber with the "smoke test." The house drainage is discharged into the public sewers through three lines of 6-inch extra-heavy cast-iron pipe, run at a uniform grade of one-fourth inch per foot, trapped just inside of the area or vault walls and having fresh-air inlets 5 inches in diameter on the inlet side of the house traps extended up flush with the sidewalk near the street curb and covered by a cast-brass grating leaded into the flagstone. The cellars are not to be connected with the house drains, but are drained into water-tight brick cesspools and catch-basins 12'x12'x18' at different points in each yard, cellar, area, and light court. These cesspools are connected to 3-inch and 4-inch pipes trapped into the house drain.

Each line of water-closets and adjacent fixtures is connected by Y's and short lengths of iron pipes to a 5-inch wrought-iron soil pipe, connecting with the house drain by a Y branch and one-eighth or one-sixteenth bend and extending in full caliber 2 feet or more above the highest part of the roof or coping. Near light shafts or ventilating opening soil pipe is extended 5 feet above it. Three-inch wrought-iron waste pipes connect the basins, urinals, bathtub sinks, and washtubs with the soil-pipe lines, and are extended to the roof, above which they are increased to a diameter of 4 inches. The fixtures are connected with them by Y branches and short lengths of 1½-inch and 2-inch iron pipes. All branch soil and waste pipes have a fall of not less than one-fourth inch per foot. For all water-closets and adjacent fixtures there is a 3-inch wrought-iron vent pipe, connecting by short lengths of 2-inch brass pipe with the branch of each water-closet trap, and by 1½-inch brass pipes with the crowns of all other traps. The pipe is enlarged to 4 inches above the roof and has an inverted 2-inch Y branch in each story. The traps of all other fixtures are vented by 2 inch wrought-iron vent pipes connected by short lengths of iron pipe 1½ inches in diameter, with the crown of each trap. The main vent pipes extend above the roof separately, and are enlarged to 4 inches in the same manner as the soil pipes, or are connected with the waste pipe above the highest fixtures. There are 11 soil and waste and 11 vent pipes extending above the roof of the building.

Each water-closet bowl and waste pipe to urinals and slopsinks is vented, by a 2 inch local vent pipe made of 18-ounce sheet copper, with locked and soldered seams, and branched into vertical lines constructed in the same manner, the area in cross-section of which at all points is equal to the total area of the several branches counted therewith. The vertical lines are connected with the attic floor with round ventilating flues, as shown in Fig. 2, constructed of No. 24 galvanized sheet iron and terminating above the roof in a cowl. Each of these flues is provided with

a Blackman air propeller, operated by an electric motor specified to be of sufficient capacity to insure a current of at least 10 feet per second at the inlet of each branch. All vent pipes are graded so as to discharge water collected by condensation to a single point at the bottom, where it is connected with a drain, soil, or waste pipe. The inside diameter of traps is as follows: For water-closets, 4 inches; urinals, 2 inches; slopsinks, 3 inches, sinks, 2 inches; basins, 1½ inches; baths, 1½ inches; washtubs, 2 inches. Brass floorplates are used with the water-closet traps and the joints made permanently secure and gas-tight by means of bolts and red lead.

The enumeration of the total number of fixtures in the house shows their distribution as follows:

	Cellar.	Basement.	First Floor.	Second Floor	Third Floor	Fourth Floor
Water-closets............	..	4	2	6	6	3
Slop hoppers............	..	4	..	2	2	1
Urinals.................	..	..	1	..	..	..
Washbasins.............	..	4	..	6	6	3
Bathtubs...............	..	4	..	3	5	1
Washtubs...............	..	3	..	..	..	..
Sinks..................	1	2	..	..	..	..
Pantry sinks............	..	3	2	2	..	..
Pumps.................	2	..	..	..	..	..
Refrigerators...........	..	1	..	..	..	..

All water-closets are syphon-jet closets. In the boudoir bathroom on the second floor the bathtub is built to occupy the whole octagonal end of the room, and is constructed of a solid block of marble with a gracefully curved and molded front and a slightly projecting top, above which the wainscoting panels are made continuous for several feet in height, and contain in the center of the tub, at the back side, a large beautifully carved shell, in which a cupid is seated between two graceful dolphins, from whose mouths the hot and cold water supplies gush, being controlled by wheel-handled valves conveniently set in the wall at the end of the tub. In the corresponding bathroom in the other part of the house the porcelain tub has a silver needle-bath canopy frame with rubber curtains and a special silver-plated stool to set in the tub beneath the canopy. The servants' basement water-closets are plain with hardwood, round-cornered tanks with polished-brass brackets, hardwood seats, with polished-brass brackets and polished-brass flush pipes. The mezzanine toilet-room closets have nickel-plated trimmings. The second-floor closets in Mr. and Mrs. Astor's rooms, guests' rooms, and salon toilets have gold-lined closets and plated trimmings. The nursery toilet-closet has silver-plated trimmings. The maids' toilet-closet has nickel-plated trimmings. The closets in guests' and public toilets on third floor have silver-plated trimmings. The closets in the servants' toilet-room on the fourth floor have nickel-plated trimmings. The bathtubs in Mr. and Mrs. Astor's, salon, the guests', public, and the nursery toilets are porcelain roll-rim baths, glazed inside and out, set on marble feet, with trimmings plated to match the water-closets. The six servants' bathtubs are roll-rim enameled iron baths, with polished-brass trimmings in basement and nickel-plated trimmings on the fourth floor. Shower baths are provided in Mr. and Mrs. Astor's rooms and in the guests' toilet-room.

All washbasins are 19x15 inches with ground rims. The waste pipes are vertical to the floor and the trap vent branches are horizontal to the wall. The supply pipes are provided with finished valves and air chambers, and the slabs are supported on plated ornamental brass legs and apron holders. The basins

WATER SUPPLY IN THE HOUSE OF MR CORNELIUS VANDERBILT, NEW YORK CITY.

(PUBLISHED IN 1895.)

PART I.—FILTERS, SUCTION TANK, PUMPS, AND COLD-WATER DISTRIBUTION.

THE new residence of Cornelius Vanderbilt, Esq., Fifty-seventh Street and Fifth Avenue, New York City, is one of the largest and most costly private

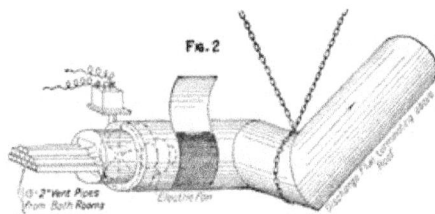

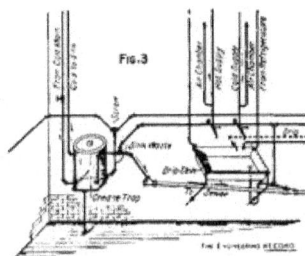

PLUMBING DETAILS IN THE RESIDENCE OF JOHN JACOB ASTOR, NEW YORK CITY.

in Mr. and Mrs. Astor's rooms, guests' salon, and public toilets are specially decorated. In the basement laundry the set of eight porcelain roll-rim wash-trays is fitted up with bronzed iron standards, marble backs, and nickel-plated fittings. These trays are divided into two sets separately trapped. The four 37½x24½-inch porcelain roll-rim sinks in the kitchen and scullery are set on galvanized-iron frames fastened to the wall, and have polished-brass legs. These sinks stand 4 inches clear of the finished wall, and waste through No. 2 polished-brass Tucker grease traps, and have all pipes and fittings of polished brass. These grease traps are connected up in the cellar below in the manner shown in Fig. 3. The two 25½x18½-inch porcelain roll-rim basement pantry sinks are fitted up in the same manner. The two German silver first-floor pantry sinks have German silver drain boards and backs, ½-inch brass hot and cold water pipes placed in the wall with air chambers and swing pantry cocks. They waste through No. 2 Tucker grease traps, with brass waste and vent connection, and all exposed brasswork is nickel-plated. There is in the butler's pantry a No. 6 nickel-plated "Perfection" combined filter and cooler with supply and waste connections. There are in the cellar four 16x24-inch enameled iron sinks trapped and connected with soil and vent pipes, and supplied with hot water through ½-inch brass pipe. These sinks are to receive the drip from safe and refrigerator wastes, and are all connected up essentially like the one shown in Fig. 3.

The plumbing above described was executed by John Tucker according to plans and requirements of the late Richard M. Hunt, architect.

city residences in the world, and the unlimited care and expense devoted to its construction and decoration have also governed its equipment with modern mechanical and sanitary apparatus which for power, heating, ventilating, illuminating, water supply, and drainage is of the most improved and complete nature. It is chiefly constructed under special supervision and specifications intended to secure above all the utmost superiority of workmanship and materials and efficient operation regardless of cost. The

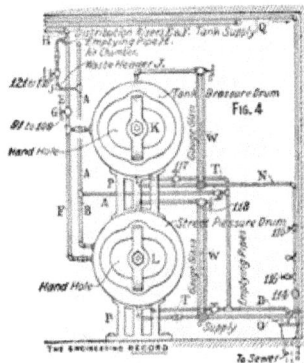

END VIEW OF COLD-WATER DRUM.

WATER SUPPLY IN THE HOUSE OF MR. CORNELIUS VANDERBILT, NEW YORK CITY.

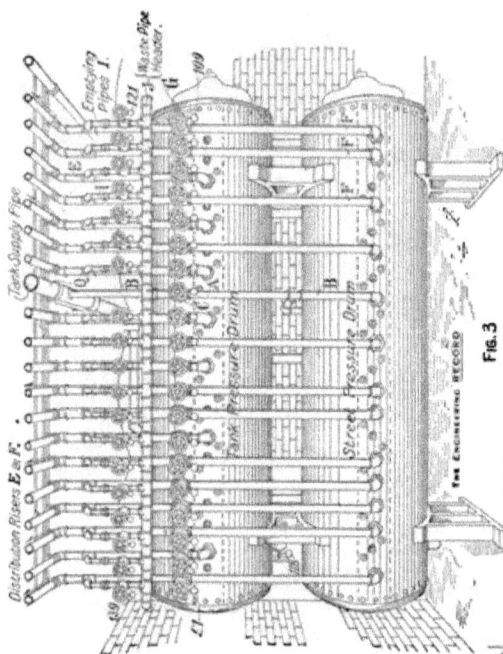

Emptying Pipes I.
S.121 (Master Pipe
Register.
Tank Supply Pipe
Distribution Risers E & F.
Tank Pressure Drum
Street Pressure Drum
B
E
F

Fig. 3

ELEVATION OF TANK AND STREET-PRESSURE CLEAN-WATER DISTRIBUTION SYSTEM

THE ENGINEERING RECORD

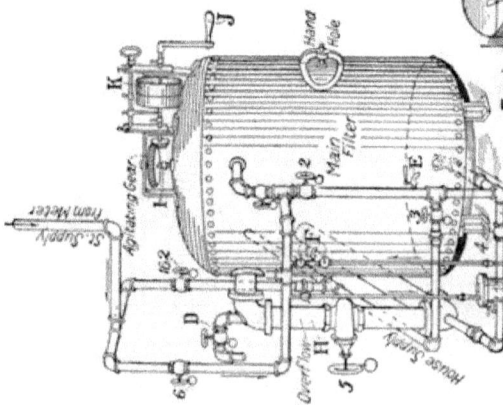

To Tank
Waste to Tank
Waste
Side Elevator
Oil Engine Pump
Electric Pump.
Supply Pipe
Basement Pipe
Suction Pipe
Suction Tank
Overflow Pipe
Sink
SUCTION TANK FOR PUMPS

Fig. 2

THE ENGINEERING RECORD

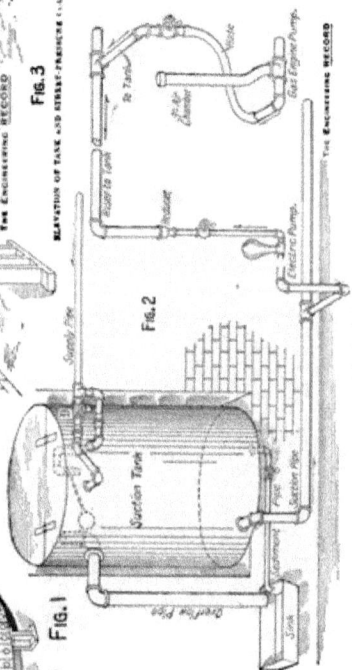

St. Supply
from Meter
Alternating Gear
Hand
Hole
Main
Filter
Wash Out
Tank
Alarm
House Supply
Rotating Supply
Overflow
J
K
I
D
H
G
E
F

Fig. 1

THE ENGINEERING RECORD

FILTER FOR ALL THE WATER SUPPLIED

WATER SUPPLY IN THE HOUSE OF MR CORNELIUS VANDERBILT, NEW YORK CITY.

plumbing system for the varied requirements of the extensive establishment includes complete service of hot and cold water, filtered, unfiltered, and double-filtered, under street and tank pressure, and the drainage, drip, and waste for the domestic establishment, besides the service for numerous general and private bath and toilet rooms and the servants' quarters. Water pipes are of galvanized iron or tinned brass, and waste and drainage pipes are of screwed galvanized iron up to 1½ inches, and of extra heavy cast iron for larger sizes. No lead pipes are used, and all are tested to a maximum water pressure of from 160 to 210 pounds. Gate valves are used throughout on all pipes above 1½ inches, and all hot-water lines have return circulation. Sets of complete standard apparatus are established in the kitchen and laundry, where steam is supplied from the boilers for the power and heating service and cooking and scullery uses.

There are several miles of water and drain and back-air pipes in the house, and all the main hot and cold water lines are interchangeable from tank to street pressure, are independent of each other and other parts of the system, and have separate branches with valves commanding each set of fixtures. The main or riser lines themselves are all commanded by sets of valves concentrated at one point in the cellar and controlled by the engineer, who has charge of a complicated labyrinth of pipes. This installation is more extensive and elaborate than the installations in many large hotels and public buildings, and presents many special and interesting features of construction, arrangement, and operation. The features and details shown have been sketched by a member of our staff, who received a general explanation of the work from Mr. George B. Post, of New York, the architect of the building, under whose requirements and superintendence the work was executed by James Muir & Co. The foreman in charge of the work for the Messrs. Muir explained the mechanical details and operation.

The water supply is taken from the city mains on both Fifty-seventh and Fifty-eighth Streets through two 2-inch connections, which deliver it through two Worthington meters in the cellar to the suction tank. From this point it is pumped to the attic tank, and to the steam boilers and the No. 33 Jewell's filter, Fig. 1, which delivers it eventually to the distribution drum, where all the house supplies are controlled. The filter is designed to purify 50,000 gallons of water per diem without appreciable diminution of pressure head, valves 1 3, and 6 being open and the others closed normally. It is intended to wash the filter daily, for which operation valves 1 and 3 are closed and 2 and 4 are open for a few seconds, when valve 4 is closed and 5 open, valve 2 being set so as to admit only enough water to liquefy the bed sufficiently to promote the revolution of the agitator. The water is forced upward through the screens and perforated diaphragms that confine the filtering materials until it overflows into the sewer through valve 5. This operation is to cause the separation of the grains of quartz, and by increasing the sizes of the interstices between them to detach the accu-

mulated particles of impurities and carry them upwards and outwards with the flowing water, while the heavier quartz, being continuously agitated, remains in semi suspension and scours itself by the rubbing of its particles together. The revolution of the agitator is intended to break up any films or lumps and to thoroughly wash the sides of the filter. About five minutes is required for the washing, which is continued until water runs clear from the try cock D. When the filter is clean it is quickly rewashed to remove the unfiltered water left therein. To do this valves 2 and 5 are closed and 1 and 4 are opened until the discharge from try-cock E is clear and bright. The machine is then set to filtering by simply closing valve 4 and opening valve 3.

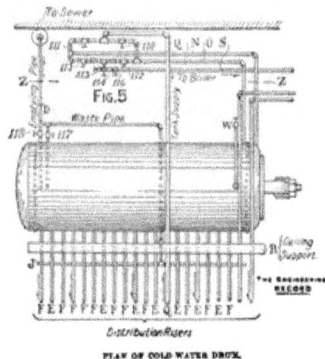

FIG. 5

PLAN OF COLD WATER DRUM.

In filtering, the water from the street mains is received on top of the quartz filtering bed, and percolates downward through its rugged interstices and the screens and pipes at the bottom. The filter requires from 10 to 15 minutes' attention daily, and can be thoroughly washed with less than 1 per cent. of the amount of water filtered. When a coagulant is used its cost varies from 1 cent to 10 cents per 10,000 gallons of water, according to its quality. The overflow and discharge pipes are 5 inches in diameter, the other pipes shown are all 2 inches except the ½-inch ones connecting the alum tank, which is controlled by valves F F. The shaft I extends vertically downward into the filter and carries, a little below the overflow outlet, a crosspiece from which a set of beveled rakes parallel to it extend about 2 feet down into the filter bed, and thoroughly cut it up and loosen it when revolved by the hand crank J. Tight and loose pulleys are also provided to drive it by a belt if it is wished to operate it by power. Just above the dished bottom piece an internal horizontal diaphragm-plate makes a false bottom, beneath which the water is collected as filtered, and upon which the filtering bed (about 2½ feet of two sizes of White's machine-crushed quartz, claimed to be 99 per cent. pure silicon) is supported. The diaphragm is perforated by numerous round holes, which are capped above with inverted conical aluminum bronze strainers that distribute the washing water in small jets in every direction, and prevent the passage of the quarts through the diaphragm.

Figure 2 shows the 4-foot suction tank, about 5 feet high, which automatically receives water under street pressure direct from the meters and filter through a 2-inch pipe and ball cock, although by opening valve D the tank may be independently filled by hand. The overflow and waste pipes discharge freely into an adjacent trapped sink, and the 3-inch pump suction pipe is connected to a Crocker-Wheeler Electric Company's one horse-power pump, which is driven by an attached motor with a speed of 1,050 revolutions per minute, and to a two horse-power Rider gas engine pump, the delivery of which is connected up with a section of rubber hose inserted just beyond the air chamber to diminish the transmission of noise, vibrations, etc., through the house by means of the riser pipes.

Figures 3, 4, 5, and 6 show the arrangement and construction of the cold-water distribution drums in

and prevent water hammer in the pipes. The amount of air in the drums is indicated by the gauge glasses, and if its volume becomes diminished it can be increased by shutting off the riser lines, emptying the drums and waste pipes, admitting air, and refilling them with water. All the valves are consecutively numbered and marked by attached brass labels, and the service thus commanded is recorded on a printed chart or key numbered to correspond and framed and hung up conveniently near. The pipes are symmetrically arranged in a regular and mechanical manner, and are so connected up with unions that any one can be taken off for alteration or repairs without interfering with the others.

PART II.—HOT-WATER SYSTEM, BOILERS, SUPPLY LINES, RETURN CIRCULATION, STEAM CONNECTIONS, AND AUTOMATIC REGULATING VALVE.

To secure an estimated maximum consumption of 500 gallons of hot water an hour for all culinary,

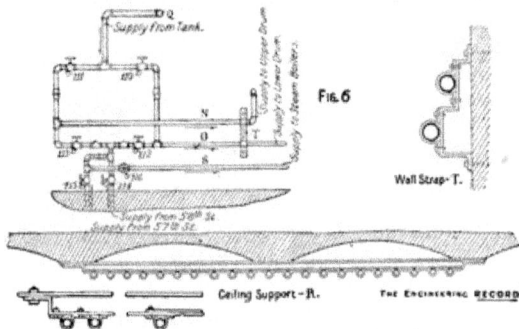

Fig. 6

WATER SUPPLY IN THE HOUSE OF MR. CORNELIUS VANDERBILT, NEW YORK CITY.

the cellar. Figure 3 is a front perspective, Fig. 4 is an end elevation from X X, Fig. 5 is a plan from Y Y, and Fig. 6 is an elevation from Z Z of the pipes on the wall. The drums are of ¼-inch galvanized steel with flanged ends and tested to 200 pounds per square inch. They are supported solidly on heavy cast-iron chairs and are about 21'x8' long, with a 6x1-inch longitudinal bar riveted on inside to provide reinforcement for the screwed pipe connections. The upper drum is supplied with filtered tank water and the lower one with filtered street-pressure water, each through 2-inch pipe, while the riser lines to different parts of the house above and below the second floor are respectively supplied from the branches E E, etc., and F F, etc., most of them 1¼ inches in diameter. Each line E or F has a ¼-inch emptying pipe H connecting it with a 1½-inch waste pipe J, through which it may be emptied into the sewer and the line left free for disconnection at any point by closing the main valve G and opening the small one I.

The upper portions of the drums are designed to be filled with air, forming a cushion to absorb shocks

toilet, and domestic purposes two boilers of ¼-inch galvanized steel with flanged heads were provided. They are about 36 inches in diameter by 6 feet long and are compactly and symmetrically arranged in a narrow space between two massive foundation piers in the cellar adjacent to the cold-water drums and near the post of the engineer, who controls their operation.

Figure 7 is a view, nearly in elevation, from a photograph, of the front end of the boilers. Figure 9 is a similar view of the rear ends of the boilers after the hot-water pipes and valves were in place, but before the steam connections had been made.

Figure 8 is a general isometric drawing of the boilers and piping as seen from the front.

Figure 10 is an isometric diagram of the rear after the connection of the steam pipes.

Figure 11 is a plan from above the boilers.

Figure 12 is a vertical section and elevation at Z Z, Fig. 11.

Figure 13 is a diagram of the connections for drip, waste, and return circulation to the underside of one boiler.

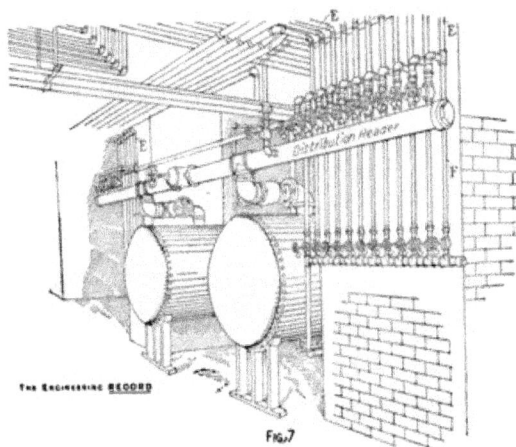

Fig. 7

FRONT OF HOT-WATER DOUBLE-PRESSURE BOILERS.

Figure 14 shows the steam connections only, omitting the water pipes shown in the preceding figures, and Fig. 15 shows the construction and details of the automatic steam regulating valve to control the amount of steam required to maintain the water at a given temperature under varying demands. Tank and street pressure cold water is delivered in 2-inch pipes so connected as to deliver either kind to either boiler without danger of backing into the other one. Water enters the bottoms of the boilers, and being heated by the interior steam coil is delivered from the tops through valved 5-inch pipes which connect

with a 5-inch horizontal header which has a valve in the center between the boilers to separate them when they are operated as is usual under different pressures. The header distributes the hot water to 1-inch risers E E, etc., that supply the different groups of fixtures throughout the house, and are each connected at their highest points with ¼-inch return-circulation pipes that enter corresponding 2½-inch headers J at the rear of the boilers. From these headers 2-inch pipes K K connect with the cold-water inlet so that the water that has been cooled in circuit enters with the fresh supply and is continually reheated. All the

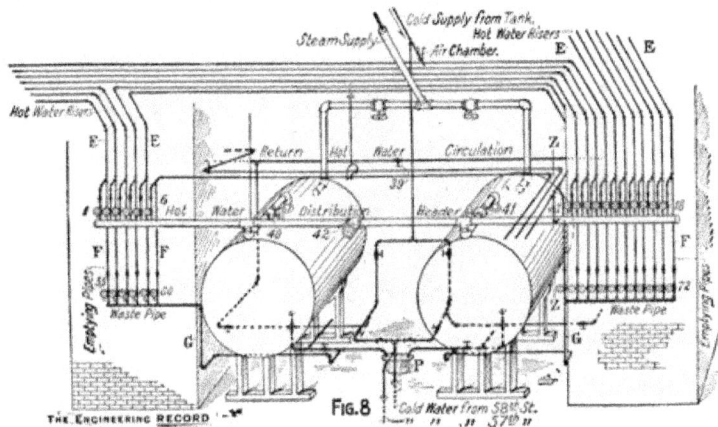

Fig. 8

HOT-WATER DELIVERY AND CONTROL OF RISER LINES FROM HOT-WATER BOILERS.

WATER SUPPLY IN THE HOUSE OF MR. CORNELIUS VANDERBILT, NEW YORK CITY.

risers are valved at the headers, and just above are tapped or bled by emptying pipes F F, etc., also valved, and wasted into an open bowl H, contents of which are trapped into a sewer pipe. This system is essentially the same as that for the cold-water drums, Figs. 3, 4, and 5, and similarly provides for the control or emptying of any line by the reversing of its two valves. As with the cold-water distribution drums and in all other places each valve is tagged and its number and corresponding service is printed on a key hung up near by. Some idea of the arrangement and extent of the system is given by the following copy of the hot-water board:

TANK PRESSURE, HOT WATER.

1 and 34, boys' bath, Fifth Avenue bath, third floor, and servants' bath Fifty-eight Street, fourth floor.

2 and 35, Mrs. Vanderbilt's bath. 3 and 36, Miss Vanderbilt's bath and Fifty-seventh Street bath, third floor.

4 and 32, west bathrooms, second, third, fourth, and fifth floors.

5 and 33, slopsinks, second, third, fourth, and fifth floors.

6 and 31, Miss Vanderbilt's bath.

STREET PRESSURE, HOT WATER.

7 and 30, Miss Vanderbilt's bath.

8 and 19, boys' bath and Fifth Avenue bath, third floor.

9 and 20, Mrs. Vanderbilt's bath.

10 and 21, Miss Vanderbilt's bath, Fifty-seventh Street, third floor bath, housekeepers' bath in basement, and library toilet-room.

11 and 22, gentlemen's toilet, men's cellar bathroom, and cellar sink.

12 and 23, ladies' toilet and musicians' toilet-room.

13 and 24, smoking-room toilet, laundry toilet, and cellar sink.

14 and 25, laundry trays.

15 and 26, west bathrooms, second and third floors, and basement slopsink.

16 and 27, kitchen sink, pastry-room, butler's pantry, and two cellar sinks, southwest.

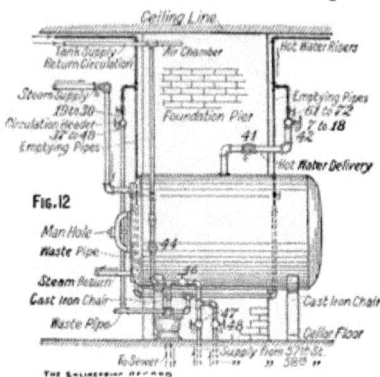

FIG. 12

ELEVATION OF HOT-WATER BOILER AT Z Z, FIG. 11.

17 and 28, second and third floor slopsinks.

18 and 29, office toilet-room, scullery sink, and basement sinks in waiter's and brush rooms.

37, on circulation distributing pipe of tank pressure boiler.

38, on circulation distributing pipe of street pressure boiler.

39, intermediate valve on distributing pipe connecting tank and street pressure boilers.

40, on tank boiler distribution pipe.

41, on street boiler distribution pipe.

42, intermediate valve on distributing pipe connecting street and tank boilers.

43, tank supply to tank pressure boiler.

44, tank supply to street pressure boiler.

45, street supply to tank pressure boiler.

46, street supply to street pressure boiler.

47, supply to boilers from Fifty-seventh Street.

48, supply to boilers from Fifty-eighth Street.

49, to empty tank pressure boiler.

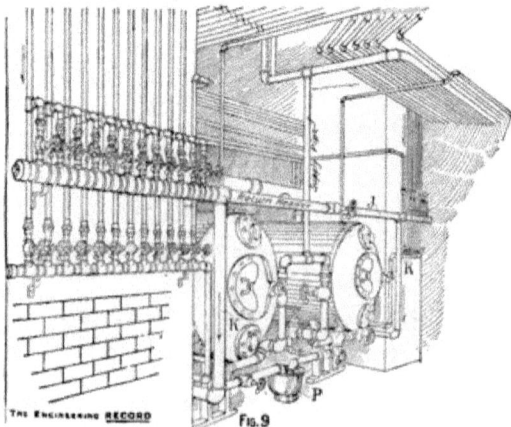

FIG. 9

REAR OF HOT-WATER DOUBLE-PRESSURE BOILERS.

WATER SUPPLY IN THE HOUSE OF MR. CORNELIUS VANDERBILT, NEW YORK CITY.

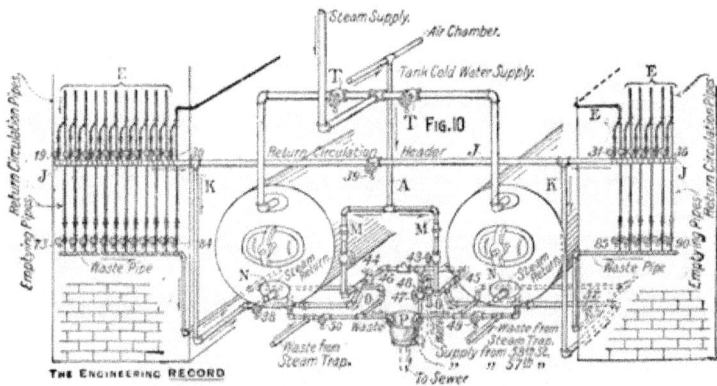

RETURN-CIRCULATION CONNECTIONS TO HOT-WATER BOILERS.

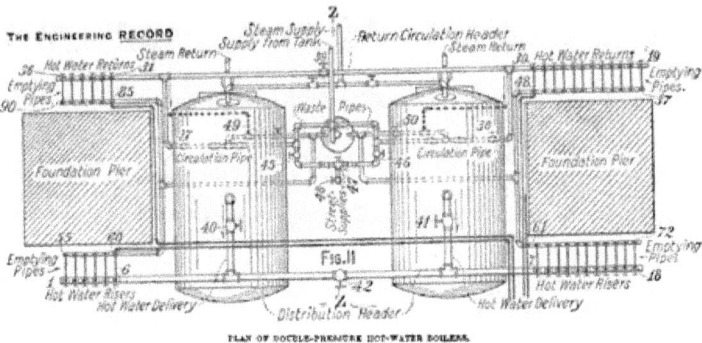

PLAN OF DOUBLE-PRESSURE HOT-WATER BOILERS.

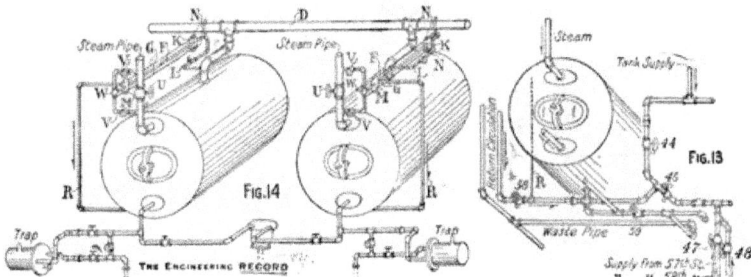

AUTOMATIC STEAM CONTROL FOR HOT-WATER BOILERS.   PIPES BENEATH HOT-WATER BOILERS.

WATER SUPPLY IN THE HOUSE OF MR. CORNELIUS VANDERBILT, NEW YORK CITY.

50, to empty street pressure boiler.

51 and 52, supply to steam regulator on tank boiler.

53 and 54, supply to steam regulator on street boiler.

55 to 90 inclusive are emptying valves.

Although it is intended to use one boiler exclusively under street pressure for the lower-floor service, and the other one under tank pressure for the upper-floor service, they are arranged so as to be independent and interchangeable and either or both can be operated from either street or tank supply. Ordinarily, however, the left-hand boiler, Fig. 10, is used for street and the right-hand one is used for tank pressure, and valves 44, 45, 42, and 39, and all emptying and waste valves are closed and all the other water valves are open. Closing valves 46 and 43 and opening 45 and 44 would admit street pressure to the tank boiler and tank pressure to the street boiler, and opening valves 39 and 42 would equalize the pressure between them and connect each boiler with all the hot-water lines, while closing 38, 41, 44, and 46 would cut out the street pressure boiler and allow it to be emptied for cleaning and repairs. By opening valves 39 and 42 in the delivery and return headers all the pipe lines would be served by the tank pressure boiler. Similarly, by closing valves 37, 40, 43, and 45 the tank pressure boiler would be cut out.

The steam pipes supply steam at about 40 pounds pressure to a 60-foot coil of 2-inch brass pipe in each boiler, which is estimated to be capable of heating 300 gallons of water per hour up to 200 degrees. The Kieley traps, drip, and return, etc. are arranged in the usual manner and each boiler is independently supplied through valves T T, Fig. 14. Between valve T and the coil each boiler has a throttle valve U and two by-pass valves V V, so as to permit the

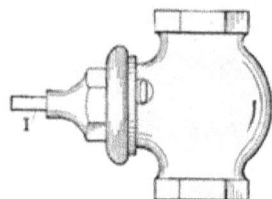

Fig. 15

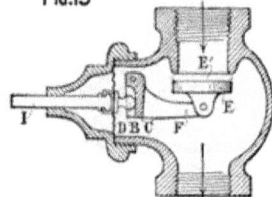

AUTOMATIC SPECIAL STEAM VALVE.

steam supply to be automatically proportioned to the amount of cold water heated. To effect this U is closed and V V are opened, admitting steam through regulating valve W, but by reversing valves V V and U the automatic arrangement is cut out and a full head of steam under boiler pressure is constantly freely admitted.

The pipe F receives the hottest water from the top of the boiler and returns it in a continuous stream through pipe N to the bottom of the boiler, thus always maintaining itself at the maximum temperature of the water in the boiler. The variations in

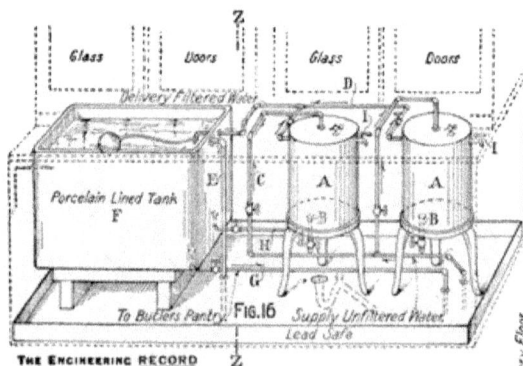

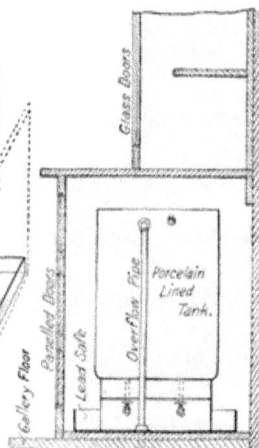

WATER SUPPLY IN THE RESIDENCE OF MR. CORNELIUS VANDERBILT, NEW YORK CITY.

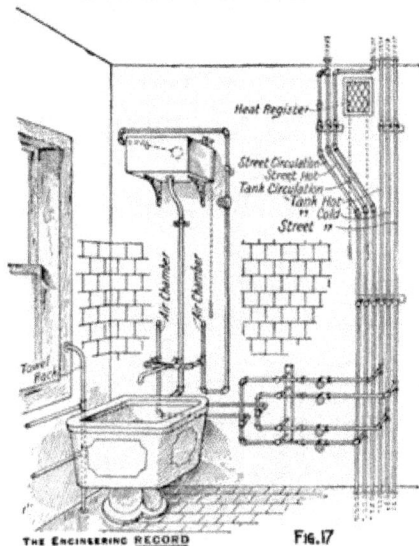

Heat Register

Street Circulation
Street Hot
Tank Circulation
Tank Hot
" Cold
Street "

Air Chamber

Air Chamber

Towel
Rack

THE ENGINEERING RECORD   FIG. 17

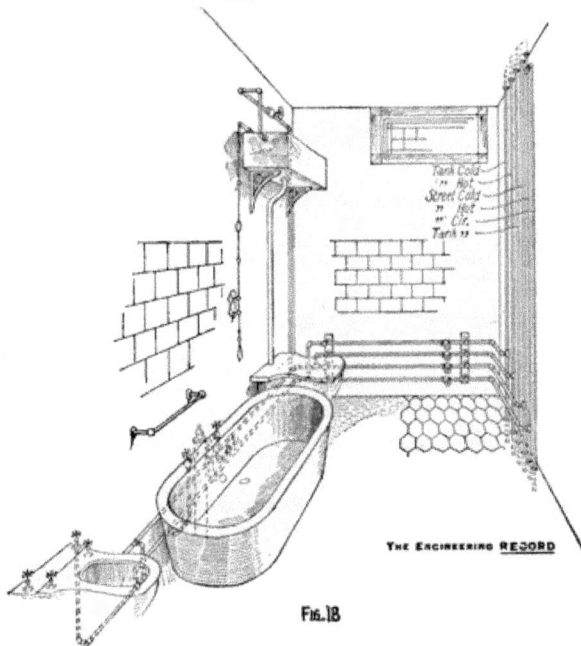

Tank Cold
" Hot
Street Cold
" Hot
" Cir.
Tank "

THE ENGINEERING RECORD

FIG. 18

WATER SUPPLY IN THE HOUSE OF MR. CORNELIUS VANDERBILT, NEW YORK CITY.

this temperature produce small but perceptible changes in the length of the upper horizontal portion of the pipe F (about 4 feet long), and as one end K is relatively fixed the other end that is supported by a loose head M sliding on guide rods G G vibrates longitudinally and actuates the valve stem I. By means of a multiplying device this stem opens and closes the valve W. Screwing up the adjustment nuts N N one revolution each shortens the distance to valve W by an amount equal to the contraction of the pipe F between N and W produced by a fall of temperature of 5° Fahr. and causes a slight motion in the elbow joints L L, which permits the pipe F to move slightly in a longitudinal direction towards valve W. It may thus be made to operate valve W at any required degree of temperature. The instrument is set to open and close the valve at 170° and 175° Fahr. respectively.

Figure 15 shows the details and operation of valve W. Its stem I is fastened to pipe F or head M, Fig. 14, and moves with it so that its expansion by increasing temperature pushes corrugated diaphragm D to the right and makes shoulder B engage the short arm C of the lever L, whose long arm F throws the valve E into its closed position E'. This regulator was invented and manufactured by Timothy Kieley, New York City.

PART III.—REFILTERING SYSTEM IN BUTLER'S PANTRY, TYPICAL ARRANGEMENTS IN SLOPSINK CLOSETS AND BATHROOMS.

THE cold-water supply for the butler's pantry is received under street pressure from the cellar filter, and is refiltered through Pasteur filters conveniently placed in a cupboard under a dresser in the balcony or gallery of the butler's pantry. Figure 16 shows the arrangement of the two separate and independent filters, each about 15x18 inches high, and having a rated capacity of 50 gallons per hour. They are supplied through a ¾ inch pipe, and deliver through a ball cock into a rectangular porcelain-lined tank about 36x18 inches and 24 inches high, with a 1¼-inch overflow pipe emptying into a trapped sink that also receives the discharge from the drip pipe of the lead safe. The dotted lines in the illustration indicate the position of the inclosing cabinet-work which forms below a table and cupboard with sliding doors and top, and is a glass case above, thus inclosing and protecting the filters and tank, while leaving them perfectly accessible.

There are numerous washbowls, sinks, toilet, bath, and dressing rooms throughout the house that are very completely and carefully fitted up. The fixtures designed for the use of the family and guests are remarkably elegant and costly. Those in the boudoir bathrooms are luxurious, with specially designed rich metal-work and large carved bathtubs hollowed out of solid blocks of marble, but there are no remarkable features in the mechanical details or in the arrangement and system of piping and connections.

The general methods of arrangement and connection and exposed valves and piping are shown in Figs. 17 and 18, which are typical of the distribution of street and tank pressure, hot and cold water, position of fixtures, etc. Figure 17 shows the interior of the maid's closet and slopsink on the third floor. Valves V V V V are introduced instead of a cut-off to utilize street or tank pressure at will. Offsets are made by one-quarter and one-eighth L fittings. All metal-work is heavily plated. The sink is porcelain, and the floors and walls are covered with marble or white ceramic tiles.

Figure 18 shows the piping in the third-story family bathroom, the work being similar to that shown in Fig. 17.

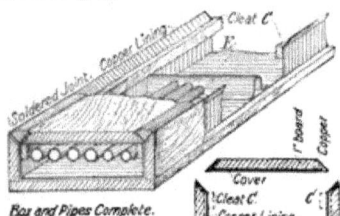

FIG. 19

Box and Pipes Complete.

Cross Sectional Box only.

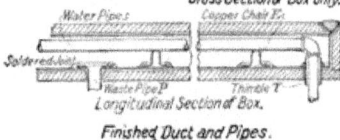

Longitudinal Section of Box.

Finished Duct and Pipes.

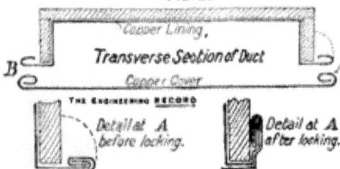

FIG. 20

Transverse Section of Duct.

Detail at A before locking.   Detail at A after locking.

Figures 19 and 20 show the methods adopted for the protection of walls and ceilings from any possible leakage in the pipe lines, which, in the remote contingency of its occurrence, would be caught in the sealed boxes and discharged by their waste pipes through numbered flap valves into cellar sinks, where the drip would be immediately evident, and would be readily identified so as to indicate the line requiring repairs.

All horizontal water pipes are run beneath the floors and over the ceilings in hermetically sealed boxes as shown in Fig. 19, about 6 inches deep and 18 inches wide, made of 1-inch boards. From the boards simple troughs were first constructed as shown at B, with beveled upper edges. Over these the 12-

ounce copper lining was folded and nailed. The joints at the ends of each section were locked and soldered, and the ends soldered tightly to make a water-tight receptacle. Then at intervals transverse copper ridges or chairs E E, etc. were soldered to the bottom so as to form supports, elevating the pipes about 3 inches from the bottom. Where the pipes enter the box, vertical sleeves or thimbles T pierce the bottom of the box (to which they are soldered tightly), and extended up to the tops of the chairs so as to allow the pipes to enter freely and expand and contract without any danger of leakage, even if the box should become filled with water nearly 3 inches deep. Each box, however, is provided with a waste pipe P, through which any leakage would be carried directly to a slopsink in the cellar without any opportunity to collect in the boxes. After the pipes are laid in them the boxes have cleats C C nailed along each side, and the beveled cover piece, also lined on the bottom and ridges with 12-ounce copper nailed on, is supported on the cleats, and the V-shaped space between the beveled edges is filled with a slightly rounded wiped joint that completes the sealing of that portion of the pipes. The vertical sections of them are run through somewhat different vertical ducts, as shown in Fig. 20. These ducts, having to carry 3-inch pump and tank risers, also are larger than the horizontal ones, and are about 30 inches wide and 6 or 7 inches deep. They consist of a wooden trough nailed vertically to the wall and lined with 12-ounce sheet copper, bent at the edges as shown in the transverse detail, and provided with corresponding cover sheets. The pipes are run in the open box and secured by ordinary jointed saddle hangers H H, etc., which were screwed

into the wood and white-leaded and tightly drawn up to the copper lining. Then the copper pieces, in convenient sections, slipped into place, their bends engaging and locking with those of the lining at A and B, and turned up against the outside of the duct and finally flattened tightly there, but not soldered. The different sections of the cover sheets were soldered together, and the bottoms of the ducts sealed and provided with waste pipes to the drip sink. The pipes are thus left practically inaccessible and most carefully provided against possibility of doing damage by leaking.

A SPECIAL ROOF TANK.

(PUBLISHED IN 1889.)

In a recent search for plumbing details, a member of our staff sketched the construction shown in Figs. 1 and 2 of a roof tank in a new residence on West End Avenue, New York.

Figure 1 is a general perspective. A is a 1-inch pipe affording an independent supply to a slopsink on third floor. B and C are ¾-inch relief pipes from the hot-water supply and heating systems. D is the strainer-plate covering safe waste pipe.

Figure 2 shows the tank broken at Z Z, Fig. 1, and shows pipes and details not visible in Fig. 1. A is the safe with lead cover B. C is a relief pipe. D is a 1¼-inch pipe supplying tank water to the house; it is connected with the city pressure, which, when sufficient, forces street water into the tank through branch E and ball cock G. F is a check valve that closes towards the right and thus prevents street water from overflowing the tank when ball cock G is closed. As the street water is likely to rise above

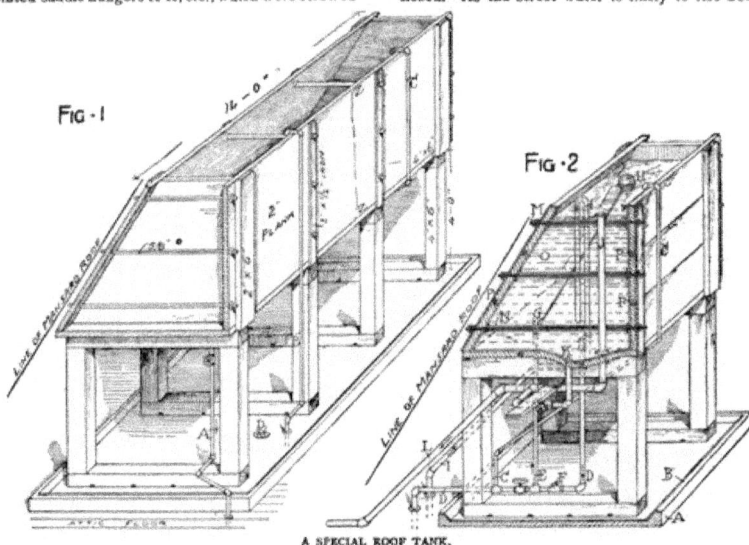

FIG. 1

FIG. 2

A SPECIAL ROOF TANK.

the bottom of the tank sooner and oftener than it would rise to the top, the ball cock G is set very low and its float H works on a long arm so as to operate whenever possible. I is the 1½-inch force pipe from the pump. J is the overflow and K the emptying pipe, both discharging through the 2-inch pipe L to the roof gutter. The tank is reinforced by twelve ⅝-inch tie-rods (see Fig. 1) M; the six inside rods are jacketed by 1-inch copper pipes N, that are soldered to the copper tank lining O. P P, etc., are 2x1-inch tongues set with white lead in the grooves of the tank boards at all joints.

The plumbing in this house was done by Byrne & Tucker, of New York, who were at that time completing the work in an apartment-house on Eighty-fourth Street, where the hot-water pipes from the kitchen boilers to the bathrooms conspicuously crossed the dining-room walls. All the plumbing in this building was exposed, and as the position of these pipes could not be changed they were bronzed, supported on frequent brackets, and served for tasteful picture rods.

All distribution pipes were pitched up a little continuously all the way from the main pipe, and so arranged that when the water supply was turned off from any apartment all the pipes in its system could be emptied by one drip cock, thus avoiding the danger of freezing when the place chanced to be untenanted.

KITCHEN BOILER ARRANGEMENT IN A NEW YORK RESIDENCE.

(PUBLISHED IN 1891.)

THE accompanying illustration shows a kitchen water heater as fitted up in the residence of W. C. Andrews, 854 Fifth Avenue, New York.

The upper part of the illustration represents the boiler in the basement, while the lower part shows the water heater in the cellar. W is a hollow copper cylinder about 6 inches long, the end which is inserted into the boiler being closed up so as to make it water-tight. The end projecting from the boiler is open so as to admit a thermostat.

When the heater above is not required the valves B E and F are closed and the valves S and H are opened, and it will heat from the range. To use both heater and range open valves B E and F and close valves S H.

A designates the Croton supply to kitchen sink and butler's pantry sink; B, tank supply to hot-water heater in cellar; the valve on this pipe is to be shut when hot water is not in use, and to be opened when in use; C, intermediate pipe; T, check valve; D, Croton supply to boiler; E, inside boiler or tank circulation, to be kept open when heater is in use, and to be kept closed when not in use; F, connecting pipe for boiler and heater, to be kept open when heater is in use, and to be closed when heater is not in use; G emptying pipe for outside boiler; H, inside boiler or tank circulation, to be kept closed when heater is in use, to be kept open when heater is not in use; I, emptying pipe for inside boiler; J, Croton circulation for inside boiler; K, tank or inside circulation; L,

hot water from outside boiler to supply lower floor. M, hot water from inside boiler of tank, to supply upper floor; N, tank supply to inside boiler, or to heater in cellar; O, hot-water supply to kitchen sink; P, Croton supply to kitchen sink; Q, hot-water supply to butler's pantry sink; R, Croton supply to but-

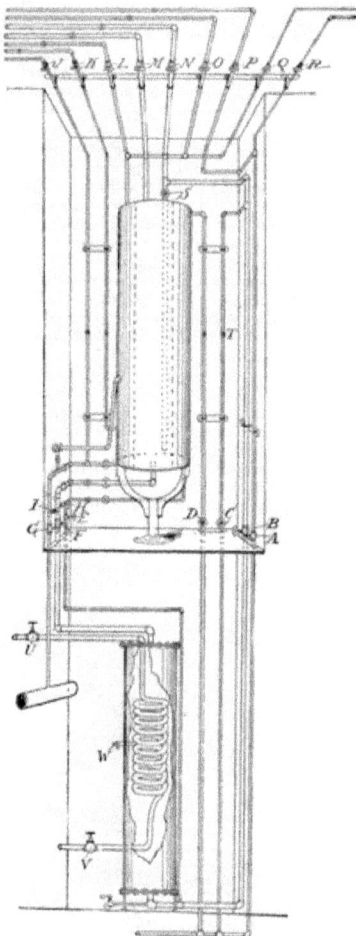

KITCHEN BOILER IN A NEW YORK CITY RESIDENCE.

ler's pantry sink; S, tank supply to the inside or high-pressure boiler, to be kept closed when hot-water heater is in use, and to be kept open when heater is not in use; U, steam supply to boiler; V, return from boiler; W, thermostat.

In this case the boiler in the basement is in a small room adjoining the kitchen. The walls are lined with Italian marble. The boiler and fittings are all nickel-plated; the pipes are of tinned brass. The pipes are hung about 4 inches from ceiling on nickel-plated plates and hangers, and where pipes go through the wall, floor or ceiling there is a nickel-plated flange, all in one piece, through which the pipes are put.

We have here a self-acting water heater (connected with the kitchen boiler), heated from steam, and regulated by electricity, and the temperature of water is said to never vary beyond 3 degrees. It is thus claimed to be a great improvement on the old hot-water heater, heated by a coal fire, and requiring constant attendance to maintain a proper temperature, besides entailing dust and danger from a fire in the cellar.

The reason that this heater was put in was that after taking one or two baths, the hot water in the boiler was used up, and consequently considerable time would elapse before another supply was available. In this case, however, the thermostat is so set that while drawing water for a bath, the cold water, which has displaced the hot water, is heated up again in a comparatively short time, because as soon as the temperature of the water drops 3 degrees the thermostat opens valves, admitting steam to the interior coil shown.

Mr. W. C. Andrews has had one in his residence at 2 East Sixty-seventh Street, New York, for two years, where it is said to have given every satisfaction.

Messrs. Lamb & Rich, of New York, were the architects, and the arrangement described was put in by Mr. George A. Pace, of New York.

PLUMBING IN A NEW ENGLAND RESIDENCE.

(PUBLISHED IN 1896.)

IN designing and executing the plumbing of the new house of Mrs. Edward Perkins, of Hartford, Conn., it was intended to make the work as complete, simple, efficient, and durable as possible, and to provide very carefully for its operation and maintenance. The accompanying illustrations and description are intended to show the general plan, methods, and workmanship adopted for this purpose. The number and location of fixtures, their style and arrangement, and the storage distribution and control of the hot and cold water supply were not unusual, but the principal features of the waste and vent systems and the materials, proportions, and workmanship of the whole installation are shown by the following drawings and description, prepared from the original designs.

Figure 1 is a basement plan showing connections of soil and waste pipes to the main drain, and indicating by dotted lines those pipes which are on the ceiling or are buried, those on the cellar floor being shown in full. The different cleanouts, traps, and air pipes, leaders and tile drains are also noted on the drawing to show the thorough provision for the flow and ventilation, and for access to all parts of the system to remove obstructions if necessary.

Figure 2 is a small-scale plan of the second story, merely intended to show the position of the fixtures, which in general were served by adjacent stacks of vertical pipes, not here shown.

Figure 3 is a partial vertical section at Z Z, Figs. 1 and 2, showing an elevation of the main stack of soil pipes and most of the fixtures.

From the street sewer a 6-inch salt-glazed Akron hub and spigot earthenware drain is laid, with continuous grade of not less than 1 foot in 10 feet, and jointed with Alsen's neat Portland cement, each joint being thoroughly swabbed out. All pipe lengths bear throughout on the body of pipe, pockets in trench being cut for hubs. No stones or rock were allowed to come within 12 inches of the pipe. The backfilling was well rammed for each 12 inches in depth filled. At the property line a running trap was set, and a cleanout and 5-inch iron fresh-air inlet and sewer vent, one on each side of the trap, were carried 18 inches above ground, and finished with one-quarter bends and brass-bar gratings screwed into the bends. Four-inch earthenware branches were connected for the rainwater leaders. Outside of the house this pipe connects with a 5-inch extra-heavy iron drain with the Y on the house sewer. In the cellar there is also a running trap, with a heavy brass cleanout on the house side, an iron plug on the sewer side, and a 4-inch fresh-air inlet, extended 30 feet or more from outside of the house, and turned up 18 inches above the ground, with a one-quarter bend and a brass-bar grating. An 8x8-inch brick pier was built at the base of each vertical column, and the pipe was supported solidly on the pier. All branch connections are Y's and one-eighth bends or TY's. Y-branch connections are well turned up. The floor openings were all cut from the top to the bottom of every column before the vertical pipe was put in. All openings in floors and ceilings about lines of pipes are entirely closed and packed with mineral wool so as to entirely seal the opening. All pipes at the roof are flashed with heavy copper, with slip-collar joint clamped to the pipe to allow for expansion. The rainwater leaders are connected in the cellar to the area drains, using an extra-heavy cast-iron ending 2 feet above grade for the leaders. The leaders and area drains have independent extra-heavy cast-iron traps inside of the cellar wall, with heavy brass cleanouts on the house side of the traps, which have not less than 3 inches of water seal. All wrought-iron pipe and fittings are galvanized. The water pipes are of "Standard" pipe, factory-tested to 300 pounds per square inch. All wrought-iron drain, waste, and vent pipes are "extra strong." Pipe up to 1¼ inches diameter is butt-welded, and the larger sizes are lap-welded. All water pipes and fittings are standard galvanized iron. Ground brass unions were used at frequent intervals to admit of ready repair.

Bedroom basins were particularly required. The entire waste system from these basins is quite independent of the drains connecting with the sewer.

The main waste from these basins discharges over an earthenware flushing rim slopsink in the cellar, provided with flush tank. The basin waste main at the sink is trapped and provided with an independent 3-inch fresh-air inlet extended 30 feet outside of the house. The basin waste branches are all vented to the roof, forming a complete vented system of basin wastes entirely cut off from any connection with the sewer or sewer-connected drains.

The cellar underdrains connect into a deep-seal anti back-pressure trap in a brick pocket below the cellar floor. The discharge from the back-pressure trap connects back of a rain-leader trap and has a weeper from the cellar slopsink flush pipe, which will keep both the anti back-pressure trap and rain-leader trap supplied with water in case of drought, and will prevent evaporation of the seal in those traps. Stoppage in the waste beyond leader trap will be announced by flooding at the area drain at the foot

of the steps leading to the cellar. Backflow into the under-cellar drains will be prevented by the anti back-pressure trap.

All threaded pipe work is put together with red lead. All lead soil, waste, and vent pipe is drawn pipe of the best quality, and of the following weights per lineal foot:

Diameters.	Weight per Foot.
½ inch	1 pound.
¾ "	1½ "
1 "	2 "
1½ "	3½ "
2 "	4¾ "
3 "	6 "
4 "	8 "

All connections of lead and iron pipe are made by " heavy " brass ferrules of the same size as the lead pipe, threaded and screwed or calked into the hub of the iron pipe. All hot and cold supply pipes at fixtures have 12 inches extension beyond faucets or branches to prevent water hammer. The following

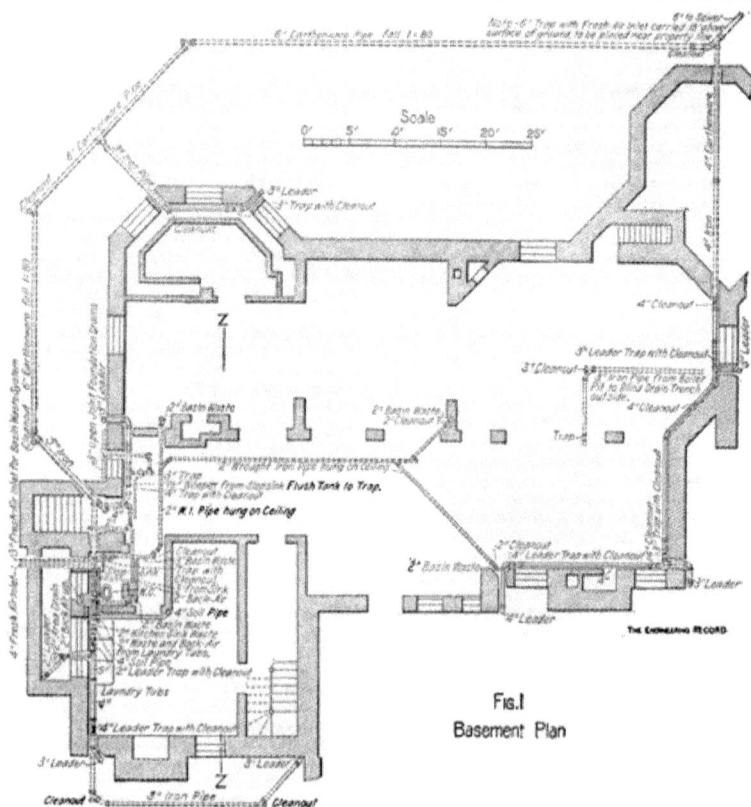

Fig. 1
Basement Plan

PLUMBING IN A HARTFORD, CONN., RESIDENCE.

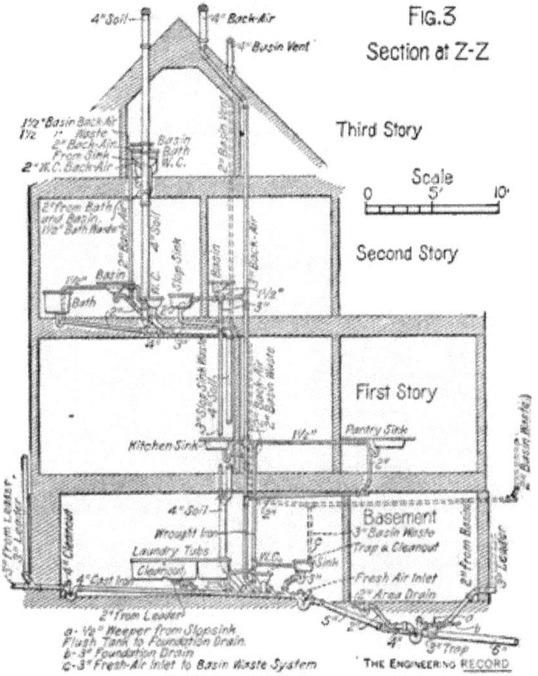

FIG. 3

Section at Z-Z

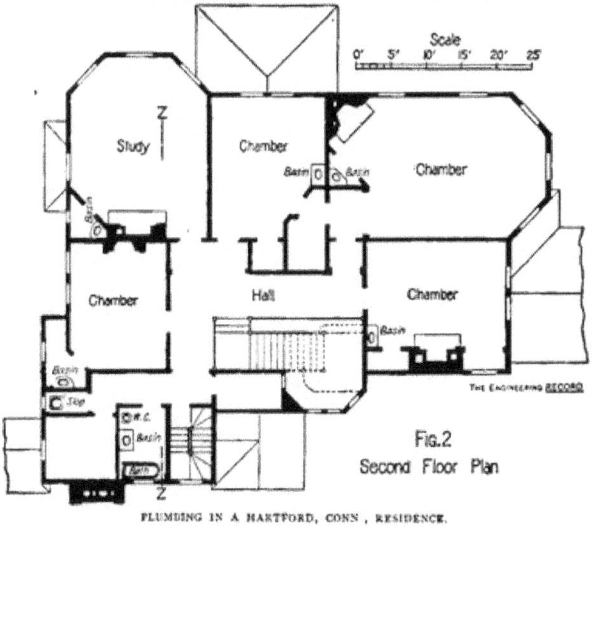

4" Soil
4" Back-Air
4" Basin Vent

Third Story

Scale
0   5'   10'

1½" Basin Back Air
1½" Waste
2" Back-Air
From Sink
2" W.C. Back-Air

Basin
Bath
W.C.

2" Basin Back Air

Second Story

2" from Bath
and Basin
1½" Bath Waste

Basin

Slop Sink

Basin

Waste
3"

Bath

First Story

2" Back Air
2" Basin Waste

Pantry Sink

2" Basin Waste

Kitchen Sink

1½"

2"

4" Soil

Basement

3" Basin Waste
Trap & Cleanout

Wrought Iron
Laundry Tubs

W.C.

Sink

Fresh Air Inlet

2" from Basin

2" Area Drain

2" from Leader

3" Cleanout

4" Cast Iron

4"-3"

4½-3"

2" from Leader

3" Trap

5'

6'

a - ½" Weeper from Slopsink
Flush Tank to Foundation Drain.
b - 3" Foundation Drain.
c - 3" Fresh-Air Inlet to Basin Waste System

THE ENGINEERING RECORD

Scale
0'   5'   10'   15'   20'   25'

Study

Z

Chamber

Basin   Basin

Chamber

Basin

Chamber

Hall

Chamber

Basin

Basin

Slop

R.C.

Basin

Basin

Z

THE ENGINEERING RECORD

FIG. 2

Second Floor Plan

PLUMBING IN A HARTFORD, CONN., RESIDENCE.

size branches were provided for fixtures: Laundry tubs, ¾-inch; water-closets, ½-inch; slopsink, ¾-inch; pantry sink, ½-inch; kitchen sink, ¾-inch; bathtubs, ¾-inch; basins, ½-inch; dark-room sink, ¾-inch; sill cocks, ¾-inch. All stop cocks and valves throughout the building have nickel-plated brass tags with their number neatly stamped on them. All cold pipes in the kitchen and elsewhere where the pipes will "sweat" are painted three coats of white lead and shellac. The photographer's wood sink and 18-inch back was made and set by the carpenter, and is 36x18 inches deep, with a shallow gutter at the back drained to the outlet, with a ¼-inch pitch in the bottom of the sink. The sink and the 18-inch back were lined by the plumber with four-pound sheet lead. The sink has a large ash drainboard and rim, and two ½-inch brass compression cocks 9 inches from each end of the sink. The cocks are provided with 18 inches of ¾ inch rubber hose wired on. All basins except bathroom basins connect with 2-inch extra-strong galvanized wrought-iron waste columns extended 3 feet above the roof, increased to 4 inches at the roof, with expansion copper roof flashing joint. The basin waste columns connect in the cellar with a 3-inch extra-strong galvanized wrought-iron waste main hung to the first-floor beams and discharging over the cellar slopsink. The waste main has a 3 inch running trap and an independent 3-inch fresh-air inlet extended 30 feet outside of the house wall, with a one-quarter bend and a grating 18 inches above ground. The basin traps on this system are

returned vented into their respective waste columns. The trap for the third-floor dark-room sink has a 3-inch water seal. At the points shown on drawings heavy brass screw-cover cleanouts are provided. All brass trap and fixture fittings and water-closet back airs have ground brass fittings. No washer couplings or fittings are used.

All of the soil, waste, drain, and vent pipes were tested by the plumber in the presence of the engineer by a water-pressure test, and upon final completion of the work and after all fixtures were set and in working order the smoke test **and** the peppermint test were applied satisfactorily **to the** entire system **of** drains.

The list of fixtures **is as follows:**

	Basement.	First Floor.	Second Floor.	Third Floor.
Water-closets	1	..	1	1
Baths	..	..	1	1
Basins	..	..	6	1
Sinks	..	1	..	1
Slopsinks	..	1	1	..
Laundry tubs	1 set of 3	..	..	..

The above work was designed and supervised by Mr. Albert L. Webster, C. E., New York.

Peabody & Stearns, of Boston, were the architects, and the installation was made by W. H. Spelman & Co., of New York City.

PLUMBING OF HOTELS.

PLUMBING IN THE WALDORF HOTEL.
(PUBLISHED IN 1893.)

PART 1.—ARRANGEMENT OF MAIN DRAINS IN THE BASE-MENT.

PROMINENT among the hotels which have lately been erected in that region of New York City extending from Twenty-third Street to the Central Park is the Waldorf Hotel, now approaching completion. This structure, which is a 14-story building of iron, stone, and brick, occupies a corner of Fifth Avenue and Thirty-third Street. Its general drainage scheme is indicated on the accompanying plan, Fig. 1, which shows the run of the horizontal pipes and the location of risers. About two-thirds of the area of the plan shown has a subbasement beneath it, and in this part of the building all the pipes are suspended from the subbasement ceiling, and are of screwed wrought iron. Where there is no subbasement the sewer pipes are of cast-iron and are bedded in trenches in the earth. With the exception of one piece of leader drain on the wall this is the only cast pipe used in the plumbing of the building. All the hot and cold water supply lines are so connected that every branch and fixture may be removed without interfering with the others on the same line.

The floor drains are connected to special wastes which are trapped into the sewer and are flushed at daily intervals by special automatic flush tanks. All drainage below the level of the basement floor is received in a cesspool and automatically pumped out. In the general toilet-rooms there are three cisterns, each automatically flushing the urinals, and so connected that each cistern flushes all the urinal traps at each discharge.

The trap vent risers are of large diameter, most of them being 4 inches or 5 inches at the foot and larger at the roof. In some instances. where practicable, several of them are connected at the foot by an open horizontal pipe intended to maintain an equilibrium between them, and to afford an air supply to each cluster of fixtures from each of the connected risers.

The lines of sewer, soil, waste, and vent pipes have been tested in sections from time to time as the work progressed, so as to give frequent checks on the quality of the work and prevent the possibility of serious defects being developed by the ultimate test. The final test was made upon every cluster of fixtures served by each set of risers, and so arranged as to be connected in one circuit and to receive air pressure from a single point. The pipes tested were

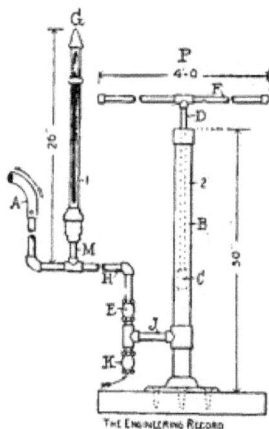
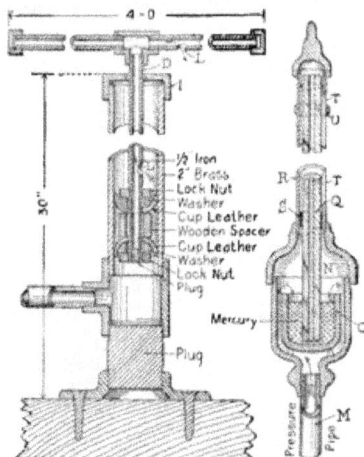
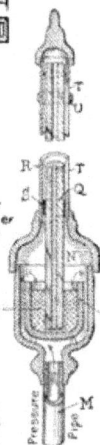

½ Iron
2" Brass
Lock Nut
Washer
Cup Leather
Wooden Spacer
Cup Leather
Washer
Lock Nut
Plug

Mercury

Plug

Pressure Pipe

THE ENGINEERING RECORD

Fig. 3   Fig. 4   Fig. 5

PLUMBING IN THE WALDORF HOTEL, NEW YORK CITY.

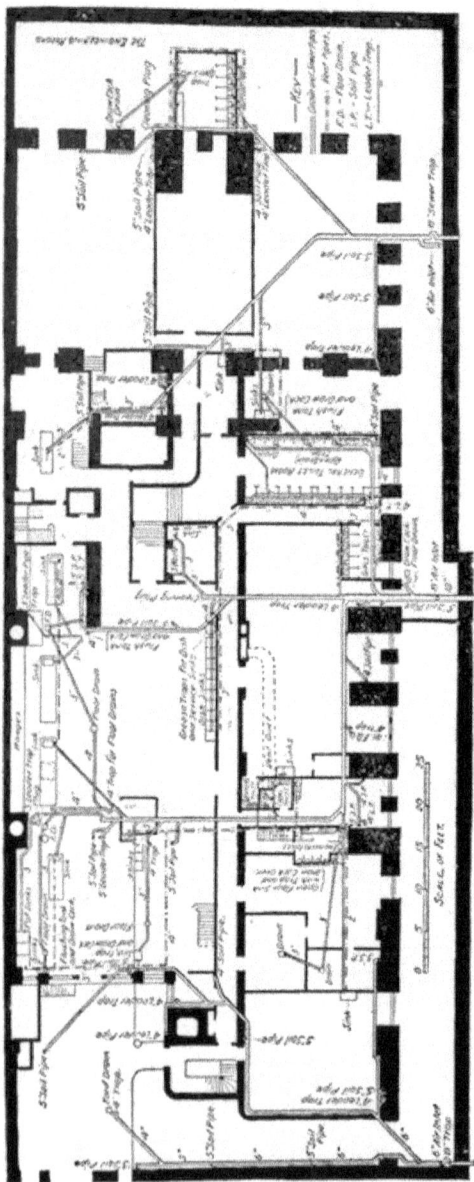

FIG. 1.

PLUMBING IN THE WALDORF HOTEL, NEW YORK CITY.

from 1½ to 8 inches in diameter, and aggregated over 15 miles in length. They satisfactorily sustained an air pressure of about 20 pounds per square inch, which was maintained for half an hour.

The difficulties of designing and arranging the drainage system were much increased by the impossibility of carrying any part of it through large areas of the lower floors which were required to be unobstructed for corridors, rotunda, halls, etc. Especial care was necessary in proportioning the lowest branches of the soil-pipe risers which served many clusters of fixtures and were obliquely carried for long horizontal distances. Their flow was carefully computed and then the pitch was made excessive enough to provide for the delivery of much more than the total volume of water that could come from the simultaneous discharge of all the fixtures, so as to prevent all possibility of backing up at the lowest connections. Their capacities were demonstrated by connecting three leaders to one of the waste pipes designed to serve only one, which freely discharged their combined flow without backing up at all.

The estimated cost of the plumbing and gasfitting was about $200,000, and the contracts for it were executed by Messrs. Byrne & Tucker, of New York, in accordance with the plans of H. J. Hardenberg.

PART II.—TESTS OF SOIL AND VENT PIPES.

THERE are in the hotel 29 sets of soil and ventilation riser pipes extending above the roof, and each serving clusters of fixtures in vertical tiers in most or all of the 15 stories. These pipes, all of screwed wrought iron, were tested from time to time as they were extended from the bottom up, and again when completed ready for the connection of the fixtures. The average height of each stack was about 170 feet, and an idea of the number of joints, 1,800 to 3,000 for each set of two pipes, connections and fittings thus tested and retested, may be derived from the statement that nearly every one of the 29 sets of risers served 10 bath and toilet rooms arranged as indicated in Fig. 2, where, however, only the system is shown and the necessary offsets and extra pipes are not indicated. A special trap screw, not shown in the figure, was put just beyond the vent pipe on the soil-pipe side of the bathtub trap, which was originally designed to be accessible through the overflow connections.

Two important features of the arrangement are the unvarying location of the vent-pipe branches above the overflows, so as to avoid the possibility of their acting as overflows should the waste be obstructed, and the provision of what is substantially a double ventilation of the water-closet trap, which may be effected either through branch A or connected branch I. In testing, the open ends were all closed by screwed caps and air pressure applied through the flexible hose A, Fig. 3, one end being connected by screw coupling to a point near the foot of the soil pipe which was connected to its adjacent vent pipe, or *vice versa*, and the other end to the discharge of an air pump P, Fig. 3, on which was set a mercurial gauge G. The commercial portable air pumps not proving of sufficiently rapid action, and being con-

sidered not adapted to this work, it was found convenient and economical to construct pumps on the spot, which was very easily and simply done, as indicated in Fig. 4. The barrel of the pump was simply a piece of 2 inch brass pipe B, which was so true and straight and of so exact a bore as manufactured as to need no finishing whatever to receive the plunger C, the rod of which D was constructed of a piece of ½-inch gas pipe, and furnished with a gas-pipe crossbar F long enough for four men to grasp. A little piece of ¼ inch iron pipe J served both for suction and delivery, and the inlet and discharge were respectively controlled by the two ordinary check valves E and K, both opening upwards. On the upstroke of the plunger valve K opened and admitted outside air to the partial vacuum formed in the lower part of barrel B. On the downstroke of the plunger this air was compressed, and closing valve K opened valve E and was discharged through it, pipe H and hose A into the closed system of pipes. When plunger C again ascends the pressure in the barrel becomes diminished and the pressure in pipe H closes valve E. Atmospheric pressure opens valve K, more air is drawn into the pump barrel and forced into the pipes, and so on. Just above check valve E there is in the pipe H a stop-cock, which is not shown in Fig. 3.

Figure 4 shows the details of construction of the plunger, etc., and like Figs. 2, 3, and 5, is not drawn to exact scale or dimensions, but is intended merely to show the arrangement, construction, and operation. Rod D passes loosely through cap I, but as the clearance did not allow the air to escape freely enough on the upstrokes, L L were cut in the rod and handle and allowed free discharge as indicated by the arrows. The apparatus was secured by a bottom flange to a 2-inch base plank, 12 inches wide by 6 feet long, on which the men stood to operate it. The entire weight was about 100 pounds, and it proved very efficient and satisfactory. Four of these pumps were made for this job at a nominal cost, as all the materials were taken from ordinary stock, and they required no repairs or alterations.

In the beginning of a test, first one and then two men were required to pump. The average time required to produce a pressure of 10 pounds was 20 minutes. The Board of Health Inspectors specified that a test would be considered satisfactory if a mercury gauge-column 20 inches high did not fall more than one half inch in 20 minutes after the pumping ceased. In the final tests the greatest leakage discovered caused a fall of only one-half of an inch in the 20-inch mercury column in 22 minutes after the pumping ceased, and in many instances there was no appreciable fall whatever. When it is considered how slight an aperture will permit the escape of air and that the pressure is uniform at all points, it will be seen how thorough this test is, and how much more positive it is than the usual water column, especially for the upper part of the system, where the hydrostatic pressure diminishes to zero.

Figure 5 shows the ordinary mercurial gauge G, Fig. 3, used to indicate the test pressures. Air pressure from the pump discharge pipe H is admitted by pipe M to the cast-iron chamber N, in which is an open iron cup O filled with mercury. A plain glass tube Q, open at both ends, dips beneath the surface of the mercury nearly to the bottom of the cistern, and is protected by a brass tube or case R, which is capped at the upper end and screwed to the iron case

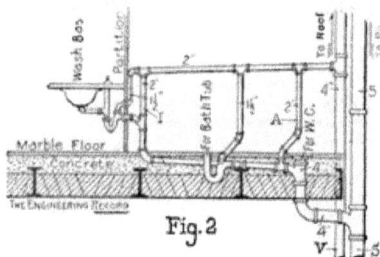

Fig. 2

containing the cup O, the packing S making an air-tight joint. The pressure is in the pipe system being communicated by pipe M to chamber N, acts downwards on the surface of the mercury in the cup and forces it up in tube Q, where its height is observed through a slot T T, in case R, and can be marked by the index U, which is a close-fitting sleeve, which can be set at any position to measure variations from. About 100 tests were made by Foreman John Hanson, who devised the pump, and was in charge of the work.

PART III.—PIPE SYSTEM AND TANKS FOR DOMESTIC AND FIRE SERVICE, SPECIAL BACK-PRESSURE VALVE, AND PNEUMATIC MERCURY GAUGE AND HOT-WATER BOILER, AND CIRCULATION LOOPS.

ALL the cold water used above the first floor is pumped to the tanks A and B, Fig. 6, in the attic by the two 8-inch and 12″x8½″x10″ compound duplex Worthington steam pumps C C in the basement.

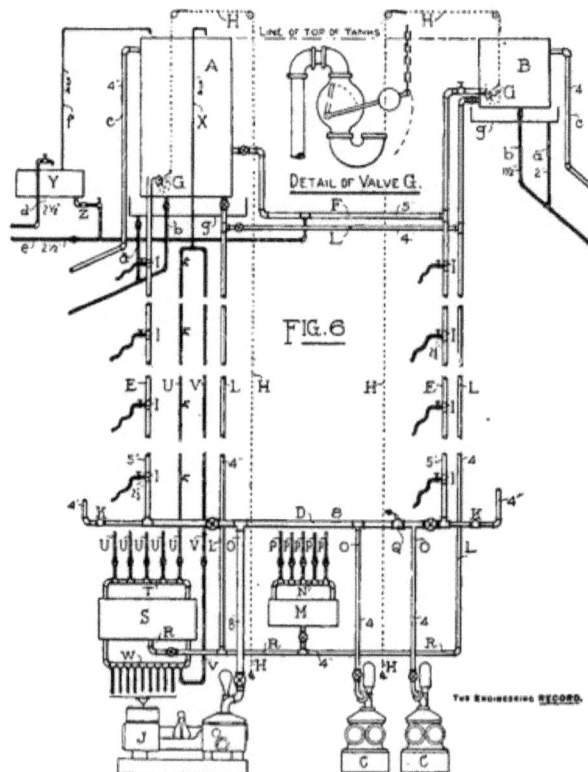

FIG. 6

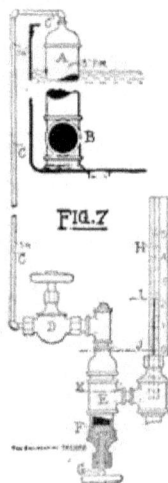

FIG. 7

PLUMBING IN THE WALDORF HOTEL.

These pumps connect through pipes O O O with the 8-inch horizontal pipe D, from which two 5-inch pipes E E rise, connecting to the two attic tanks. They are cross-connected by the 5-inch pipe F at the level of the bottom of tank A. Pipes E E discharge into the tanks through the 5-inch back-pressure valves G G, set so that the disk of the valve may be held open by the lever and chain connection H extending to the engine-room, where the ends are secured in such manner that they may be released at the will of the engineer. The details of valve G are shown in this figure. The tanks are built of ⅜-inch wrought iron, and each has a wrought-iron safe pan with a 2-inch waste *a a*. The tanks may be emptied through the 1½-inch emptying pipes *b b*, and *c c* are the 4-inch overflow pipes. All these pipes may empty into the rainwater leader pipes.

Upon each floor there is taken off from pipes E E a 2½-inch connection to which is attached valve I for fire service. Each of these valves has connected 100 feet of linen rubber-lined hose and a hose pipe, all of which are housed upon a swinging rack ready for immediate use. For fire-extinguishing the 16,000 gallons of water in the attic tanks is always available, the water passing back to the rising lines E E through the back-pressure valves G G which are held open by the lines H H. By letting go either line H the flap of the valve which it controls is closed and by starting the 20'x11'x15' Worthington steam fire pump J, which is connected to the pipe D, its full pressure may be directed into the lines being used for fire purposes. At the ends of pipe D are located the 4-inch check valves K K. The discharge ends of these connect with 4-inch pipes passing through the walls and turning upward, where they end with standard fire-department coupling connections. In the event of a fire calling the city department to the building the firemen would couple their hose to this pipe, using the pipes E E as stand-pipes, and availing themselves of the hose already in position upon the several floors. Safety valve Q on pipe D is intended to relieve excessive pressure in the event of the closing of valves G G while the steam pumps are working.

The height of the water in the attic tanks is recorded in the engine-room by means of a pneumatic mercury gauge especially designed for this work by Mr. Hanson, the foreman plumber engaged upon it. It is illustrated in Fig. 7. In each of the tanks an air compressor is erected, formed of the 3-inch wrought-iron pipe A, which is of equal height with the tank. The base, a cast-iron flange, is fastened to the tank bottom, and close to it is the open centered tie B. The upper end of the compressor is reduced to three-eighths of an inch in diameter and is connected to the ⅜-inch pipe C, which, descending to the engine-room, is there connected with the mercury column H, Fig. 7. It is controlled at the point of connection by the valve D. The mercury column is of ordinary construction, except that at the base of the reservoir E is inserted the close-fitting piston head F, which may be raised as required by the screwed stem G to allow of adjustment when, by reason of reduction of the quantity of mercury in the column, the scale fails to accurately register the height of water in the tank. The tank being empty mercury is introduced into the reservoir and columns until it shows at O. The valve D is opened and the water as the tank is filled closes the opening B, preventing the farther escape of air. As the water rises it gradually compresses the air, which pressure, transmitted through the pipe C to the mercury in the reservoir E, drives it up in the graduated glass tube H, which is long enough to indicate a full tank. The full marks on the glass agree with even feet of water in the tank, and the one-twelfth graduations correspond to immediate inches of depth. As the water is lowered in the tank and the air pressure is lessened the mercury drops in the gauge glass.

From each of the attic tanks the 4-inch wrought-iron pipes L L, Fig 6, which, descending to the cellar, are connected through 3-inch pipe R with the 30x84-inch heavy wrought-iron tank M. On the top of this tank is connected the 4-inch horizontal drum N, from which radiate the 2-inch cold-water pipes P P, etc. leading to the several riser lines supplying cold water to the plumbing fixtures on all the floors above the first. All connections to tank M and the pipes from drum N have valves for use in case of repairs. These risers are cross-connected by a 4-inch horizontal pipe so valved that either tank may be cut out of service if necessary. Pipe R also serves to connect the cold-water pipes L L with the wrought-iron boiler S in the basement. This is built of ⅜-inch iron, and is strongly made to withstand the pressure from the attic tanks. The water in this boiler is heated by a brass coil, to which steam from the power boilers is admitted. There are also arrangements for heating this coil by exhaust steam from any of the several steam pumps or engines in the cellar. On the top of this boiler is built the double-connected 4-inch drum T, from which is laid the 2-inch galvanized-iron pipes U U, etc. supplying hot water to plumbing fixtures above the first floor. Pipes T rise to a point above the highest fixtures and then return by the 1½-inch galvanized-iron circulation pipes V to the 3-inch drum W, which is double-connected to the bottom of the boiler S, thus forming a circulation loop which gives hot water at any point of draft on its line. The tops of all these loops are relieved of air or steam, which would interfere with the circulation, through the ¾-inch galvanized-iron relief pipes X.

Upon the same floor as tank A is the 30x84-inch wrought-iron tank Y, Fig. 6, which is used for heating the water for the laundry located on the floor immediately beneath. Cold water enters the boiler from the pipe F through the 2½-inch pipe *e* having the check valve Z to prevent the return of hot water into the tank A. The water in boiler Y is heated by a brass coil to which steam is supplied from a live-steam pipe in the laundry service, or by the exhaust steam from the 40 horse-power engine in the laundry which furnishes power for all the laundry machinery. From the top of this boiler Y is laid the 2½-inch galvanized-iron pipe *d* to the laundry, and from the highest point of this pipe is carried a ¾-inch relief pipe *f* into tank A.

PART IV.—DRIP BOXES AND RUNNING TRAPS ON HOUSE DRAINS.

ALL of the floor drains shown in the ground plan, Fig. 1, and those in the cellar have square cast-iron drip boxes, as shown in Fig. 8, which are also used as cleanout boxes. Some of them are large enough to allow of the assembling in them of several drains, which are then carried to the main drainage system through a larger size pipe, having a full-size running trap. Each of these traps had a cleanout G set in a manhole with cast-iron cover to allow of ready inspection. On long lines of drain pipes and at some of their intersecting points these cast-iron junction boxes are used in order that they may be readily cleaned or inspected. On these cases a flat cast-iron cover takes the place of the brass perforated top A. So far as possible two or more of these floor drains were connected with each run of the drainage pipes, and at the highest ends of each of them were set automatic flushing tanks supplied by the waste from the cold-water drinking-fountains. The cooling coils for these fountains are made a part of the general refrigerating and icemaking plant, and a current must consequently be continuously maintained to prevent freezing. Floor-drain bell traps, which are liable to become clogged when subjected to the severe service to be expected in a job of this magnitude, are thus avoided, and a cleanout arrangement is provided. All of the brass covers A are arranged for ready removal. The inlets B and the outlet D' in the cast-iron box C are cast flush with the inside bottom, which is pitched to form canals through which the water may flow unimpeded.

The cast-iron running traps A, Fig. 9, on the main house drains were set sufficiently beyond the inside lines of the area walls to give a vertical fall to any

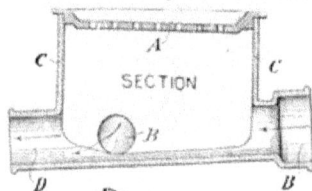

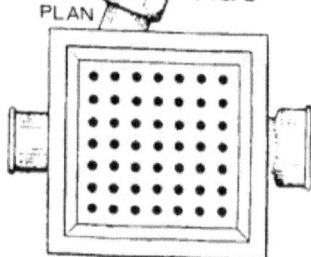

dirt, water, etc. which gained entrance to the fresh-air inlet B on the house side of the traps in order to guard against a common cause of failure of fresh-air inlets to operate. All matter or water so falling passes directly into the trap and is readily washed away. The cast-iron round bottomed bowls C were made especially for this job. They were set in the brickwork of the sidewalk arches, the wide flanges D allowing of a water-tight cement joint. The stone walk above the bowls was pierced and fitted with the oblong brass bar plates E. Ample clearing space is left between the bars, and as the widest part of the opening is at the bottom they are practically self-clearing. On the bottom of the bowls were cast collars upon which were tap-screwed the 6 and 8 inch flanges to which the upright wrought-iron pipe F

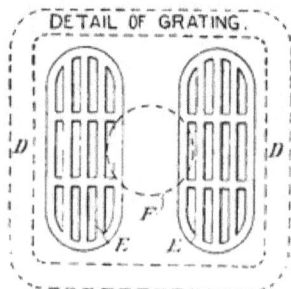

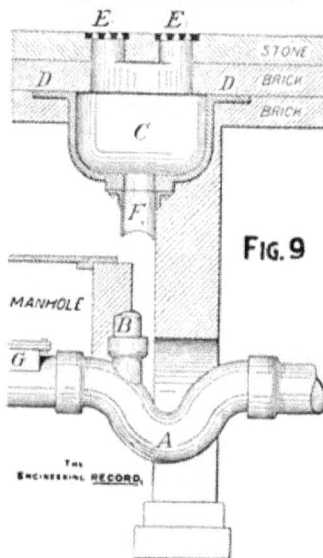

DETAIL OF GRATING.

FIG. 9

FIG. 8

PLAN

DRIP BOXES IN THE WALDORF HOTEL.

connecting with the standing collar B was screwed.

There are separate sinks for washing the silverware, dishes, and cooking utensils, each having a separate cast-iron Tucker grease trap 23 inches high and 28 inches in diameter, which is the largest size made, conveniently located for cleaning out. All the cold water used in the kitchen and pantry is passed through the water chambers of these traps, the proportion being divided according to the service.

PART V.—WATER FILTERS AND CONNECTIONS.

All the water used in the hotel is filtered. The plant for this purpose comprises two Cummings filters made by the Cummings Filter Company, of Philadelphia. Through one is passed the water for the icemaking apparatus, the boilers, the culinary department, and for all other uses in the cellar, base-

is perforated with $\frac{1}{8}$-inch holes and has soldered upon its under side a fine meshed brass wire cloth. Three inches below this plate is a similar one, and between them is a 3-inch bed of clean-washed pea-size gravel H. At the bottom of the cylinders at joints S S are placed perforated plates, gravel, joints, etc. similar to those described at the top. Upon the top of the lower diaphragms is placed 4 feet of loosely filled-in animal charcoal, shown at I. The cock B being set for delivery to the filter water turned on from the pipe A, it passes up C C and into the cylinders passing through the perforated plates and gravel bed. It fills the water space J and passes down through the charcoal I and through the lower plates and gravel to the pipe K leading to the pumps. The animal charcoal is heavier than water and consequently remains sufficiently compacted for filtering

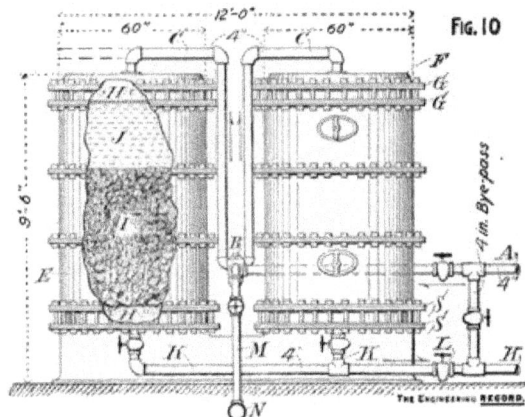

FIG. 10

WATER FILTERS IN THE WALDORF HOTEL.

ment, and first floor. Through the other is passed all the water used on floors above the first. Each of these filtering tanks is composed of a 60-inch sectional cast-iron cylinder with bases and tops as shown in elevation and broken section in Fig. 10. The filters stand 9 feet 6 inches high. All the internal metallic surfaces are treated by the Bower-Barff anti-rusting process.

Water is delivered from the 4-inch Thomson meters through the 4-inch pipe A to the six way cock B. The line is then branched into the 4-inch pipes C C, which enter the tops of the separate cylinders E E. The cast-iron top plate F is dished so as to allow a 3-inch space between its upper surface, and a $\frac{1}{16}$-inch galvanized-iron diaphragm plate which, with its packing, is inserted at the upper joints at G. This plate

without compression. By closing valve L which controls the flow of filtered water to the pumps and turning the six-way cock B, filtered water from one of the cylinders may be passed upward into the other through pipe K for the purpose of washing the filter. This water passes in the reverse direction through the bed of charcoal and down through pipe C to the six-way cock by which it is directed into the waste pipe M leading to the sewer N. A single movement of the cock B will make the effluent of either cylinder wash the other.

Each of these cylinders occupies a ground space of 72x150 inches. The weight of each charged and ready for service is about 14 tons, of which three tons is animal charcoal. The plant is designed to filter 150,000 gallons daily.

PLUMBING IN THE LAKEWOOD, N. J., HOTEL.

(PUBLISHED IN 1891.)

PART I.—ENGINE-ROOM, PUMPS, SUCTION TANKS, AND
LAUNDRY BOILER.

THE Lakewood, at Lakewood, N. J., is a four-story
brick and iron building, with a frontage of 465 feet
and a total depth of 408 feet.

Figure 1 is a general view of the engine-room in
the basement. Lake water is received in suction
tank A, through two ball cocks in the supply pipe B.
C is a 4-inch suction pipe to the (Blake) house pump
D, which delivers through 3-inch pipe E to the roof
tanks. F is the live and G the exhaust-steam pipe

H is the fire pump, with a 4-inch suction I, from the
tank, and a 3-inch suction J, from an artesian well.
K is an air chamber. Pump H delivers through pipe
L, in which a short section of heavy rubber M is set
to prevent the transmission of vibrations through the
house by the riser and branch pipes.

N is a branch by which the tank A may be filled
from the artesian well if lake supply fails. O is a
4-inch fire line, with hose cocks and reels in every
corridor. P is a branch connecting with delivery E
from pump D to the house tanks. Q is a pressure regulating valve. R and S are branches from the exhaust
and live steam pipes G and F respectively. T is the
400-gallon hot-water boiler supplying the wash-

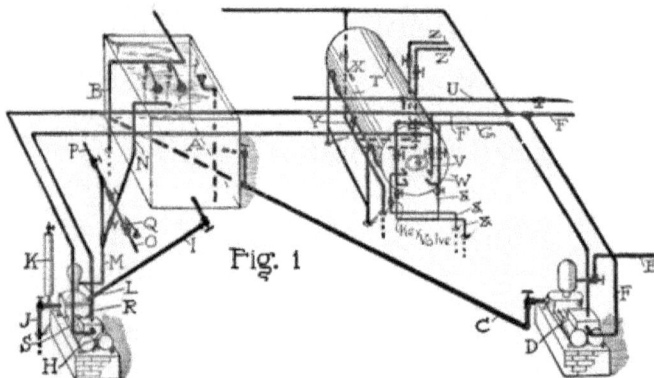

Fig. 1

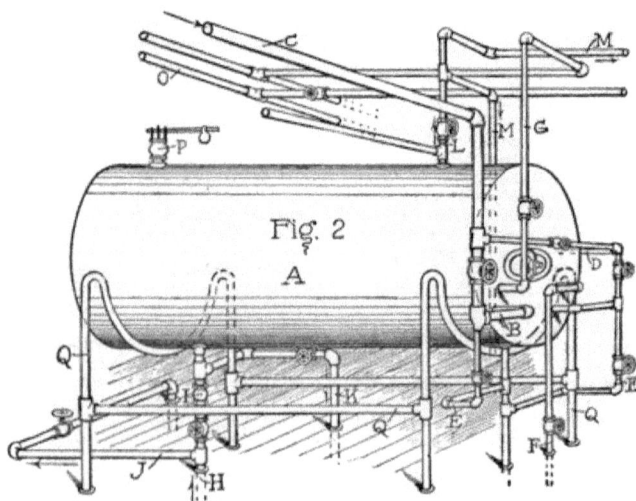

Fig. 2

PLUMBING IN THE LAKEWOOD, N. J., HOTEL.

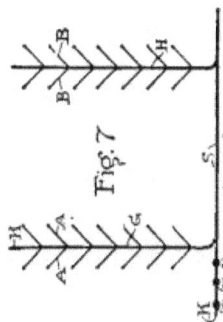

Fig. 7

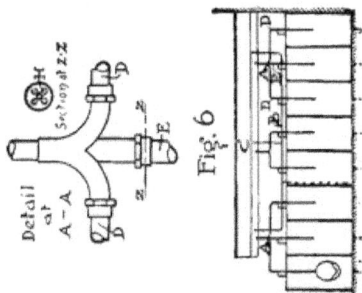

Detail at A—A

Fig. 6

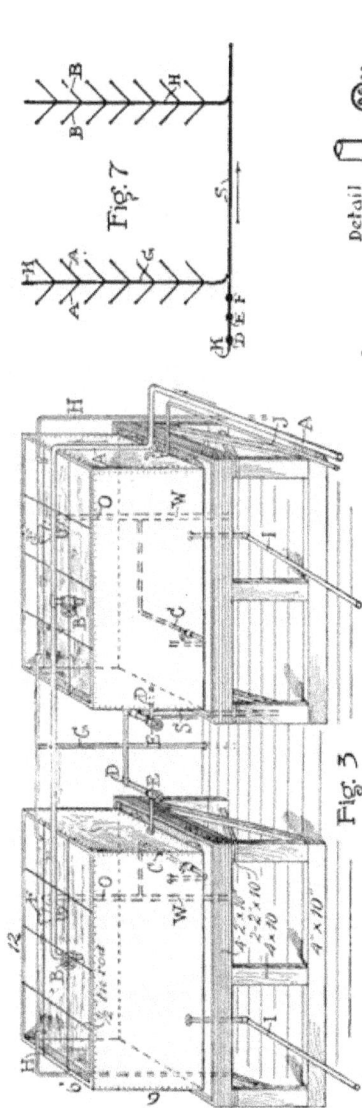

Fig. 3

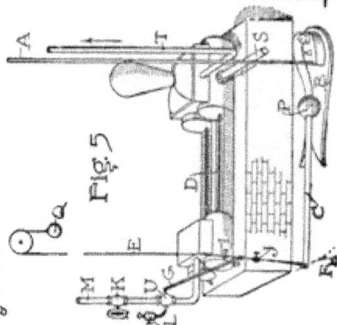

Fig. 5

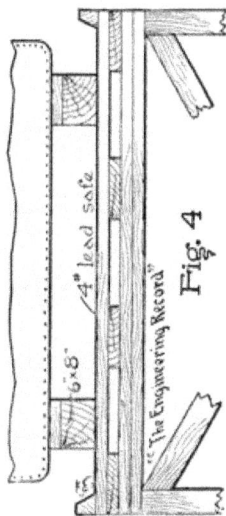

Fig. 4

basins and bathtubs throughout the house and the kitchen through 2-inch pipe U. It is supplied through branch X and may be emptied through Y.

The coils receive live and exhaust steam through pipes V and W respectively. & & & are condensation and drip pipes and Z Z are emptying pipes for riser lines.

Figure 2 shows the laundry boiler A, about 8 feet long by 3 feet 6 inches diameter, and containing a 100-foot coil of 1½-inch copper pipe. This receives steam through pipe B, connected with live supply C and exhaust supply D.

E E are 1-inch return pipes, and F is a drip and emptying pipe for the coil. G is a special cold-water supply from the roof tank, but the regular supply is from city mains through pipe H and check valve I that prevents tank water escaping to the street. J is the 1½-inch cold-water distribution to the laundry, and K is the 2-inch emptying pipe. L is the 1½-inch hot-water supply with 1¼-inch branches to servants' toilet and bath room, and laundry tubs. O is a steam pipe, and P is the safety valve set at 75 pounds. Q Q, etc., is a 1½-inch pipe frame supporting the boiler.

PART II.—ROOF TANKS, PUMP GOVERNOR, URINAL AND SOIL PIPE.

FIGURE 3 shows the construction and arrangement of the house storage tanks in the attic. They are built of ¼-inch wrought iron, and have a united capacity of about 15,000 gallons. They are filled through the 3-inch pump pipe A and separate valves B B on 2½-inch branches, and may be emptied through the 2½-inch pipes C C that discharge through the waste pipes W W of the 4-inch overflow stand-pipes O O. The latter terminate in copper funnels F F, about 8 inches in diameter and 8 inches high.

The house supply is through the 4-inch pipe S, with 3-inch branches D D, which serve as equalizing pipes between the two tanks and valves E E. These last enable either tank to be cut out for emptying, etc., without interrupting the house supply.

G is a ¾-inch pipe from the hot-water boiler T, Fig. 1, and H H are ¾-inch relief pipes from the hot-water circulation pipes of the east and west wing systems.

I I are 2-inch safe wastes that discharge into a sink in the pump-room. J is the ¾-inch pressure pipe to the pump governor.

Figure 4 is a section through the bottom of the tank, showing details of safe and support. Figure 5 shows the automatic governor for pump D, Fig. 1. S is the suction and T the tank pipe. A is a ½-inch pressure pipe from the tank connected to a diaphragm damper regulator B, which has a slide weight P, and suspended weight Q, so adjusted that the lever arm C will only rise when the pressure is just equivalent to a full tank head on pipe A. E is a flexible copper wire cable tightly strained by weights P and Q, and having two adjustable clips I and J, set so that when lever C rises, clip J engages the lever G of valve U, and raising it, closes it and shuts off steam supply to the pump through pipe M. When tank pressure falls, lever C also falls, and clip I engages lever G, depressing it and opening valve U so as to admit steam and start the pump; when tank is again full, lever C rises and shuts off the steam and so on. K is a valve for operating the pump by hand, and L is a petcock.

Figure 9 is a view of the urinals in the principal toilet-room. The cistern C is in three compartments, each supplied through double ball cocks through water pipe B and flushing through ¾-inch curved, silver-plated brass pipes that have special connections at A A to equalize the flow through pipes D D and E, in order that the latter may not receive more than one-third of the total amount. The diaphragm H is introduced at Z Z (see detail) to obstruct pipe E, and divert equal amounts of water through D and D.

Adjacent to the above toilet room is another containing three sets of water closets and one row of basins. The waste and soil pipes from all these fixtures are carried exposed just below the ceiling of the basement room underneath, and this arrangement is shown in the diagram, Fig. 7. In this A A and B B are the branches to fixtures, each being connected with double Y's to the main soil pipes G and H, that Y into the 6 inch sewer-pipe S. The latter goes direct to the main sewer. D is a rainwater leader flushing out pipe S, and E and F are connections to washstands and urinal wastes. All parts of the pipes are commanded by the cleaning screws K K K.

PART III.—KITCHEN AND SCULLERY SYSTEMS.

FIGURE 8 is a view of the kitchen arrangement. K K, etc. are boilers, and L L, etc. are steamers, to all of which live steam is supplied through pipe S, and returns through pipe E. F F are wooden vegetable tanks, and D is a wooden soaking tank. Q Q are iron safes, and Z Z their drip pipes.

G is a trap vent from the waste of tank F, and I is a trap vent from trap on safe wastes. C C C are cold, and H is a hot-water supply pipe. A is a steam, and B a hot-water pipe for flushing traps for tank D.

J J, etc. are 2-inch ventilation pipes for taking the vapor away from the cooking food. They are received in 3-inch branch N, which discharges through 6-inch pipe O into an exhaust flue, which conveys it away from any danger of penetrating the hotel rooms.

Figure 9 shows a cross-section through one of the copper boilers K; the meaning of the same reference letters is the same as in Fig. 8. A chamber is formed between the outer and inner shells, and receives live steam which is used for cooking.

W is a perforated strainer cap, and X an emptying cock. U is an iron supporting frame, and I the cold-water cock. R is a vent pipe controlled by valve M.

Figure 10 shows the dishwashing sinks G G, in the scullery. The waste traps are commanded by screws as at S. E E are draining boards. H is a hot, and C a cold-water supply, and A is a live steam pipe with branches D D, etc., and perforated heads K K, etc., for blowing steam into the water to heat it rapidly in the sinks. The 2-inch trap pipes V V are connected by branches M and N respectively with the steam and cold-water pipes, so as to clean out the wastes in the following manner: Close valve O and open P; then steam will be forced through lower

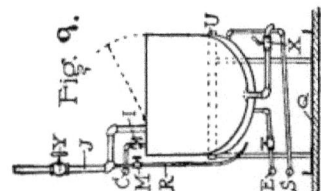

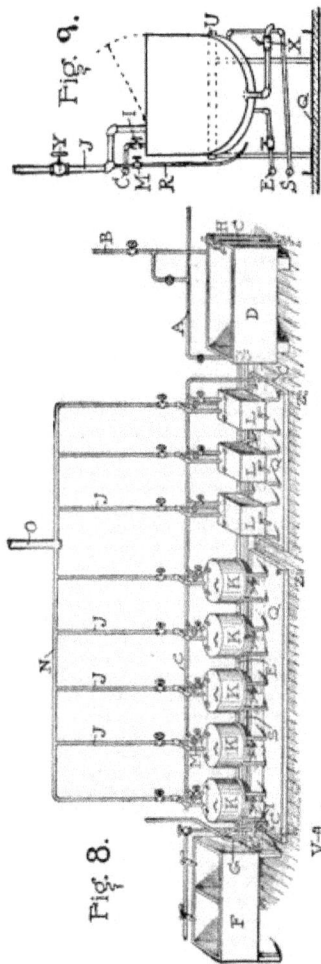

Fig. 9.

Fig. 8.

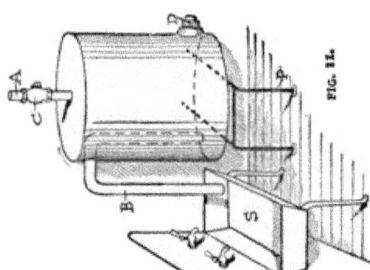

FIG. 11.

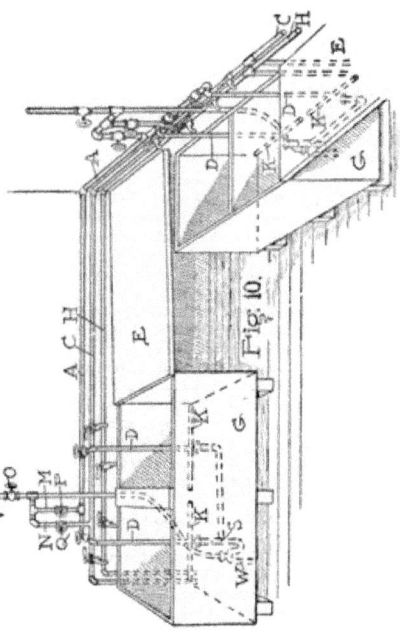

Fig. 10.

PLUMBING IN THE LAKEWOOD, N. J., HOTEL.

part of vent pipe V and through waste pipe W to the sewer, blowing out all grease, sediment, and other obstructions; then close P and open Q, and the pipe will be flushed with hot water, and the trap seal restored if broken; finally close Q and open O and then the operation is completed. This system has been applied to all the sinks in the kitchen.

Throughout the kitchen waste system, which is separate from main building, all pipes are cast iron with rust joints. A full diameter, open-way valve is set below the sink strainer in each waste pipe, permitting discharge from above and preventing the escape of steam into the sink.

Figure 11 shows a large grease trap, through which all waste water from the kitchen is discharged to the subbasement sink S. A is the inlet, and B the outlet pipe, both 2½ inches in diameter; C is a key valve, and D a screw plug for emptying by means of a hose. P is a supporting-pipe frame. The top is removable. All water pipe in this hotel is of galvanized wrought iron, except where plated brass pipe is exposed in the toilet-room, etc.

All waste, soil, and vent pipes were tested to about 40 pounds hydraulic pressure after being set. William Schickel & Co., of New York, were the architects, and John Tourney & Son, of New York, executed the plumbing.

PLUMBING IN THE NEW COATES HOUSE, KANSAS CITY, MO.

(PUBLISHED IN 1891.)

PART I.—GENERAL DESCRIPTION, BASEMENT AND MAIN PIPE PLAN.

THE following illustrations and description of the plumbing-work in the new Coates House, at Kansas City, Mo.; of which Messrs. Van Brunt & Howe were the architects, have been prepared from photographs of the completed work and blue-prints of the plans under which the work was done. These were prepared by Messrs. E. D. Hornbrook & Co., plumbers, of Kansas City, and submitted with their proposal.

The source of water supply for the new addition, as well as the old portion of the hotel, is from the house tanks located on the roof of the old part of the building. The hot-water supply is connected with the hot-water boilers located in the basement of the old part. The main galvanized-iron supply pipes suspended from ceiling, and which are connected to hot and cold-water headers in the old part of the building, are 2 inches in diameter.

All rising lines throughout the building, hot and circulation pipes, are of brass. Cold water and safe waste pipes are of galvanized iron. At the base of all risers there is a valve placed in the hot, cold, and circulating pipes for controlling each line separately, with tees and drain cocks for draining the lines. Each private bathroom is cut off independently by valves located in each room.

All underground house and rainwater drainage, as well as rising lines, including ventilating lines, with closet traps, vent and waste connections all connected in place with ends closed up, were tested

by water test, by filling the entire system with water to 5 feet above the roof, which is 100 feet high.

The method of making the soil-pipe joints was by special tools. The entire job was perfectly tight with but three exceptions, and the plumbing inspector was called upon only twice to pass this entire job.

Underneath all fixtures in private bathrooms there are Italian marble floor slabs, fitted with 6-inch base all round, and polished-cherry wainscoting, 5 feet high all round the rooms. Water-closet partitions, back and sides are all Italian marble 7 feet high. Urinal and closet tanks are also cased with marble. Urinal stalls, backs and sides are Italian marble, backs 7 feet high, and partitions 6 feet. The marble-work is put together and supported in a special manner, without the use of brass or nickel-plated clamps, bolts, etc., such as are ordinarily used.

Figure 1 is a basement plan, showing the arrangement of the underground drainage, soil and waste pipes of the different fixtures throughout the building, as well as the roof drainage. The main roof drainage is 8-inch extra heavy cast-iron pipe with connections leading to the various down-spouts (D D, etc.) and area drains throughout the building. The bottoms of the areas are 5 to 6 feet above the basement floor. This 8-inch pipe is connected to a 12-inch main drain outside of the main house trap, with back-pressure valves as shown, so that if at any time the main sewer becomes stopped it would not back up into the roof or down-spout drainage pipe.

The main house drainage is 10-inch extra heavy cast-iron soil pipe, fitted with house trap and 6-inch fresh-air inlet. I is a section showing the method of connecting the branch soil pipes with the main 10-inch iron drain ; J is a detail showing method of connecting the wastes from six bathtubs in the Turkish bath department ; K is a detail of the overflow and waste from the plunge pool in the Turkish bath department, which is fitted with 4-inch gate valve and 4-inch Barrett back-pressure valve and trap.

This pool is also fitted with a 3-inch polished-brass nickel plated standing overflow pipe, which is connected outside of the gate valve. The top of the standing overflow is funnel-shaped, 8 inches in diameter, and is so arranged that it can be lifted out of its socket at the bottom of pool when it is desired to empty the pool in a hurry, by the use of both valve and the removal of standing waste. L is the trap under each bathtub, which is fitted with a 4-inch polished-brass trap screw flush with marble floor, from which the vent is connected by a union joint. This method of trapping the bathtubs is carried out throughout all the bathrooms. M is a 4-inch extra heavy branch drain fitted with 4-inch Barrett back-pressure valve and trap, which is connected to the various polished-brass floor strainers for draining the marble floors under needle baths, rubbing and shampoo slabs.

PART II.—GENTLEMEN'S TOILET-ROOM, PLAN, ELEVATION, SECTION AND DIAGRAM OF URINALS AND CLOSETS, SECTION AND DESCRIPTION OF PUBLIC TOILET-ROOMS.

ALL the public toilet-rooms are in the same vertical line in the center of the house. The back-air pipe is

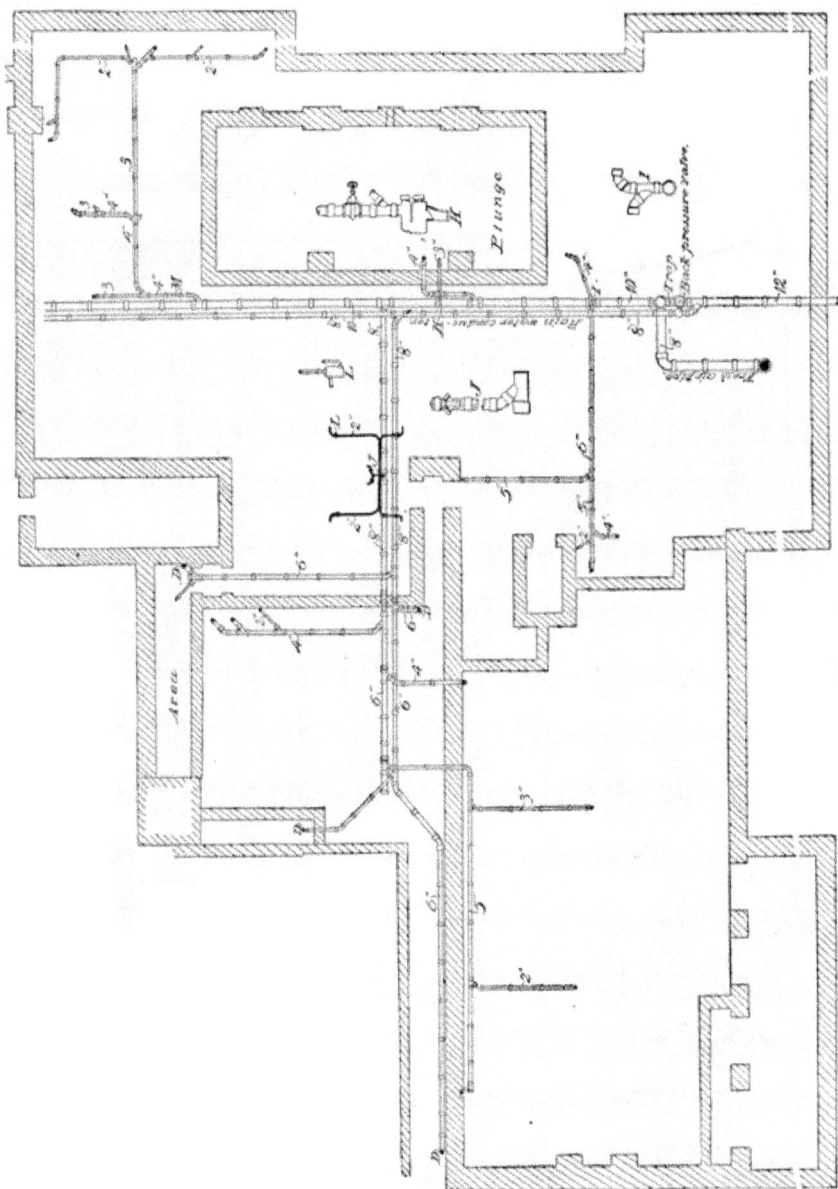

FIG. 18

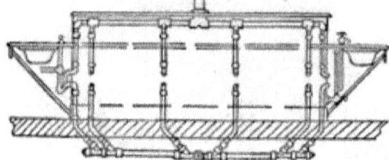

FIG. 17

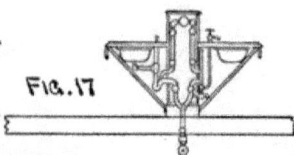

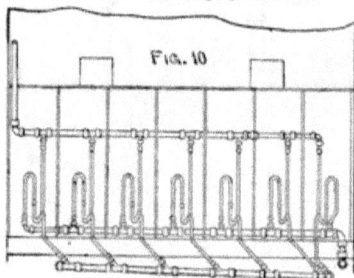

FIG. 10

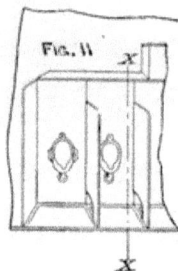

FIG. 11

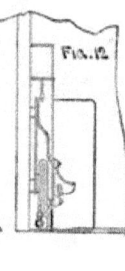

FIG. 12

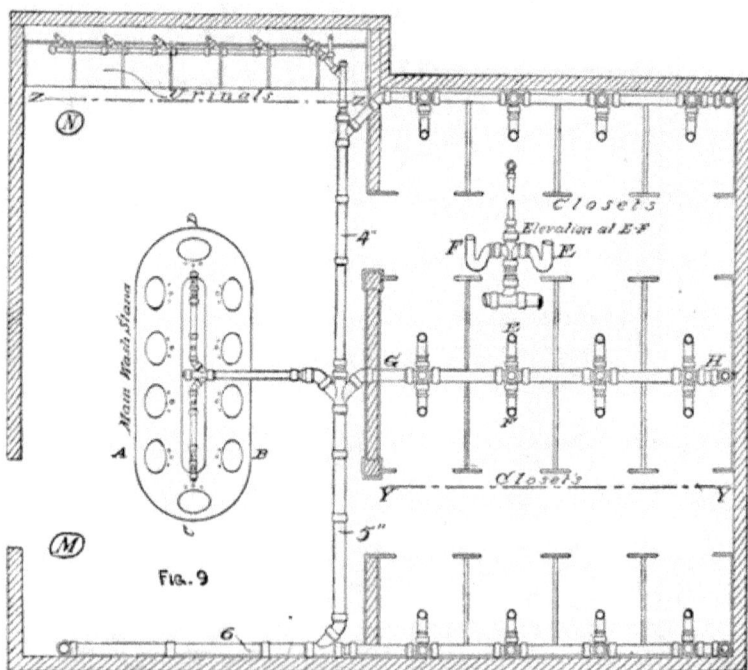

FIG. 9

PLUMBING IN THE NEW COATES HOUSE, KANSAS CITY, MO

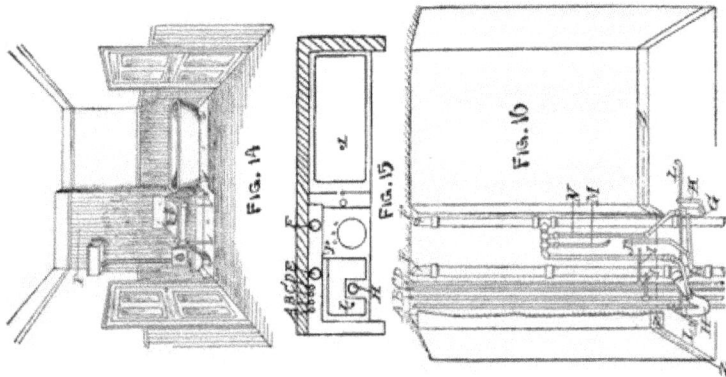

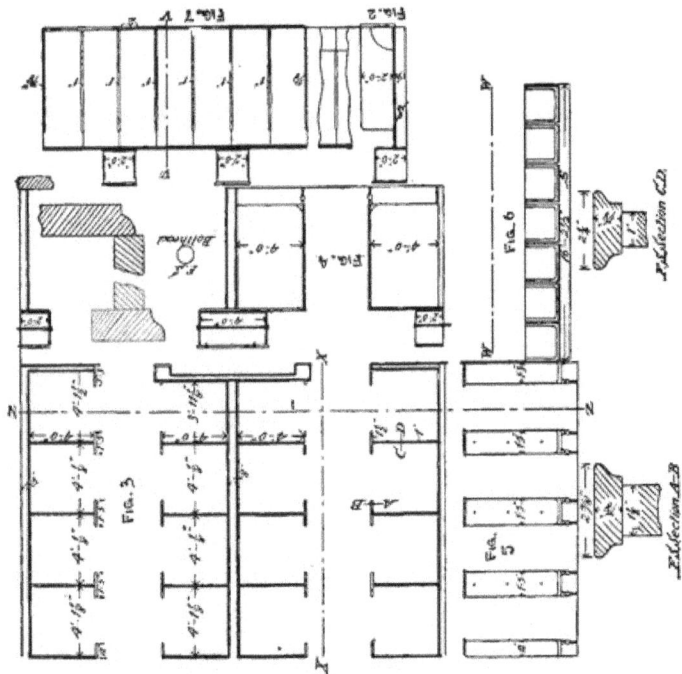

PLUMBING IN THE NEW COATES HOUSE, KANSAS CITY, MO.

3 inches from basement, where it is connected to the traps of six bathtubs and one lavatory, increasing on the third floor to 4 inches, and on the fourth floor increasing to 5 inches, and so continues through the roof. On the third floor there are four water-closets, supplied with a flush tank over each closet. On each of the fourth, fifth, and sixth floors there are three water-closets, two bathtubs and one sink, the latter not shown on the plans. There are no fixtures located on the first and second floors on this line.

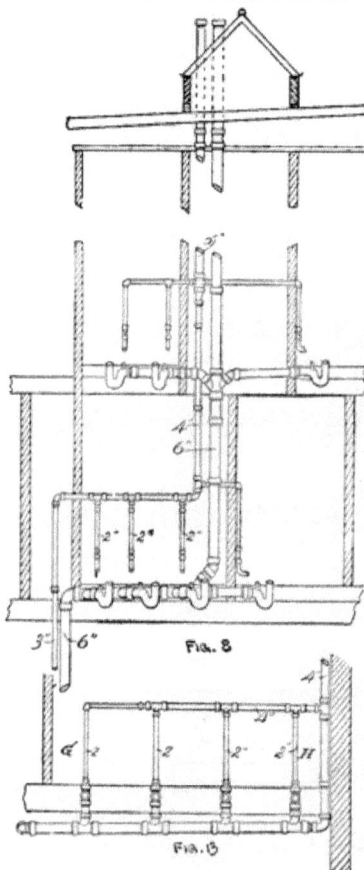

FIG. 8

FIG. 13

Figure 8 is a section through the closets showing the arrangement and connections of the soil and trap vent pipes. Figure 9 is a plan of the gentlemen's toilet-rooms on the first floor, showing arrangement of soil pipes. Figure 10 is an elevation at Z Z, Fig. 9, showing the connection and arrangement of the urinal waste, vent and soil pipes. The 3-inch local

vent pipe is connected to the main ventilating shaft, which is 6 feet in diameter, and in which there is an exhaust fan with connections leading to Turkish bath for ventilating the same.

Figure 12 is a section of urinals at X X, Fig. 10. Figure 11 is a general view of two urinals standing at position N. Access to urinal traps and connections is obtained through an 8 and 10-inch opening left in the marble slab and covered by urinal. By unscrewing the supply and waste cap the urinal can be removed. Figure 13 is an elevation from Y Y, Fig. 9, showing the soil and vent pipe connections for the water-closets.

Figure 17 is an end section at A B, Fig. 9, of the main washstand, showing arrangement of traps, waste, vent, and supply connections, also brackets for supporting stand. Figure 18 is a longitudinal section at C D, Fig. 9.

The top slab of main washstand is 1¾ inches thick polished Italian marble, and is made in four pieces; length over all is 13 feet 6 inches; width 5 feet 6 inches, fitted with 10 15x19 oval basins.

PART III.—CONNECTION DETAILS AND PLAN OF MARBLE-WORK, PRIVATE BATHROOMS.

FIGURES 2 to 6 inclusive show the diagrams by which the marble-work was ordered and set. Figure 3 is a plan of the main public toilet-room water-closets. Figure 4 is a section at Z Z, Figs. 3 and 5. Figure 5 is an elevation at X X, Fig. 3. Figure 6 is a plan of adjacent urinal stalls with inclosed pipe chamber S. Figure 7 is an elevation from W W, Fig. 6, and Fig. 2 is a section at V V, Fig. 7.

Figure 14 is a general view of one of the 35 private bathrooms. Figure 16 shows the arrangement of pipes behind movable paneling P, Fig. 14. Figure 15 is a floor plan and section at Z Z Z, Fig. 16. The different pipes are designated by reference letters as follows. A, safe waste riser; B, hot-water supply; C, hot-water circulation; D, cold-water supply; E, soil pipe; F, back air; II, basin supplies; J, waste from basin; M, vent from basin trap; K, bathtub waste; N, vent from bathtub trap; LL, safe wastes; G, bathtub trap; H, water-closet trap.

PART IV.—TURKISH BATH, MAIN WASHSTAND, DRINKING-FOUNTAIN, BARBER SHOP, ETC.

FIGURE 19 is a view taken of the work underneath the main washstand from position M in the gentlemen's toilet-room, Fig. 9. The supports for carrying the main slab are 1¾ inches polished brass, the bolts passing through the flanges on same at base of marble run through from side to side of base. Waste fixtures, traps, wastes, and vent connections, as well as supplies, valves, and all exposed work, are polished brass. It will be noticed that the back-air pipe passes up through the center of the slab and is 3-inch polished brass pipe, and connects above the ceiling of toilet-room with 4-inch extra heavy cast-iron vent pipe from closets and urinals, which extends up through and above the roof full size.

Figure 20 is a general view of the toilet-room from the same position as Fig. 19. Figure 21 is a general view of the arrangement of supply pipes and valves

over the Turkish plunge bath T, which is 20'x50'x6'
deep to the average surface of the water. Hot and
cold water is supplied to the plunge through 2-inch
pipe M, discharging at the bottom of plunge. The
two lions' heads A A are supplied with hot and cold
water through two 1¼-inch supply pipes, which flow
through the mouths of the lions' heads into the pool,
thus constantly keeping the water in motion.

There is also just over center of pool a large shower
B, 4 feet in diameter, made of 2-inch polished-brass
pipe, nickel-plated, with a large douche in its center.
Surrounding the douche or center shower is
a circle C, of incandescent electric lamps. This
shower is also supplied with hot and cold water
through two 1¼-inch pipe connections. The effect
of the electric light on the water when the shower
is in use is considered very pretty. The valves for
controlling the various supplies to the pool are all
placed overhead on one side of the pool.

The different supplies, etc., are controlled by
valves; D, hot to plunge; E, cold to plunge; F and
valve opposite, hot and cold to special shower; G and
valve opposite, hot and cold to lions' heads; H, I, and
J, hot circulation and cold supplies leading to a tier
of private bathrooms up through the building; K, K,
and K, drain valves; L L, etc., pipe-hanger supports.

In the barber shop the four center columns are
surrounded by a large washstand, which contains 10
15x19 oval basins. The 3-inch galvanized-iron vent
passes up through a large cup case in center of the
stand to above the ceiling, and there connects to a
4-inch extra heavy cast-iron vent pipe from the slopsinks, which extends up through the building and
above the roof. There is also a large shampoo stand
in one end of the room; its supplies are suspended
from ceiling, with branches leading across to a large
center stand, which pass down through the large cup
case, supplying the basins and have drains on their
lower ends.

These supplies are all finished in silver bronze to
suit the decoration of the room. The woodwork
around the large stand is polished cherry with 10-inch
marble base. A marble drinking-fountain in the
rotunda of the main office is a very handsome piece
of carving on Italian marble, and is supplied with
ice-cold water. The waste leading from the fountain
does not connect direct with the house drainage. It
discharges in an open sink located in the basement
room below. This sink also receives all the safe
waste pipes from fixtures on upper floors throughout
the building. The discharging ends of these are
fitted with check valves. The waste from this sink is
a 2-inch pipe, properly trapped and ventilated, and
the trap receives an almost constant flow of water
from the drinking-fountain waste, as this sink is used
for no other purpose.

All the gas and electric-light fixtures on the first
and second floors, including a chandelier, 6½ feet
diameter, in the office dome, are finished in light
steel, and were also furnished by E. D. Hornbrook
& Co., together with the total amount of plumbing
fixtures, as follows: One carved Italian marble
drinking-fountain, 60 water-closets, 49 lavatory
basins, 50 porcelain-lined French bathtubs, 10 large

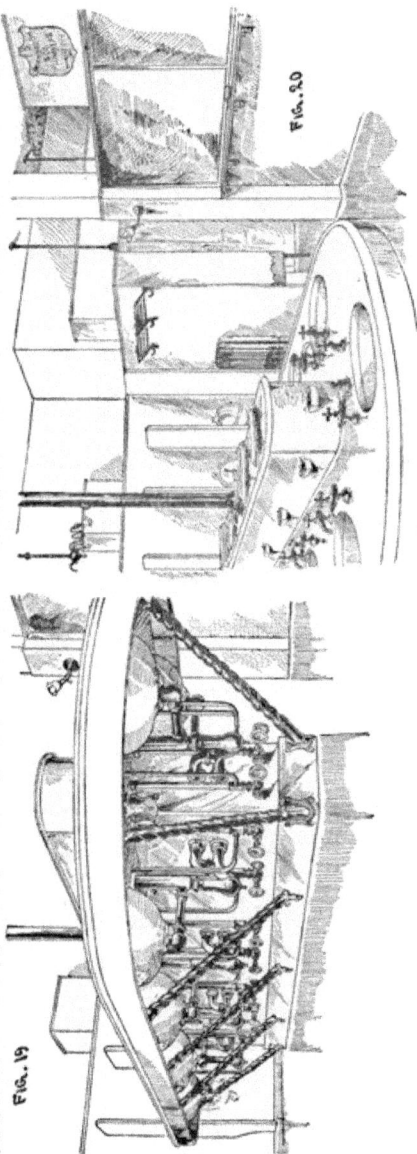

Fig. 20

Fig. 19

PLUMBING IN THE NEW COATES HOUSE, KANSAS CITY, MO.

FIG. 21

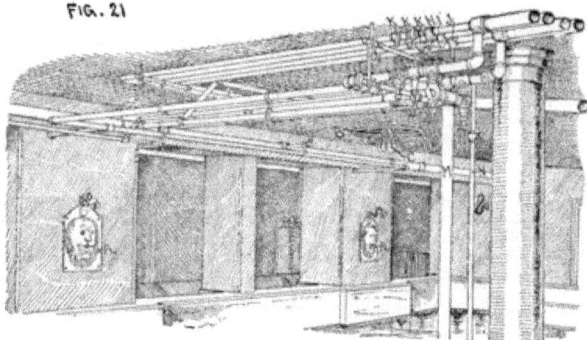

PLUMBING IN THE NEW COATES HOUSE, KANSAS CITY, MO

lip urinals, six slopsinks, two large nickel-plated needle baths, fitted with needle spray, shower, liver spray and douche attachments, cased in Italian marble; nine rubbing slabs with hot and cold showers, two vapor showers, one large shampoo slab, one plunge, and one fountain in reception-room of Turkish bath department.

HOT AND COLD-WATER SYSTEM IN A MIL-WAUKEE HOTEL.

(PUBLISHED IN 1893.)

THE Pabst Hotel, now called the St. Charles, Milwaukee, Wis., was built in 1860, and was refitted and the plumbing extended and modified in 1891 by Halsey Brothers, Milwaukee, who arranged the work so as to control all hot and cold-water supplies from the engine room in the cellar, as shown in the accompanying view of the valve board, heater, and main connections, the pipes and valves being set to meet previous conditions, and conveniently and compactly arranged with symmetry on the walls and low ceiling. The street supply is through a 4-inch main, which ordinarily delivers through a 4-inch Worthington meter, which is by-passed as shown for repairs or emergencies. The cold supply pipe from the meter has independent branches to the 4-inch house distribution drum Z from 1-inch galvan-

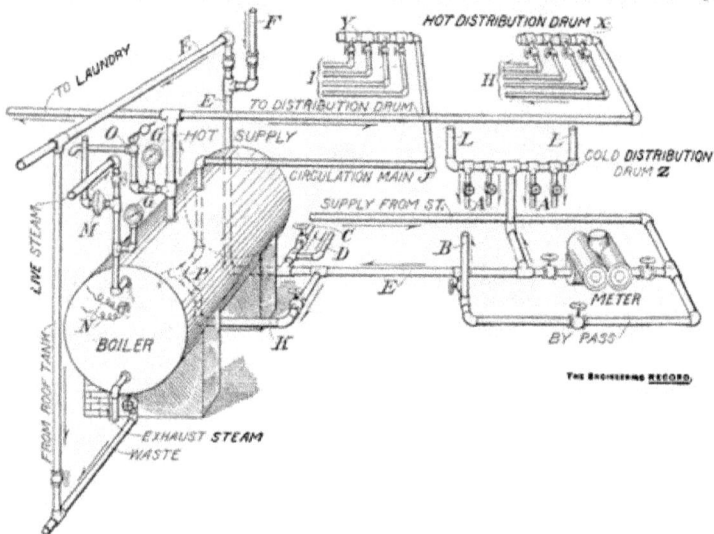

HOT AND COLD-WATER SYSTEM IN A MILWAUKEE HOTEL.

ized-iron pipes A A, etc., to supply the different
upper stories; B to boiler, fire and elevator pumps;
C to the barber shop; D to the basement fixtures;
E to the roof storage tank, and K to the hot-water
boiler, which contains an interior brass steam coil,
whose action is automatically regulated by the
operation of the pneumatic diaphragm M, which
closes the attached steam valve as soon as the temperature of the water in the boiler reaches a fixed
point (usually required to be 200° Fahr.), and causes
the thermostat (not here shown) to complete the
electric circuit through wires N and actuates an
electromagnet that admits pneumatic pressure to
the diaphragm M. Hot water is distributed to the
various lines H H, etc., from the 4-inch drum X.
Each line is vented and has a 1-inch circulation pipe
I returning to the 3-inch drum Y, which discharges
continuously into the boiler through 2-inch pipe J,
whose outlet is connected by a Y (marked L) with
the feed pipe K, so that the discharge of the latter
may cause a suction promoting the return circulation.
The drum Z is provided with two air chambers L L,
4 inches in diameter and filled with rubber balls,
which are said to satisfactorily absorb the shocks of
a water hammer in the risers.

PLUMBING IN THE HOLLAND HOUSE, NEW YORK.

(PUBLISHED IN 1891.)

PART I.—SUCTION TANK, FILTERS, PUMPS, AND PUMP
GOVERNORS.

The Holland House is an 11-story marble hotel at
Fifth Avenue and Thirtieth Street, New York City.

The architects are G. E. Harding & Gooch, and the
plumbers, James Muir, Sons & Co., both of New
York. The plumbing-work is extensive and handsome; opportunity having been given to design and
execute to the best advantage the complete supply,
waste, and vent system, which includes the fixtures
and provisions usual in such recent work in this
metropolis. Standard details, methods, etc., have
been adopted, and although care and skill have been
exercised throughout, our description will consider
only some features of the design and operation of the
general hot and cold-water supply. Other characteristics of the work follow more or less closely the illustrations of methods and details which have appeared
from time to time in THE ENGINEERING RECORD.

Water is taken from the street mains by a 4-inch
pipe A, Fig. 1, and passing through the meter B and
pipe E is delivered through branch C to the distribution pipe D, and through its branches at F F F F to
the two filters G G. The filtered water then returns
through pipes H H, etc., to header I, and thence
through a 3-inch branch J to pipe K, which fills the
main suction tank through ball cocks on the 1½-inch
branches L L L L. M is a 1-inch supply to employees' basement toilet-room, and N is 1½-inch supply direct to the boilers. O is a 1½-inch waste pipe
to empty pipe D into the drip sink P. Each filter
has two valves R R, and two valves S S, commanding its four connection pipes H H and F F respectively. These gate valves are connected by links
(omitted here to avoid confusion), which are arranged
to command them all by a single handle which operates them simultaneously, closing all or opening all
at once. Ordinarily, valves Q, T, U, and U are

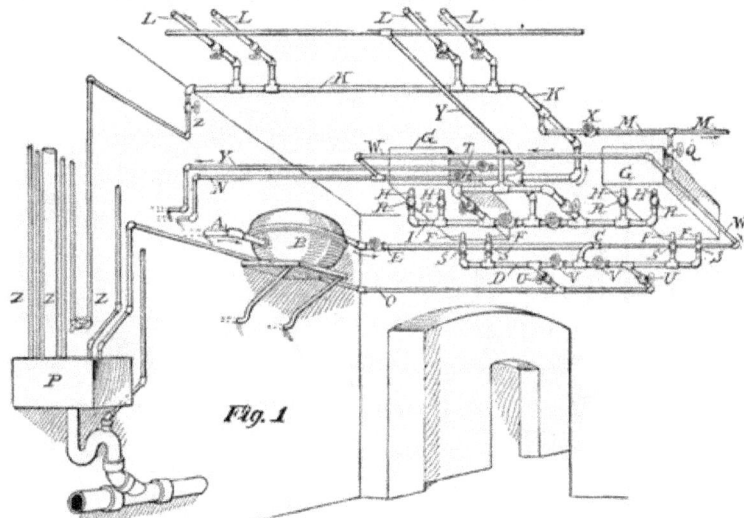

Fig. 1

closed and all others are open, but to wash out the filter these valves are opened and V V and X are closed. Unfiltered water is then supplied from pipe E, through by-pass pipe W W and pipe J to the header I, and entering the filters through connections H H H H escapes to header D through connections F F F F, and is discharged through waste pipe O. If it is only desired to wash one filter at once while the other is in use, valves Q T and X are left as usual, and the valves V and U belonging to this filter being respectively closed and opened, the flow of water will be reversed in it and it will be washed as before described, only filtered water will be used. Y Y are separate supplies to the elevator tanks and hot-water boilers, and Z Z Z Z, etc. are drip pipes.

In Fig. 2, A is the open suction tank about 10'x25'x3' deep, supported at the ceiling of the corridor by the 6-inch rolled iron beams B B, etc. It is filled through four ball cocks on branches L L, etc. of 3-inch pipe K (see also Fig. 1), and overflows through the 4-inch pipe C; D is a ¼-inch telltale, and E is the 6-inch suction pipe to the two Worthington pumps F F for the house service. They deliver through the 5-inch pipe G to the roof tanks, about 140 feet above them, and have a capacity of about 500,000 gallons in 10 hours. H H are hose cocks and I is a pipe to the boiler-room. J J are steam pipes, K K and L L are patent valve regulators, which automatically cut off steam from the pumps when the tanks are full. O is the usual hand throttle valve. P is a ¼-inch telltale from the roof tanks. When they are nearly full water escapes through P and the partly closed valve Q, and producing a pressure in branches J J forces down the piston in the cylinder M, so that its rod R closes the steam valve in S.

PART II.—ROOF TANK CONNECTIONS, SUPPORT AND PRO-
TECTION.

FIGURE 3 is an isometric general view of the twin 60,000-gallon roof tanks T T, with the tank-house and all pipe connections removed for clearness. Figure 4 is a plan of the tanks and tank girders, and Fig. 5 is an elevation at Z Z, Fig. 4. The tanks are made of 7/16-inch tank steel, single-riveted, on frames of 2"x2"x¼" angle iron, about 24'x10'x6' deep. Each tank has two longitudinal tie-rods and two sets of eight transverse tie-rods, all one-half inch in diameter, and hooked into angle clips. The tanks stand on five rolled-iron I beams 12 inches high, which are framed into similar longitudinal girders B B, resting on wooden cushion sills C C, laid on the main outer wall W, and on special bearing wall V. Beams A A, etc. are connected by 32 small crossbeams D D, etc., which may have been designed to support wooden cushion pieces on their top flanges, or a mass of concrete bedding to afford support for the bottom of the tanks throughout. Light vertical angle and beam posts E E, etc. are riveted to the beams A and B, to support the roof and side frames, which are covered with corrugated iron, and all connected by angle iron. This covering, built after the tanks were finished and connected, is thought to be sufficient, together with the frequent fluctuations of its level, to prevent the water in them from freezing, but the position is very much exposed, and a steam radiator will be put in the house if necessary.

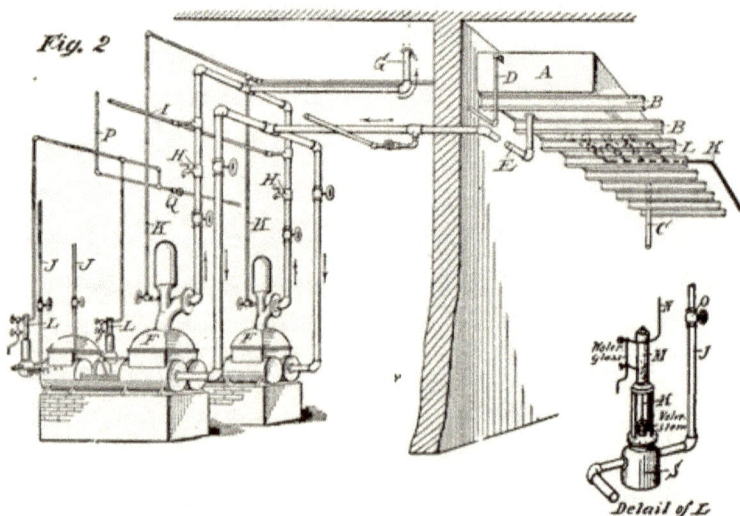

Fig. 2

Detail of L

PLUMBING IN THE HOLLAND HOUSE, NEW YORK.

Figure 6 is a view with the supporting beams, floor, and house removed, so as to show the pipe connections clearly. G is the 4-inch delivery from pumps. H the 3-inch house supply, and I the 4-inch fire line. K is a 1½-inch supply to the hot-water boilers. L is a 1¼-inch supply to the main toilet-room. M is a 1¼-inch supply to the kitchen, and N is a 1¼-inch supply to the laundry. O is a ¾-inch telltale and pressure pipe to the pump governors. The horizontal offsets in the small service pipes were necessitated to avoid the obstructions caused by beams A A, E E, etc.. Figs. 3, 4, and 5. P P are 3-inch overflow stand-pipes discharging directly to the roof gutter. V V, etc are ¾-inch vent pipes to facilitate the emptying of the risers when the upper valves are closed. It will be noticed that the valves are arranged so as to permit any service pipe to be connected or disconnected with either or both tanks.

PART III.—HOT-WATER BOILERS, FLUSHING AND VENTILATION OF URINALS AND ARRANGEMENT OF SOIL AND VENT LINES.

FIGURE 7 is a view of the basement battery of three hot-water boilers A A A, each of which is about 4 feet diameter by 10 feet long. B is a live and C an exhaust pipe, from either of which each boiler can receive steam through its branch D, which connects with a 200-foot interior coil of 1¼-inch brass pipe, whose drip, etc. discharge through branches E E E, and return main F. Cold water under street and tank pressure respectively is supplied through 2-inch pipe G and 4-inch pipe H, and the 1½-inch branches I and 2-inch branches J. Hot water is delivered at M M M, and is admitted to the 4-inch tank pressure main K, or the 2-inch street pressure main L, according to whether that boiler is supplied with

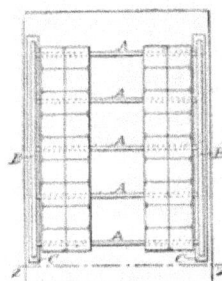

Fig. 4

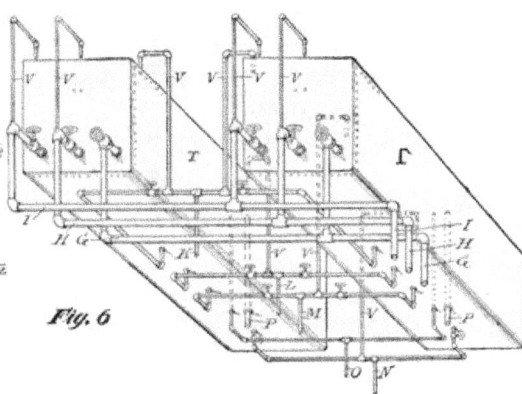

Fig. 6

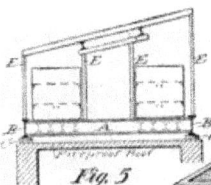

Fig. 5

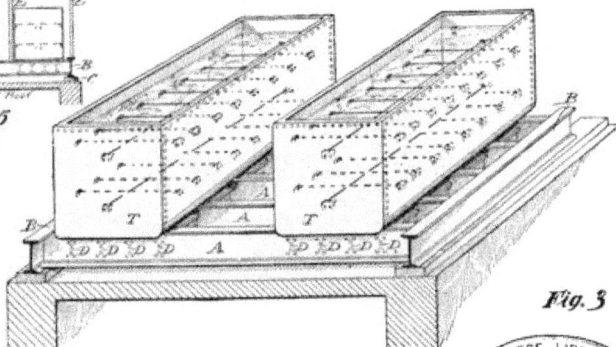

Fig. 3

PLUMBING IN THE HOLLAND HOUSE, NEW YORK.

cold, street or tank water. N is a 2-inch circulation pipe, with 1¼-inch branches O O O to each boiler. P P P are safety valves, and Q Q Q are check valves, closing with a current away from the boiler, so as to prevent possible escape of tank water into the street mains. R is an emptying pipe from rising lines to the sewer.

In the public toilet-room a group of five urinals automatically flushed by one cistern, arranged as shown in Fig. 8, the header B being designed to have a capacity in excess of the combined draft of all the branches B B, etc. to each bowl. A grated floor drainer S is set under every urinal and its ventilation, together with that of the trap and the bowl,

is shown in Fig. 9, where full arrows indicate the directions of air currents and dotted arrows indicate that of the waste. Galvanized-iron duct V connects with a vertical conduit extending through a light shaft to above the roof, and containing at the bottom 32 gas flames to promote the circulation. Figure 10 shows the arrangement of all the hot and cold-water risers which supply the different lines of fixtures throughout the house. Immediately above the distribution main C an air chamber D is branched off from the riser A, which is continued above the supply B, of the highest story, to form a second air chamber E. Beside this an air chamber is provided at every fixture. F F are petcocks.

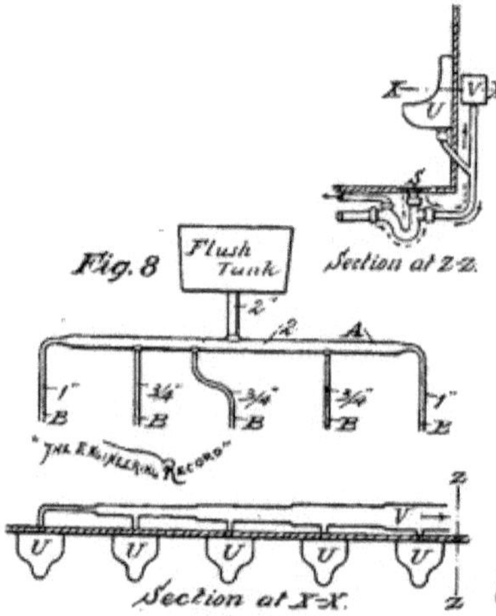

Fig. 8

Section at Z-Z.

Fig. 9　　　Section at X-X.

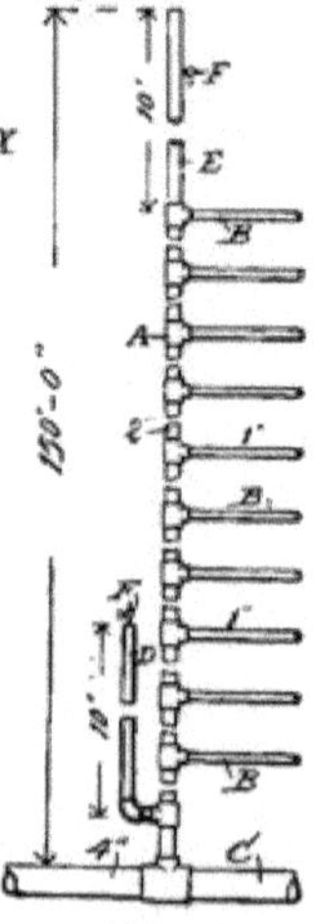

Fig. 10

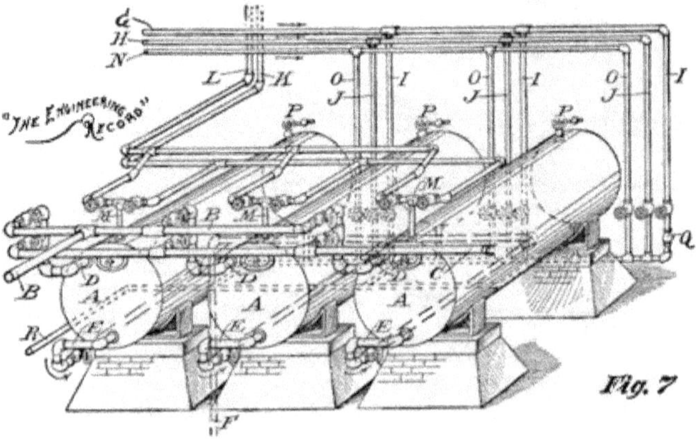

Fig. 7

PLUMBING IN THE HOLLAND HOUSE, NEW YORK.

PLUMBING DETAILS IN THE PLAZA HOTEL, NEW YORK.

(PUBLISHED IN 1891.)

PART I.—STEAM PUMP CONNECTIONS, AUTOMATIC PUMP REGULATOR AND SUCTION TANK.

In designing and executing the plumbing in the Plaza Hotel it was necessary to conform to the plans and conditions of the building, and to meet the requirements of the successive changes in construction and system that accompanied the changes made in its owners, architects, builders, and contractors. William Paul Gerhard, C. E., of New York City, was consulting engineer for the drainage and ventilation, and S. & A. Clark, also of New York, were the contractors for the plumbing, various detached features of which are illustrated in this and succeeding parts.

The water supply for the hotel is received from the city mains, passes through Worthington meters, and is delivered through 3-inch pipes A and B from the Fifty-ninth Street and the Fifth Avenue mains respectively. These pipes deliver through branches may be supplied from the main A directly through branch G. The pumps are also similarly connected with pipe B.

Figure 2 shows the connections of the pump suction and delivery pipes. A A are two Worthington pumps, whose suction pipes B B are connected with the 6-inch header D, which may be supplied from any or all of the branches, E, F, and G. E is a 6-inch pipe to suction tank D, Fig. 1. F is a 3-inch pipe to the cistern which receives the rainwater from the roofs, etc., and G is a 6-inch pipe connected direct to the city mains. The pump-delivery pipes C C are connected to the 4-inch headers H H, which have the branches N N, etc., with valves L L, etc., so arranged that either or both pumps can deliver through any one or more of the pipes O, P, Q, R, and S, four of which lead to the different roof tanks, and one to the elevator tank. All the pipes are jacketed to prevent the condensation of moisture on their surfaces.

Figure 3 shows the pipe A supplying steam to the pumping engine cylinders B B. When the tank is full its supply pipe is closed and the action of the pump develops pressure in it. This pressure operating

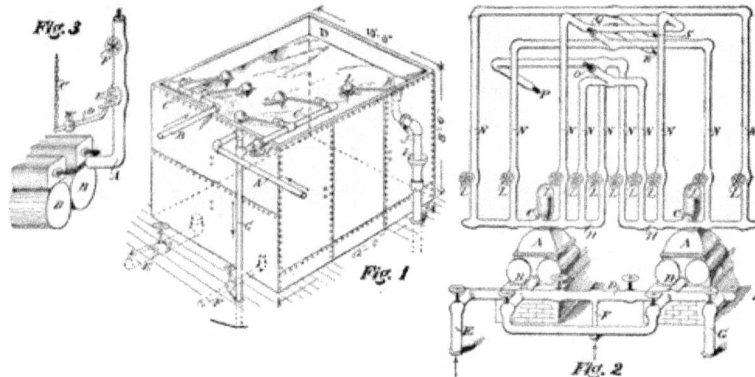

PLUMBING IN THE PLAZA HOTEL, NEW YORK CITY.

C C and double ball cocks into the boiler-plate tank D, which rests on the basement floor and affords some storage in case the city mains are shut off, besides preventing a draft on the meters.

E is a 6-inch suction pipe connected to all the pumps, and F is a 2-inch emptying pipe; J is a 4-inch overflow pipe which empties freely into the bell-mouthed pipe I. The latter is trapped into the sewer. The top of pipe J has a hinged flap plate K, to which a heavy ball float L is connected by a long arm. When the water in the tank is *below* the overflow level this float acts as a weight to hold the cover down, tightly closed, and to prevent any discharge of gas, cellar air, etc. above the water. When the tank is nearly full of water the ball L acts as a buoy to open the flap and permit overflow through J. Ordinarily the valve H is open, and all the street water is first received in tank D, but by closing H the pumps

on the diaphragm of a pump regulator, produces a pull on chain C which raises the weighted lever D and closes valve E, thereby shutting off the steam and stopping the engine. When the water level is lowered in the tank the weight of W opens valve E, and the pump starts up and so on. F is a throttle valve for independently controlling the steam.

PART II.—ROOF TANK AND HOT-WATER TANKS.

Figure 4 is a view of one of the four roof tanks which, with a united capacity of about 60,000 gallons, store the house water supply. The tanks are built of angle and plate iron, and are supported about 3 feet above the floor on broad wooden frames A A, which rest on the tops of the iron floor beams. K K K are 6-inch iron beams under the tank. They are tied together by the ¾ inch rods L L.

B is the 2-inch pump delivery pipe, through which the supply is automatically controlled.

D is the 2-inch pipe which serves both for the house-supply riser and the pump-delivery riser. It is connected to the tank by branch C, in which a check valve, not here shown, permits the flow of water from the tank, but prevents any flow *into* it.

E is a branch to a cock F, for a roof supply, and G is an air chamber. H is 2-inch emptying pipe, and I is the 3-inch overflow discharging on the roof. The tanks are covered by galvanized-iron houses, just large enough to inclose them, and containing steam coils on two sides of the tanks to protect them from freezing.

Figure 5 is a view of the rear end of the hot-water tanks. A and B are 3-inch supplies from Fifty-ninth Street and Fifth Avenue respectively. They are connected to the header C, and have valves arranged so that either or both A or B can supply either or both tanks. D D D D are the tank connection pipes, made 2 inches in diameter, for convenience of tapping into the tanks, and put on in pairs to secure equivalent area of cross-section.

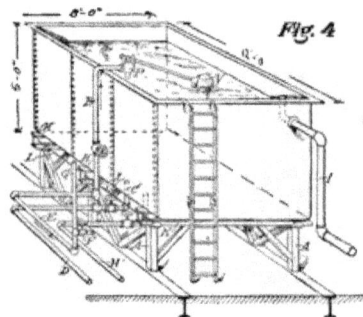

ately or together by operating the valves F and G. The drum is supplied by the branches H H, etc. These were made 2 inches in diameter for convenience in coupling up, and four of them were used to secure ample area of cross-section.

I is an independent 2-inch supply to the basement water-closets; K K, etc. are 1½-inch supplies to different parts of the house; L is an emptying pipe, and M is a hose cock; N N are 1-inch drip pipes for emptying the rising lines; O is a 1-inch safe waste pipe, and P is a 1½ inch refrigerator waste pipe.

Figure 7 is a view of another cold-water distribution drum A, adjacent to the hot-water tanks M M, Fig. 5. It receives its supply through the 3-inch mains G and H. These mains are connected to the 24-inch by 5-foot galvanized steel drum A by four branches B B B B, made of 2-inch pipe for convenience of connecting

D D, etc., are 1½-inch supplies to different lines of washbowls and bathtubs; K is a continuation of the 1-inch drip pipe shown in Fig. 5, and empties the rising lines; F is an emptying pipe, controlled by an

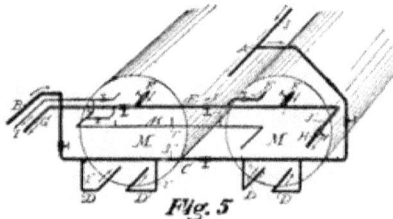

PLUMBING IN THE PLAZA HOTEL, NEW YORK CITY.

equivalent area of cross-section. E is the 3-inch hot-water header connected to the tanks by 2-inch branches F F F, and controlled by valves, so as to receive water from either or both tanks. Its 2-inch branches G and H, to different parts of the house, can be put in communication or separated by opening or closing valve J.

I is a 1½-inch special hot-water supply direct to the kitchen; K is a ½-inch drip pipe to empty various riser lines. Another direct house supply, not here shown, is taken from the opposite ends of the tanks.

PART III.—HOT AND COLD WATER DISTRIBUTION DRUMS.

FIGURE 6 shows a distribution drum A for the cold-water house supply. It is made of galvanized steel and is about 30 inches diameter and 16 feet long. It is suspended by the flat iron hangers B B, etc. from the iron floor beams above.

C and D are 3-inch supply mains from Fifty-ninth Street and Fifth Avenue respectively, and are connected by branch E so that they may be used separ-

ordinary valve F, with the handwheel removed; I is a ¾-inch cold supply to sink J; K is a 1-inch safe waste pipe; L is a 1½-inch refrigerator drip; M M, etc. are iron hangers supporting the drum from the iron floor beams above.

PART IV.—VENTILATION AND FLUSHING OF URINALS.

FIGURE 13 is a general view of the principal gentlemen's toilet-room, which is fitted up throughout with Italian veined marble and nickel plated metalwork.

Figure 14 shows the arrangements for flushing the seven urinals by the three independent automatic flush tanks F F F. That in the center has a special arrangement to operate equally for the three connected urinals. The branches A A each receive half of the flush water and are so placed that the distance D to pipe B is only one-half as great as the distance E to pipe C, thus offering less resistance and securing a greater flow to B, while C has its smaller supply doubled by the flow from the other side.

Figure 15 shows the waste and ventilation pipes of the same set of urinals; D D, etc. are 2-inch waste to the main soil pipe A; B is the 3-inch trap vent pipe and C C C are floor drains with plated strainer tops; E F and G are local vent, screwed iron pipes of 1½-inch, 2-inch and 3-inch diameters respectively. They all enter the closed galvanized-iron box H, to which they are screwed by lock nuts. Box H is 2'x2'x8", and contains a gas flame I, accessible through a handhole with close-fitting glass door. Box H is exhausted through a 6-inch copper pipe J, to a 24x 30-inch galvanized-iron ventilation stack K, which discharges above the roof and contains an exhaust steam pipe L to promote the circulation.

Figure 16 shows the urinal in the bar-room toilet-room. The paneling is of Italian veined marble; exposed metal-work is nickel-plated and the floor of mosaic. The front panels are continually washed by fine streams of water from the perforated branches A A, controlled by the valve B. Valve C controls hose coupling D for use in washing down the floor. The washbowl in this room is provided with a special overflow and emptying valve controlled by handle F. By twisting it, a peg turning in a spiral slot raises stand-pipe E from the valve seat and empties the basin. Releasing the handle F, the stand-pipe returns to its seat and acts as an overflow.

PART V.—SERVANTS' WATER-CLOSETS AND CELLAR, DRAINAGE TANK.

Figure 17 is a diagram of basement toilet-room with six water-closets and one slopsink A. This is adjacent to the laundry, and is intended for the exclusive use of the female employees of that department. The plastered walls have a hard finish, the floor is of cement, partition slabs are slate, and the only woodwork is in the half-doors and the oak seat boards. The partitions are raised from the floor, as shown in Fig. 18, to facilitate circulation of light and air, and cleaning.

The seat boards E are instantly removable by being lifted out of the open brass hinge sockets D, which are bolted to the partitions. The front support C is also brass, bolted to B.

The water-closet seats throughout the house are of a similar pattern, though of more expensive and highly polished wood in the guests' rooms. General bath and toilet rooms are provided in the servants' quarters on the upper floor. The men servants' general water-closets are in two adjacent basement rooms, intended for the separate use of the white and colored employees.

Figure 19 is a diagram of these rooms, which have cement floors, plastered walls, slate paneling with ash trimmings, and cabinet-work. A A. etc. are wardrobe cupboards; B is a washstand with marble slab and table, and plated legs and fixtures; S is a slopsink, D is a dripsink receiving the discharge from safe wastes, overflows, etc.; E is a pair of steps to the raised floor of the toilet-room; C C, etc. are water-closets; and U U are urinals.

Figure 20 is a view from Z, Fig. 19, showing the details and piping. The slate urinal trough U is constantly washed by fine streams of water from the

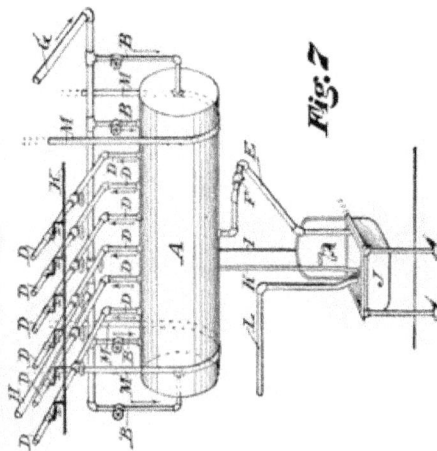

Fig. 7

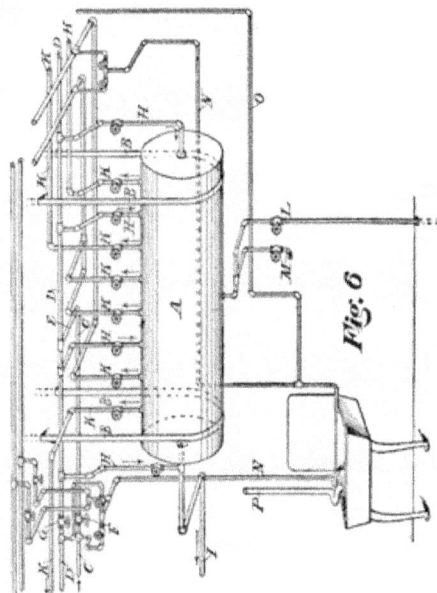

Fig. 6

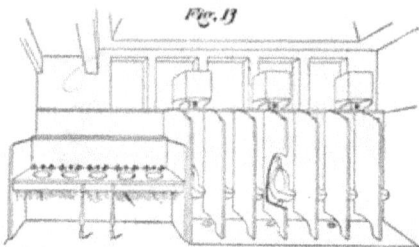

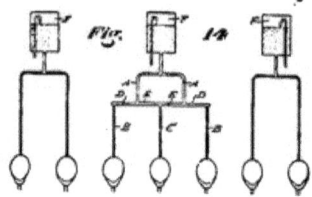

Fig. 13

Fig. 14

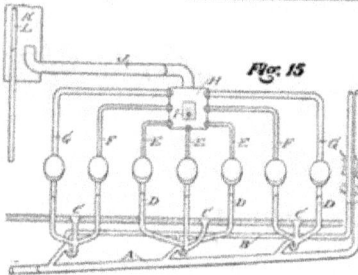

Fig. 15

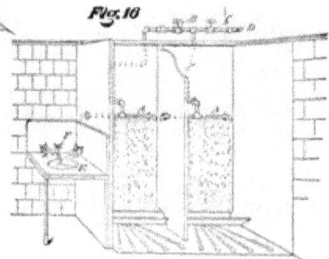

Fig. 16

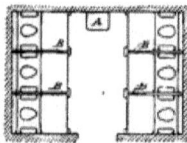

Fig. 17

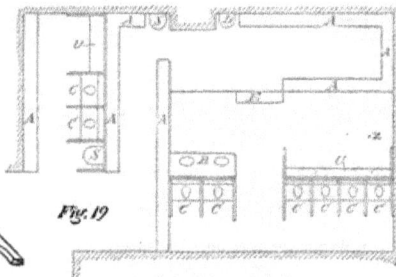

Fig. 19

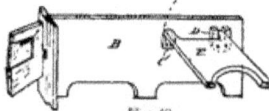

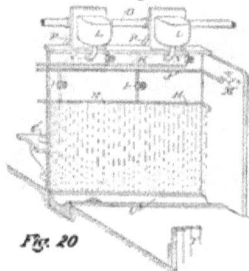

Fig. 18

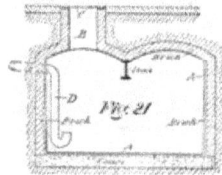

Fig. 21

Fig. 20

PLUMBING IN THE PLAZA HOTEL, NEW YORK CITY.

perforated pipes H H, commanded by the cocks I I, by which the flow can be regulated at will. J and K are 1-inch cold-water supply pipes, the former terminating in hose cock M for washing out the room, and the latter having branches to afford an independent supply to each of the automatic flush tanks L L, etc. for the water-closets. The local vent pipes P P of these are connected to main 3-inch vent flue O, which opens into a ventilation flue extending above the roof, and heated by an exhaust steam riser.

Figure 21 is a section, not to scale, showing the construction of the tank below the cellar. This tank receives the drainage from the refrigerator engine-

water-closets; E E E, washstands, F F, bathtubs, and G, a chambermaid's slopsink.

The rooms have mosaic floor, marble panels and slabs, and ash cabinet-work. The pipes are all exposed and, as well as the fixtures, are nickel-plated. The soil and waste pipe branches are shown by solid black lines. H is a 3-inch fire line with hose cock and hose reel; I, J, K, L, M, N and P are riser lines; I is the 5-inch soil pipe; J, 1½-inch safe waste; K, 4-inch trap vent; L, 2-inch cold-water supply; M, 1½ inch hot-water supply; N, ¾-inch hot-water circulation pipe; and P is 4-inch vent pipe for slopsinks and washbasins.

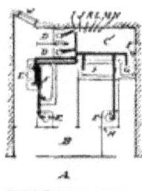

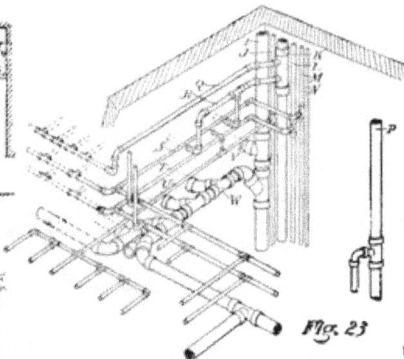

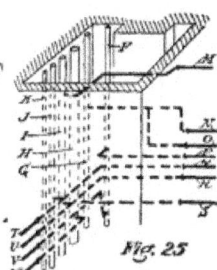

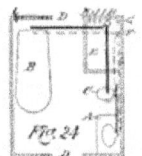

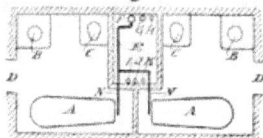

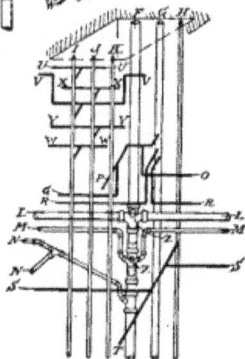

PLUMBING IN THE PLAZA HOTEL, NEW YORK CITY.

room and the laundry. It is about 12'x12'x6' high, and is lined with a 1-inch coating of asphalt. It overflows through the 6-inch pipe D, and is accessible through the manhole B, with cast-iron cover C.

PART VI.—ARRANGEMENT AND PIPING OF PUBLIC AND PRIVATE TOILET-ROOMS.

Figure 22 is a diagram of the gentlemen's toilet and bath rooms on the fourth floor, and is one of a series of six similar ones, directly above one another. A is the main corridor; B, an entrance to hall; C, a pipe and ventilation shaft, and W, a window; D D are

Figure 23 is a perspective view of pipe connections in shaft C, Fig. 22, and the same pipes are designated by the same reference letters as in that figure. Q is a 2-inch cast-iron vent pipe to three washbasins and one bathtub; R is a 2-inch cast-iron vent pipe to two water-closets; S is a ¾ inch hot-water supply to four washbasins and two bathtubs; T is a 1-inch cold-water supply to all the fixtures shown in Fig. 22; U is a 1 inch waste pipe from all the lead safes under the fixtures; V is a 3-inch soil pipe from one water-closet; and W is 4-inch soil pipe from one water-closet, four washbasins, two bathtubs, and one slopsink.

Figure 24 is a diagram of the arrangement of a toilet-room, designed to serve two suites of guest rooms, which communicate with it by doors D D. There is a set of six of these rooms in the same vertical line, on the successive guest floors, in the front of the house, and the same in the rear.

A is a washbasin; B, a porcelain bathtub; C, a water-closet; E, a ventilation, light and pipe shaft; and L, a 3-inch local vent pipe; F, G, H, I, J, and K are riser lines, designated by the same reference letters as in Fig. 25, which is a diagram of their branches to the toilet-room connections. F is a 2-inch safe waste with 1-inch branches D and W, to the water-closet and the bathtub respectively; G is a 4-inch trap vent, with 2-inch branches N, O, and T to the washbasin, water closet and bathtub; H is the 6 inch soil pipe, with 3-inch branch S to water-closet and washbasin, and 2-inch branch X to the bathtub; I is a 2-inch cold-water supply, with ¾-inch branches M, R, and V to the water-closet, cistern, the washbasin and bathtub; J is the hot-water supply with ¾-inch branches O to the washbasin, and U is the bathtub; K is the 1¼-inch hot-water return circulation pipe. In Fig. 24 none of the horizontal branches is shown except for the soil pipe, which is made solid black.

Figure 26 shows the arrangement of one of a set of six double toilet-rooms, which are in the same vertical line and communicate, through doors D D, with special suites of guest chambers on the successive guest floors of the house.

A A are bathtubs; B B are washbasins; and C C, water-closets; E is the light, ventilation and pipe shaft, about 42x36 inches square, and F, G, H, I, J, and K are the riser lines which correspond to those designated by the same reference letters in Fig. 27. The latter shows their connections for the branches serving this floor.

The soil-pipe branches N N, to the bathtubs, only, are shown by solid black lines.

F is a 5-inch soil pipe, with 4-inch branches L L to the water-closets, 2-inch branches M M to the washbasin, and 2 inch branches N N to the bathtubs; Z Z are cleaning-out screens; G is the 4 inch trap vent pipe with 2-inch branches, O to one bathtub and washbasin, P to one bathtub, Q to one basin, and R R to one water-closet each; H is a 1½-inch safe waste pipe with 1-inch branches S S, to one water-closet safe each, and T to two bathtub safes.

I is the 1½-inch cold water supply, with 1½-inch branches U U, to the bathtubs, ¾-inch branches V V to the washbasins, and ¾-inch branches W W to the water-closet cisterns, in the toilet-room next below the one shown in Fig. 26; J is the 1¼-inch hot-water supply, with 1-inch branches X X to the bathtubs, and ½-inch branches Y Y to the washbasins; K is the 1½-inch hot-water return circulation pipe.

The private toilet-rooms attached to the guest suites are all paved with mosaic tiles, have marble paneling, porcelain bathtubs, and polished natural wood cabinetwork. All pipes are exposed and nickel-plated. Throughout the house all pipes are exposed, accessible; and all waste and soil pipes are provided with numerous scrub holes for cleaning. The pipes are of extra heavy cast iron, and were tested by water pressure after the joints were calked.

Special care was taken in arranging the trap vent pipes to carry the horizontal branches always well above the fixtures so as to prevent possibility of their acting as an overflow.

There are in the house a total of about 200 water-closets, 200 washbasins, 150 bathtubs, 20 slopsinks, 20 iron sinks, and 15 urinals. Jeremiah Delaney was the foreman plumber in charge of the work.

PLUMBING IN THE NEW NETHERLAND HOTEL.

(PUBLISHED IN 1893.)

PART I.—GENERAL DESCRIPTION, PLANS AND ELEVATION OF DRAINAGE SYSTEM, PRESSURE TEST AND DETAILS OF HANGERS AND FRESH AIR INLETS.

A NOTABLE addition to the hotel accommodations of New York City was made in the construction of the New Netherland on the northeast corner of Fifty-ninth Street and Fifth Avenue, with its main entrance on Fifth Avenue opposite the "Scholars Gate" entrance to Central Park. The structure, which was constructed according to the plans of William H. Hume, of New York City, architect, is of iron, stone, and brick. It has 125 feet frontage on Fifty-ninth Street and 100 feet on Fifth Avenue. It is 17 stories high, four stories being in the mansard roof. The first or main story is 16 feet high, the others varying from 12 feet to 9 feet 6 inches. The seventeenth-story floor is 210 feet above the sidewalk. There is also a basement and cellar below grade, each 11 feet in the clear.

The plumbing, including gas and water piping and house drainage, has been done by Macdonald & Co., of New York City. The sectional elevation, Fig. 1, is a general diagram of the arrangement of four of the 19 lines of risers which serve the different vertical groups of bath and toilet rooms, only the drainage stacks being here shown, soil pipes A being indicated by heavy full black lines, basin wastes B by lighter black lines, and the main stacks of trap vent pipes C by double light lines, their branches to the fixtures being shown by broken lines.

There are 14 stacks of 5-inch soil pipe A extending from the cellar ceiling to the sixteenth floor, where they are enlarged to 6-inch, and then continued up through and above the roof to a point safely isolated from all house openings, the top of each being crowned by a copper wire hood. There are also five stacks of 3-inch basin waste B, with enlarged upper end and of the same height and general character as the soil pipes. Each of those stacks is composed of standard size wrought-iron pipe with cast-iron fittings, all tested and dipped in hot coal tar before delivery on the ground. All joints are screwed and internal burrs removed. Face joints are metallic only. All fittings have recessed threads giving a practically smooth bore to the entire drainage system; in order to prevent the lodgment of insoluble matter on or against shoulder burrs, etc. The connections at each floor are by 45 degree bends, and

FIG. 1

PLUMBING IN THE NEW NETHERLAND HOTEL, NEW YORK CITY.

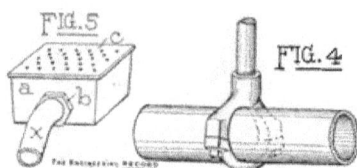

FIG.5

FIG. 4

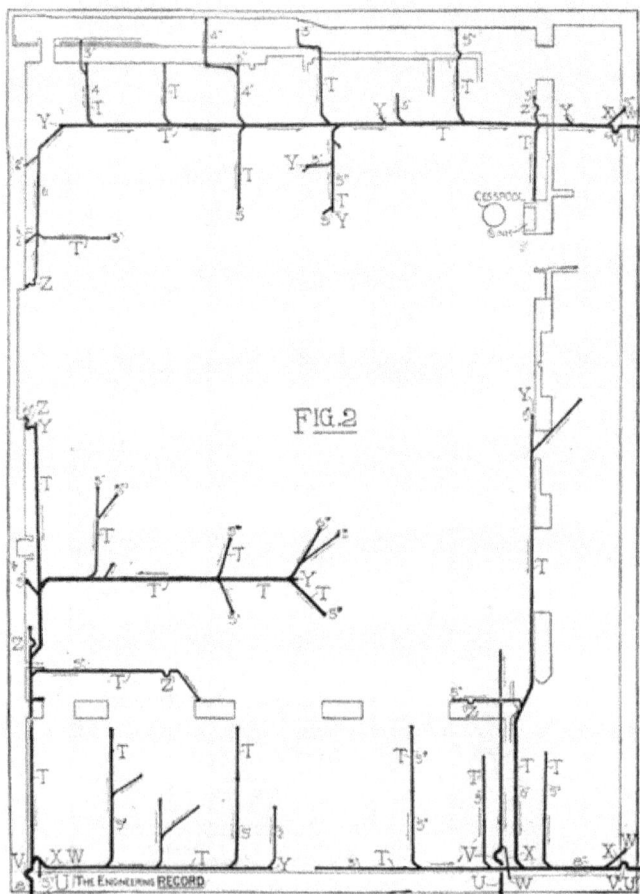

FIG.2

PLUMBING IN THE NEW NETHERLAND HOTEL, NEW YORK CITY.

the pipes are securely fastened to the floor beams, or were anchored in recesses in the walls as they were built up. Each stack of soil or waste pipes has a corresponding air vent pipe C of same quality, diameter, height, and workmanship. From these vent pipes branches are taken off at points sufficiently high to guard against a possible inflow from the fixture wastes, if from any cause these should become stopped, and are designed to afford ample protection from syphonage of traps.

The soil and waste pipes connect just below the cellar ceiling with the horizontal house drains of the same size and quality, which are hung from the cellar ceiling and shown in Fig. 2. They have a fall of not less than one-quarter of an inch to a foot, and increase in size as the service requires. The closets, sinks, etc. in the cellar are raised sufficiently to give a clear drainage into the house drain before it passes to the outside of the wall. The hangers used were specially made for this job, and are shown in Fig. 4. They are made of malleable cast iron. The saddle piece is tapped for iron pipe threads to suit the size of pipe carried, the clamp piece being held up by tap bolts. On account of the power plant and other apparatus in the cellar, direct lines could not in all cases be located for the house drains from the foot of risers to the point of exit or main drains, but they were run as directly as the condition would permit. The junctions of two lines are invariably effected by a 45-degree connection or a long 90-degree bend, as shown in Fig. 1. The same general practice is followed in laying the horizontal pipe as holds with the uprights; there are no dead ends, and ample arrangements are made for cleanouts by the full-sized screwed brass plugs Y. No short 90-degree bends are used on any of the lines. For convenience of changes or repairs, a specially designed stepped-faced union was used, which dispenses with the use of gaskets or other packing, which might in time become porous or offensive.

When the stacks had been erected to the ninth floor, it became necessary to facilitate the work to inclose some of them in the fireproof partition and to cover up much of the work in the fireproof arches so that floors could be laid. An air test was therefore made then, to the end that all joints should be proved perfectly tight before inclosure, as small leaks, which may occur in the most carefully executed work, would prove very troublesome and costly to repair after they had become inaccessible. This test was made to the satisfaction of the building inspectors.

At such points as were necessary in the suspended house drainage, 45-degree entrances T, Fig. 2, were allowed for caring for the closet, urinal, basin, and kitchen service on the basement floor. The outflow of sewage was through one 6-inch and one 8-inch line on Fifty-ninth Street, and one 6-inch and one 8-inch line on Fifth Avenue, all marked U, as shown on Fig. 2. Each line had a deep-seal running trap V with cleanout close to the inside wall line, on the house line of which there were 4-inch and 5-inch fresh-air inlets W connecting by wrought-iron pipes X to fresh-air gratings at the curb line of the sidewalks,

as shown in Fig. 5. The cast-iron box a being 10"x18"x9", a hole on one side allowed the connecting of the 4 or 5-inch pipe X, which is fastened by the two locknuts b b. These connections are close to the bottoms of the boxes, so that by flushing them occasionally they are kept clear of dirt. The brass grating c is flush on top, and by allowing the entrance of storm or other waters, helps to keep them clear.

From the running traps V, Fig. 2, extra heavy cast-iron pipes of the same size with leaded joints were laid to the city sewers. Those pipes were tested to 20 pounds pressure to the square inch, and were dipped in hot coal tar before being laid. The rainwater leaders enter the house-drainage system in the cellar through the deep-seal cast-iron running traps Z. The surface drainage of the cellar is into a 4x6-foot cement-lined brick cesspool beneath the cellar floor, and from there is automatically pumped into the house-drainage system.

When all necessary connections were made, every opening was closed and the entire system of sewer, drain, soil, waste, and air vent pipes was united by a communicating temporary pipe, so that air could be pumped into them, and an equal pressure be gotten at all points from one pumping station. A large size gas force pump was used for the purpose and a mercury column showing 25 inches of mercury recorded the pressure. The test was made in the presence of the representative of the Board of Health and the architects. When an air pressure of 25 inches of mercury was attained the pump was shut off.

For three hours there was no perceptible loss of pressure and the test was therefore declared to be satisfactory. Such a severe test as this and its result is a standing testimonial to the workmanship of those actually engaged upon the work.

PART II—GENERAL SYSTEMS OF WATER DISTRIBUTION. DIAGRAM OF VERTICAL LINES.

On account of the great height of this hotel, and in order to secure a more equable delivery to the several floors, it was decided to have three separate systems of water supply and distribution, the lower one to supply the cellar, basement, and first floor, the intermediate one supplying from the second to the eighth stories inclusive, and the upper one beginning at the ninth story and supplying all above. There are two 4 inch Croton water-service pipes, each delivering only through a Worthington meter. One of them directly supplies the water which is used in the lower system; the other discharges through ball cocks into a 5,000-gallon iron storage tank in the cellar. From this tank water is pumped to the tanks of the intermediate and upper systems. All of the water used in the building is filtered in the cellar before entering the storage tank. The entire storage capacity of the roof, intermediate and cellar tanks is over 32,000 gallons. Each tank has overflows, wastes, recording gauges and independent shut-offs on discharges. The cold-water storage supply for the intermediate system is for convenience of arrangement and economy of space distributed in two open tanks on the ninth and tenth floors. The

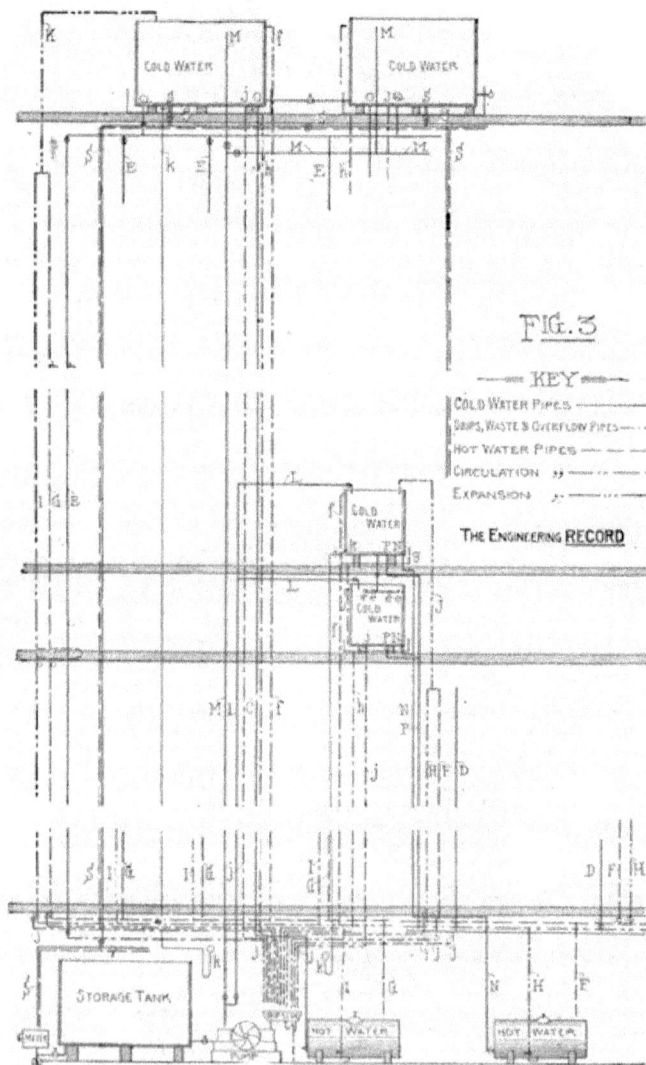

FIG. 3

KEY

COLD WATER PIPES
SOILS, WASTE & OVERFLOW PIPES
HOT WATER PIPES
CIRCULATION „
EXPANSION „

THE ENGINEERING RECORD

PLUMBING IN THE NEW NETHERLAND HOTEL, NEW YORK CITY.

tanks are of quarter-inch riveted wrought iron. Each holds 2,500 gallons, and is set in a special closet.

Three quarter-inch riveted and stayed wrought-iron tanks on the roof store about 22,000 gallons of water, and furnish a pressure head for the upper system. They are protected from the weather by two iron and brick houses, which have steam radiators to prevent freezing in winter. Each of the three pressure systems has a separate hot-water supply from its own boiler. These boilers are substantially alike and are located in the cellar. Each is a horizontal cylindrical steel tank of 550 gallons capacity, supplied with a brass steam coil which can be operated with either live or exhaust steam at will, though the former is not intended to be used when the supply of the latter is sufficient. The water is pumped into intermediate and roof storage tanks by two pumps, one for each system, has steam on constantly for fire service, and these pumps, together with a boiler feed pump, are interchangeably connected, so that all can work together, or any one can be cut out and its duty performed by any one or all of the rest.

Figure 3 is a diagram not drawn to scale or exact location and detail, but prepared to present clearly the relative position, connections and operations of the three systems and show their characteristic features. In it the filters, one of the roof tanks and the hot-water heater for the basement system are omitted to avoid confusion. Pressure gauges are shown as open circles and some principal valves are conventionally indicated by small open circles. The intermediate tanks are filled by the cellar pump through its delivery pipe L, which has branches to each tank and fills the lower one first. The roof tanks are filled by pump pipe M, delivering through separate branches, and they are connected by open equalizing pipes. The distribution of cold water in the upper system is made by horizontal pipes on the seventeenth-story ceiling and that of cold water for the lower and hot water for all three systems is made by horizontal pipes on the cellar ceiling. Vent pipes are carried from the summits of each of the hot-water risers and open freely with return bends above their respective pressure tanks. The reference letters in Fig. 3 have the following significance: S, fire system; R, safe wastes from fixtures; *e e*, etc., ball cocks; *f f*, tank overflows; *g g*, save all pans under tanks; *h h*, safe wastes from tank pans; *j j*, emptying pipes for hot, cold, and circulation lines; *k k*, water-level gauges.

In the intermediate system: D, cold-water risers to groups of fixtures; F, hot-water risers to groups of fixtures; H, circulation pipe from group of fixtures; J, expansion pipe from group of fixtures; N, tank cold supply to hot-water boiler; P, cold-water supply from tanks to distribution system.

In the upper system: E, cold-water risers to groups of fixtures; G, hot-water risers to groups of fixtures; I, circulation pipe from group of fixtures; K, expansion pipe from group of fixtures; O, tank cold supply to hot-water boiler; Q, cold-water supply from tanks to distribution system.

The highest point of every hot-water riser F or G is connected with the lower side of its hot-water

boiler by its circulation pipe H or I, and with the atmosphere by its half-inch open vent or expansion pipe J or K. This prevents the collection of cold water at any point of the system, and provides for the delivery of the hottest water in the boiler at any valve on any branch immediately on its opening, in conformity to the well-known principles of pressure head and circulation. The first is provided by the tank storage and gives sufficient uniform levels of water throughout the system, and the second insures the continual ascent of the hottest water to the summit of the system and corresponding return flow of an equal volume as fast as its specific gravity is increased by cooling. The open expansion pipe taken from the highest point of the loop, prevents accumulation of air or steam, which, if in sufficient quantities, would interfere with the circulation in those pipes, and its upper open end allows all vapors to pass over and into the tank.

Two vertical 4-inch galvanized-iron fire mains S S are supplied from the roof tanks and extend the full height of the building, connecting in the basement with the pump system and providing complete fire service by outlets taken off at each floor and controlled by a 2½-inch fire valve, which is provided with 50 feet of 2½-inch hose and nozzle housed on a swinging hose rack. By the proper manipulation of valves, the entire contents of the three tanks on the roof can be used for fire service, or, at the will of the engineer, the discharge from the fire pump in the cellar can be turned directly into the fire mains and used upon any floor. Each of the 19 sections of risers and lines of fixtures is designated by a letter which identifies every main pipe therein and locates any vertical pipe in the house, as soil D, vent G, intermediate cold E, upper hot F, etc. The save-alls of each of the 19 sections of fixtures have an independent waste. The lower ends of all of them are assembled over a large sink in view of the engineer. Each waste has a brass label lettered to show its section, so that if it shows a leak the engineer can shut off the supply valves lettered to correspond with this section. Excepting the valves of the cold water for the upper system, all valves are under his immediate control. To prevent the passage of cellar air or sound through them, the ends of each of the safe wastes are provided with a hinged flat disk, so hung as to close by its own weight, but to open out for the discharge of leaking water.

All the water pipes were made of galvanized wrought iron, those for the upper system being extra heavy throughout, although the specification only required special strength below the ninth floor, where they were subjected to very heavy pressures, reaching a maximum of about 100 pounds per square inch. The diameters of risers are cold and hot supplies, 1½ inches; return circulation, 1 inch; vent or expansion, three-quarters inch; and emptying pipes, one-half inch. The hot-water risers and circulation pipes of the upper system having a direct rise of 220 feet, their expansion was provided for by a lateral spring at the ninth floor, as indicated in Fig. 3 and shown in detail in Fig. 6. The return bends A and elbows B are extra heavy; the lower section was hung from

the crosspipe C, its expansion being downwards and
allowed for at the lower end by flexure of a horizontal connection. The upper section was made fast at
its center, and so expanded up and down from that
point.

PART III.—WATER-HEATING TANKS AND PUMP CONNEC-
TIONS.

THE water-heating tanks for the three systems are
nested as shown in Fig. 7. They are each heated by
exhaust steam through the 7-inch pipe A, which connects to each tank by the two 3-inch pipes B B with
stop valves close to the tanks. Each of the pipes B B
is connected to the 4 inch brass tube return coil
heater C, shown through the broken shell of tank D,

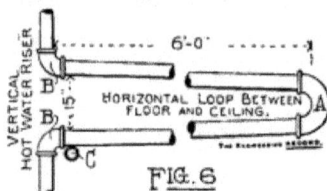

FIG. 6

The exhaust steam enters at both ends of the coil C,
acting as a surface condenser and without circulation,
the waste passing off to cesspool through the pipe H
and the valve I. Should it be necessary to heat the
water by live steam, the valves E are closed and the
valves F on the live-steam pipe G are opened. Valve
I is closed and the valve J is opened, allowing the
condensation to pass to the trap K and on to the
pump through the pipe L.

Cold water is laid on to the tank D from the upper
system cold-water tank, through the 2-inch pipe M,
and the hot water for the same system leaves the

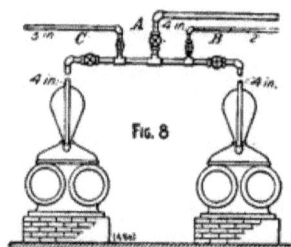

FIG. 8

tank through the pipe N. The pipes O and P serve
in like manner the intermediate system and tank Q,
while the pipes R and S serve lower system and tank
T. To guard against excessive pressure each tank
was provided with the safety valves U, the wastes of
which empty into the storage tanks. The valves V
are for emptying the tanks.

The cold water for the upper and intermediate systems is pumped to the respective tanks by two 12"x
6"x10" Worthington duplex pumps. The 4-inch pipe
A, Fig 8, is for fire service, the 2-inch pipe B is to
the tanks on the seventeenth floor, and the 3-inch
pipe C is to the tanks on the ninth and tenth floors.

PART IV.—DETAILS OF ROOF TANKS AND INTERMEDIATE
TANKS AND CONNECTIONS.

FIGURE 9 is a diagram of the arrangement of the
roof tanks conventionally shown in Fig 3. They are
filled through a 3-inch pump pipe M, and are connected by an open equalizing pipe A.

Figure 10 is a side elevation of one of the tanks,
and Fig. 11 is an elevation at Z Z, Fig. 10. The
tanks each rest on four 12 inch rolled steel beams
D D, etc., which are supported at a convenient

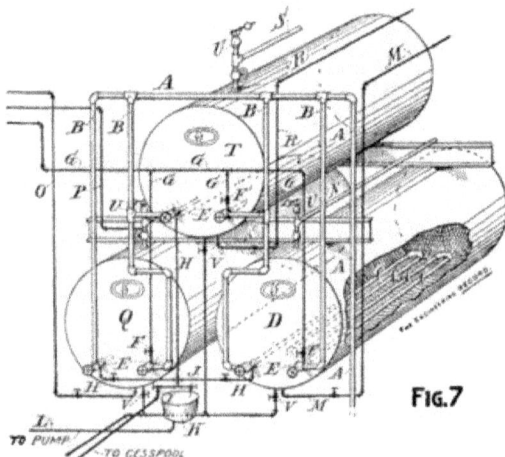

FIG.7

PLUMBING IN THE NEW NETHERLAND HOTEL, NEW YORK CITY.

height above the floor by I-beam pillars E E, etc.,
which are set on the iron floor girders below. These
pillars, as well as all pipes that pass through the roof,
are cased with copper flashings G G, etc., which have
horizontal flanges H H, etc., resting directly upon
the brickwork and covered by the cement surface I I.
The pipes are also provided with sleeves L L and
bottom escutcheons.

The water supply from pump main M is controlled
by valve B and received in stand-pipe C. Delivery
is through pipe O to the hot-water boiler in the basement, through N to the house service main for the
upper system, and through Q to the five mains. At *k*
is represented a ½-inch pressure pipe to the engine-
room gauge. J is a vent and K is an expansion pipe.
T is a telltale to a sink in the engine-rooom. P P are
a row of ¾-inch air pipes from different riser lines.
There is a 4-inch overflow *f* emptying on the roof.
S S are angle braces for the supporting beams.

Figure 12 shows the connections of the two 3,500-
gallon tanks on the ninth and tenth floors for supplying the intermediate system. Water is received
through the 2-inch pump pipe L, which is sleeved to
prevent noise and splashing, and immediately fills

the lower tank through the 2-inch pipe *d*, and five
1-inch ball cocks, which close when the lower tank is
full and permit the upper tank to fill up to the level
of its 3-inch overflow *f*. This overflow empties into
the lower tank and causes it to overflow through the
4-inch galvanized-iron emptying pipe P; at the same
time indicating the fact by a discharge of water
through the small telltale pipe *f*, the discharge of

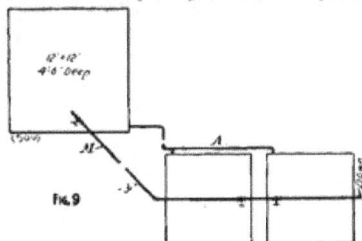
FIG. 12

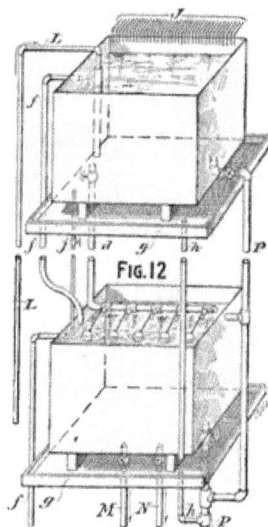
FIG. 9

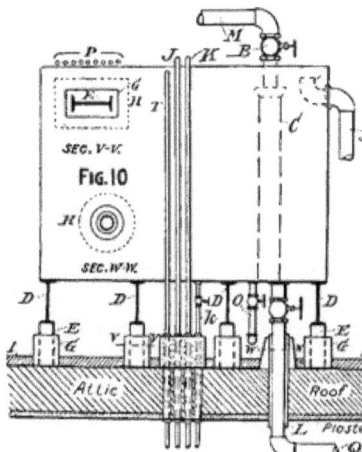
FIG. 10

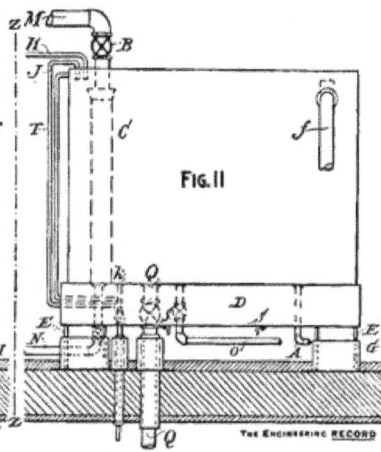
FIG. 11

PLUMBING IN THE NEW NETHERLAND HOTEL, NEW YORK CITY.

which is visible in a sink in the engine-room. At *j* is a pressure pipe to a gauge in the engine-room which indicates the height of water in the upper tank; *h h* are safe wastes. P is an emptying pipe to the sewer. M is a 3-inch supply to the cold-water distribution system in the cellar. N is a 3-inch supply to the hot-water boiler of the intermediate system. At J there are 36 ½-inch vent pipes from rising lines of the intermediate system.

PART V.—DISTRIBUTION OF COLD WATER IN THE UPPER SYSTEM, GENERAL LOCATION OF BATH AND TOILET ROOMS.

FIGURE 13 is a plan from above, looking downward, of the pipes on the attic ceiling which govern the distribution of cold water for the upper system. The valves have their handles down, but they are here shown up for distinctness. The 4-inch vertical branches C C C from the roof tanks, Fig. 9, supply the 4-inch fire line S and the pipe Q to the hot-water boiler, while a branch D connects with the 4-inch header F, from which the 16-inch pipes E E, etc. are branched to the rising lines of the different parts of the systems for the ninth to the sixteenth floors inclusive, and lines G and H are taken to supply the kitchen and laundry and the servants' quarters on the seventeenth floor. I I etc. are ½-inch vent pipes opening above the roof tank.

Figures 14 and 15 are sectional elevations looking

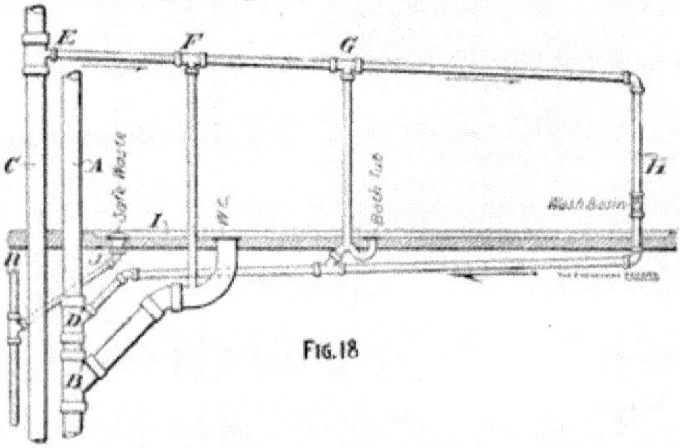

FIG. 18

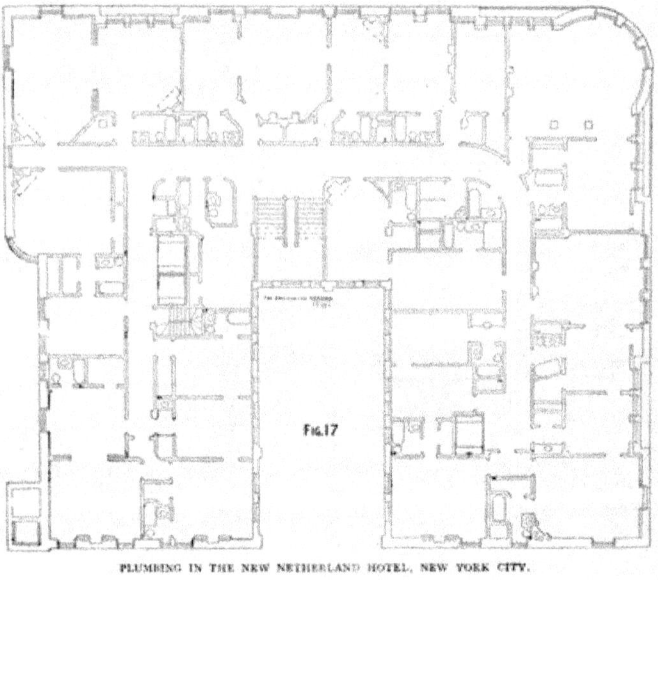

Fig. 17

PLUMBING IN THE NEW NETHERLAND HOTEL, NEW YORK CITY.

in different directions from Z Z, Fig. 13. Figure 16 is a sectional elevation at Y Y, Fig. 13. Vent pipes I I, etc. pass through the roof in 1-inch sleeves J J. with flashings above and escutcheons below. The distribution pipes E E, etc. are suspended in tees one size larger than themselves and are supported from above by vertical pipe hangers B B, etc., which either pass through the floor and have tee-connections to pipes K that lie on the iron beams or are tee-connected to 1½-inch bearing pipes A, which are supported underneath the ceiling by clamps L to the lower beam flanges.

Figure 27 is a plan of one of the guest floors showing the arrangement of the bath and toilet rooms. The position of the vertical hot and cold water risers is indicated by pairs of small black circles. The arrangement of all the other guest floors is substantially similar to that shown. Each hot and cold water riser and circulation pipe has an independent globe valve shut-off. A brass tag fastened to each valve is plainly stamped with a letter designating the particular section the valve controls. Upon the house service side of these valves ½-inch globe valves are connected to the galvanized-iron pipe I, Fig. 3, leading to a sink in the engine-room. Through these valves and pipes the lines may be emptied if necessary. All rising lines are parallel below the eighth floor, and all the pipes in the upper system are extra heavy to sustain the maximum pressure due to a head of 244 feet at the lowest point.

All of the fixtures are set on full-size marble slabs, having a ½-inch countersink for a save-all, with a 3-inch brass rose, screw-top strainer, connected to a 1¼-inch galvanized-iron waste pipe with a good incline and entering its 2-inch sectional save-all waste pipes R, Fig. 18, by a 45-degree connection. All rooms in which plumbing fixtures are located have tiled floors, tile wainscoting 7 feet high, and heavy plate-glass mirrors permanently set into the walls. All stall partitions are of marble. No toilet-rooms are built against outside walls, but to insure to each room a sufficient supply of fresh air six vertical vent shafts 2'6"x5', equipped with exhaust fans at their tops above the roof, were located in different parts of the building, and the water-closets were clustered as near them as possible, and each was connected to it by a ventilating space formed between the main ceiling and an auxiliary one hung 18 inches below it. The foul air thus removed is replaced by fresh air drawn in from the halls through wall openings just above the floor, which are screened with open bronzed fretwork.

The several soil, waste, and venting stacks, water and circulating pipes were run in the vent shafts and ducts wherever it was possible in order to make access easy for repairs or alterations. Beside the kitchen, laundry, and barber-shop fixtures and other fixtures in the cellar, basement, first, sixteenth, and seventeenth floors, the 14 guest floors have 210 bathtubs, 316 washbasins, 243 water-closets, 14 slopsinks, and two urinals. Each fixture has an independent trap which is back-vented and has cleanout holes. All water connections have a separate globe valve shut-off. The wastes are of the stand-pipe pattern

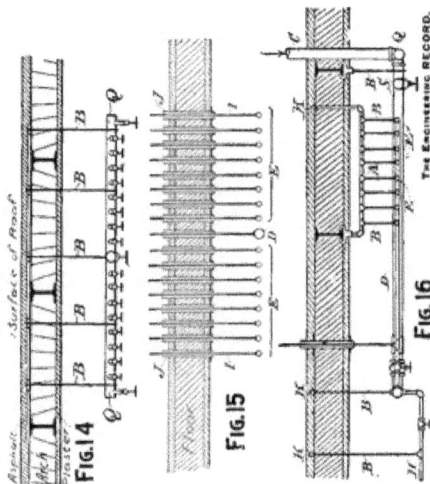

FIG. 14

FIG. 15

FIG. 16

THE ENGINEERING RECORD.

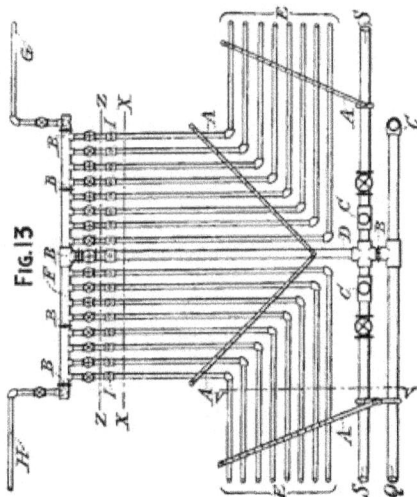

FIG. 13

PLUMBING IN THE NEW NETHERLAND HOTEL, NEW YORK CITY.

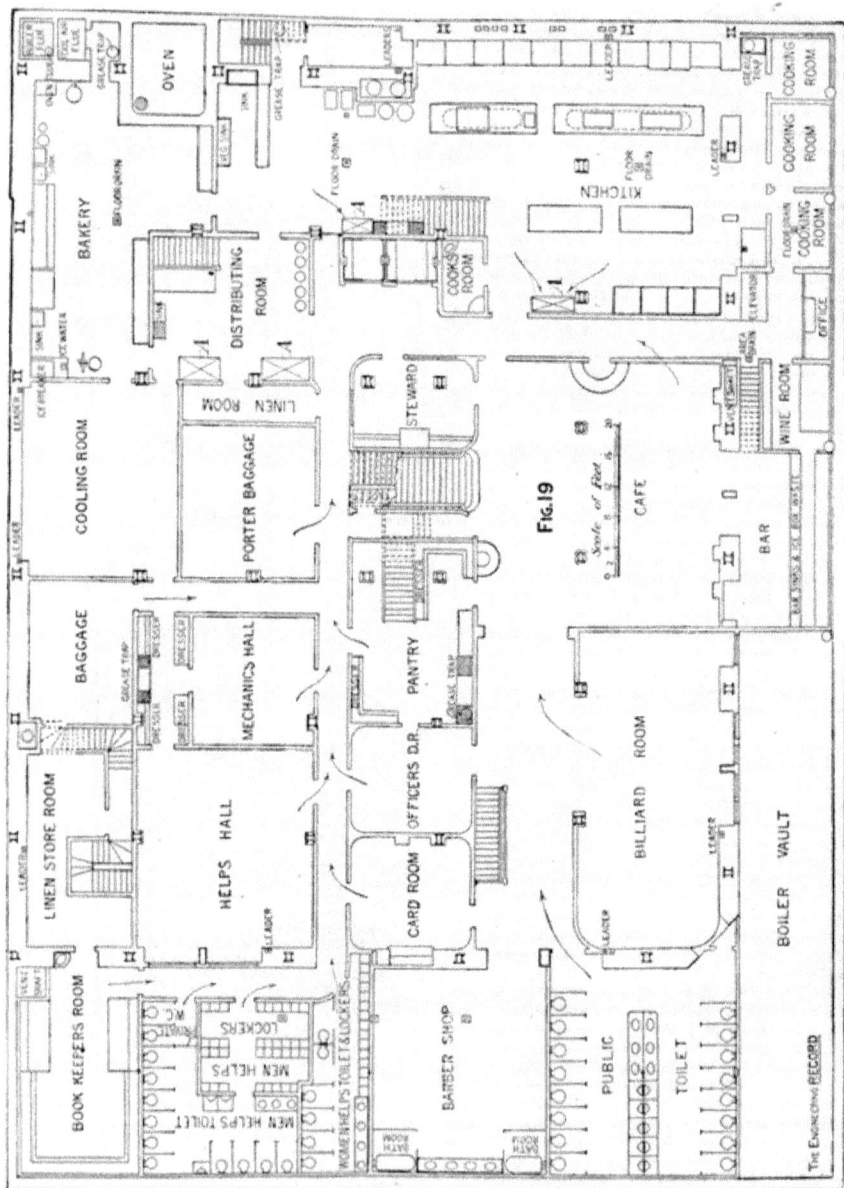

Fig. 19

PLUMBING IN THE NEW NETHERLAND HOTEL, NEW YORK CITY.

and all brasswork is nickel-plated. The marble basin slabs are 1½ inches thick and have 14-inch backs. All fixtures are exposed, the basin slabs being supported by nickel-plated brass legs and the bathtubs resting 5 inches from the floor on marble footings.

Figure 18 shows the general plan of drainage, trap vent, and safe wastes for the several bath and toilet rooms. A is the 5-inch soil pipe with the 5-inch connection B for the water-closet. From the 5-inch vent stack C at a point well above all fixtures is taken a 2-inch branch E, which has tee branches F and G to the water-closet and bathtub traps and pitches continuously to the branch H, forming the basin waste. This is intended to provide a free circulation of air through all the connected pipes. The safe waste J has a heavy pitch and is accessible for cleaning if obstructed.

PART VI.—PLAN OF BASEMENT AND DESCRIPTION OF PLUMBING IN UPPER STORIES.

Figure 19 is a basement plan showing the arrangement of the plumbing fixtures and indicating by arrows the direction of foul air exhausted into the ventilating shafts A A, etc. There are in the cellar some dripsinks and water-closets; on the sixteenth floor washbowls, water-closets, and bathrooms for the servants; and on the seventeenth floor the usual laundry apparatus and a steam drying closet. There is also on this floor the following kitchen apparatus installed by the Duparquet, Huot & Moneuse Company: 40 feet of ranges, 6 feet of broilers, two steam stock boilers, two steam vegetable steamers, one four-compartment dry steamer, one steam egg boiler, two steam bain-marie boxes, two steam tables, six steam dish heaters, two steam dish washers, 10 coffee, tea, and hot-water urns, two steam confectioners' kettles, cooks' and marble tables, freezer, ice breaker, steel canopies over ranges, kettles, etc., and copper and tin cooking utensils, etc.

REMODELING AN ELABORATE WATER SYSTEM IN A HOTEL.

(PUBLISHED IN 1895.)

After the completion of the new Hotel Pfister in Milwaukee, Wis., it was found that the hot and cold water distribution and circulation systems were complicated, confused, and unsatisfactory, and Messrs. Halsey Brothers, of Milwaukee, were engaged to make such changes and reconstruction as were necessary to secure the efficient and positive operation and control of the system. Their examination disclosed an irregular arrangement and a great multiplication of pipes, many of them not properly connected, valved, or identified. Separate lines were not always distinguishable, and it was often impossible either to locate the pipes leading to certain fixtures or to command them afterwards without involving many others, thus making satisfactory maintenance and operation almost impossible. This state of affairs illustrates the necessity of installing large work by the simplest and most carefully developed direct system with experienced practical treatment of arrangement and connections as well as adhering to the letter of specification and architects' general requirements on the plans. As it was impracticable to change the pipe lines already permanently built into the walls and floors and it was not permitted to disturb the decorations of the upper floors a series of careful measurements and plans was first made, locating and tracing every pipe line in the building down to the cellar where the engines, pumps, etc. were located. Then an accurate cellar plan was made to large scale showing the existing arrangement of pipes and connections, and on it a new system was laid out dividing the different lines into suitable groups, and arranging and connecting iron branches so as to conform as well as possible to the old arrangement and yet secure the utmost sim-

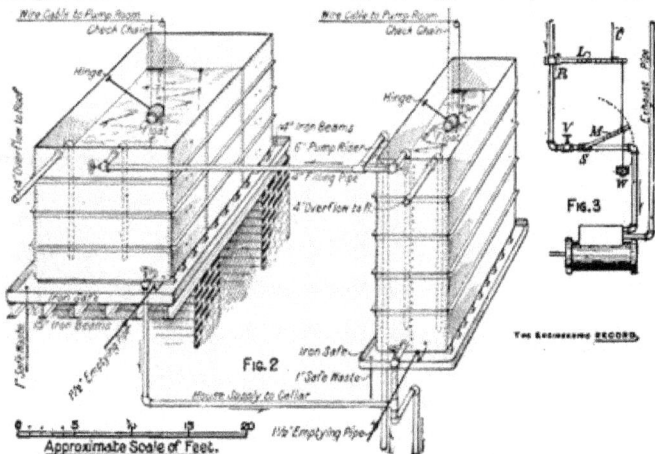

FIG. 2

FIG. 3

Approximate Scale of Feet.

REMODELING AN ELABORATE WATER SYSTEM IN A HOTEL.

REMODELING AN ELABORATE WATER SYSTEM IN A HOTEL.

plicity and directness for the new system and provide for its complete and easy control and maintenance.

Two distribution drums or headers were provided for the hot and cold water, and all the connections were gradually brought to them without interrupting the general service throughout the house, which was then occupied, and without shutting off any line except the one actually being connected at a given time.

Figure 1 shows the present arrangement of headers and distribution pipes. All the hot water is under tank pressure and is distributed from this point, but the cold water for cellar and basement is supplied from a separate 2-inch pipe under street pressure, from which branches are taken off and run on the cellar ceiling to the required points, where short risers are taken off and valved. The cold water header is supplied from the roof tank, but can be connected to the street pressure. The return circulation header is composed of 2 inch Y's and nipples. All valves have brass labels attached to them bearing the same number that is marked on the pipe lines that they control, and which is also recorded on a key provided with double references.

The cold-water supply is pumped into a 45,000-gallon pair of attic tanks built of riveted steel plates stiffened with heavy 4x4-inch angles around the sides and set on 4-inch rolled crossbeams that raise them above the bottom of 6-inch steel safe pans, which in turn are supported on heavy 15-inch girders across the top of brick walls. A large, heavy float made from a whisky keg is hinged by a 4-foot arm to the top of each tank, and its motion is limited by a check chain. A ¼ inch copper-wire cable is attached to the float and carried over pulleys to the pump-room, where the other end is secured to a lever L, Fig. 3, which is connected with another lever M that is loaded with a counterweight W, of 20 pounds, about half the weight of the float keg. When the water rises in the tank the ascent of the float keg slacks off its cable C and allows the counterweight W to pull down both levers. Lever M being attached to the stem of the steam valve, closes it and stops the pumps. As the water falls in the tank the operation is reversed, the descending float keg pulls back the cable raising the lighter counterweight and lifts lever M, opening the steam valve and starting the pump, and so on, automatically. Of course the automatic steam valve S can be fastened open, and the pump controlled at will by the hand valve V. Adjustments are made by attaching the cable to different points on line L. There are two pumps and two float cables, all similarly connected up to a shaft R, so that either float will control both pumps. There is also a gauge board in the pump-room with an index indicating the water level in the tanks.

PLUMBING DETAILS IN AN OMAHA HOTEL.

(PUBLISHED IN 1874.)

In the Millard Hotel, Omaha, Neb., a prominent feature of the plumbing is the arrangement, in the main public toilet-room, of a central cluster of eight urinals which are grouped around an octagonal marble center, which was especially designed by W. H. Spelman, of New York, then of Omaha, Neb. These are arranged so as to comprise in a compact form in the waste space between the stalls all the supply, waste, trap, vent, and flushing pipes and gas and water pipes, symmetrically arranged and effectually screened from view or disturbance while completely accessible at will, and at the same time provide for flushing without the use of a tank, which was interdicted.

Figure 1 is a front view of the stalls, which are placed in the center of a room about 20 feet square, which is lighted by the four gas jets G G, etc., whose curved branches were made from brass tubes, bent and trimmed to correspond with the other exposed pipes, and connected to the supply main in the ceiling by a brass four-way piece A. To secure rigidity, the bottom of this piece was made continuous with the vertical riser from the floor. The copper pipe B was closed at both ends, and merely braced A to the top of the air chamber C, which cushions the water in the eight copper branches D D, etc., by which the different urinals are periodically flushed.

Figure 2 is a top plan from Z Z and shows the marble cover slab H, which is cut in two on the line F F so that either side is conveniently removable for access to the pipes below. I is a handhole commanding an inside flush valve.

Figure 3 is a vertical cross-section at F F, Fig. 2. through the pipe chamber, showing the characteristic features of the most important details. Some of these are magnified or displaced to avoid confusion or obscureness. The marble top H rests loosely on the wainscot panels J, which are keyed together and mortised to the wings W, both of which fit into filleted sockets K in the 2½-inch slate floor slabs L, thus securing very firm marble-work rigidly secured without bolts or metal clamps.

The special urinal waste is commanded in front by trap screw M, and enters, just inside the chamber, a brass Y, also special, that connects with the discharge and ventilation pipes. The former are slightly trapped to break up sound transmission, and the latter communicate with an air-tight metal box made of 24-ounce copper and having a continuous outward draft through the 5-inch exhaust flue. This arrangement is designed to secure the constant removal of all vapor, etc., directly from the bottoms of the urinals, where they originate, before they can rise or become at all diffused. The supply for the flush pipes passes through the box through stuffing-boxes N N, and has a small cock which is furnished with a rubber hose and bulb, so that if any obstruction occurs in any vent pipe O a stream of water can be easily forced through it. The 3-inch main waste pipe extends above the water line and is flanged out to fit a 4-inch vertical vent pipe, which is connected to it by a wiped flange joint which takes bearing on a ring Q supported from the floor by a brass tripod.

Figure 4 is a horizontal section at Z Z, Fig. 3. Figure 5 shows the connection of the vent flue to the main chimney so as to always secure a strong up-

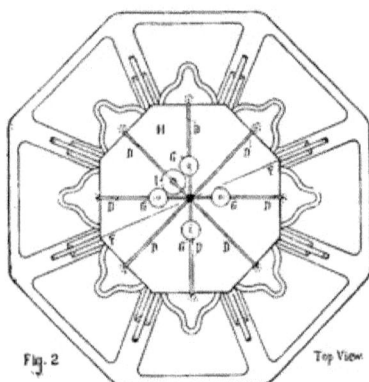

Fig. 2   Top View.

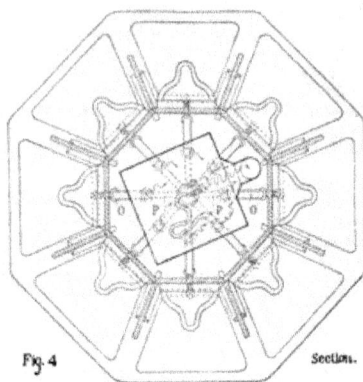

Fig. 4   Section.

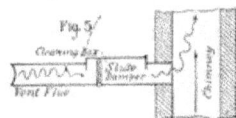

Fig. 5

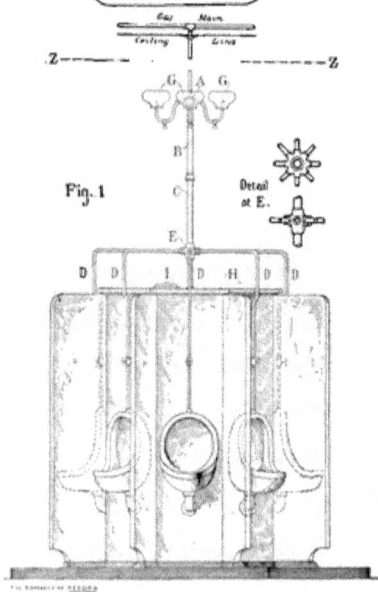

Fig. 1   Detail at E.

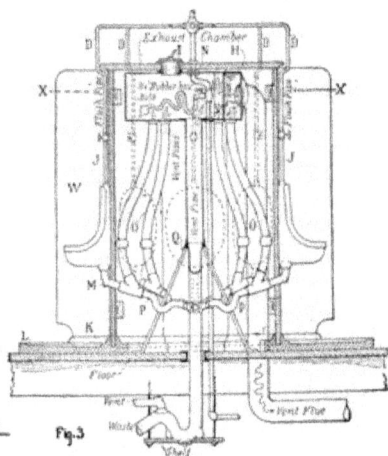

Fig. 3

PLUMBING DETAILS IN AN OMAHA HOTEL.

ward draft from the urinals. The draft is controlled by the sliding damper D, which is usually partly closed, and has a by-pass (not here shown) around it to prevent all ventilation from ever being stopped.

PLUMBING OF HOSPITALS.

PLUMBING IN A NEW YORK INFIRMARY.

(PUBLISHED IN 1894.)

PART I.—GENERAL DESCRIPTION, WATER-SUPPLY DIS-
TRIBUTION, HOT-WATER SYSTEM, PIPE FITTING,
TRENCHES, BASEMENT PLAN, **VALVE BOARD**, AND
RISER LINES.

The New York Infirmary for Women and Children
is located on Livingston Place, New York City. Its
plumbing was intended to be of the simplest character consistent with excellence, and to be especially
adapted to the hospital requirements and to the
structural conditions of the edifice. Detailed drawings were prepared showing the exact position of
every fixture and the sizes, lengths, and locations of

main supplies a 2-inch branch with a stop to the
pump in the cellar, and a ¾-inch branch with a stop
to the kitchen sink, and a 2-inch branch to supply
the steam boilers in the cellar. The hot-air pump
delivers through a 2½-inch riser to a 4,000-gallon
covered cedar roof tank, which is strongly bound
with adjustable wrought-iron bands. Six inches
below the top of the tank is a 3-inch overflow to the
roof and a 1-inch draw-off with a valve at the bottom
into the overflow. A 2½-inch house supply and fire
line is valved at the bottom of tank and runs to the
valve board in the basement. This has 2-inch
branches and screw plugs for fire purposes on each
floor. A ½-inch telltale pipe runs from the top of

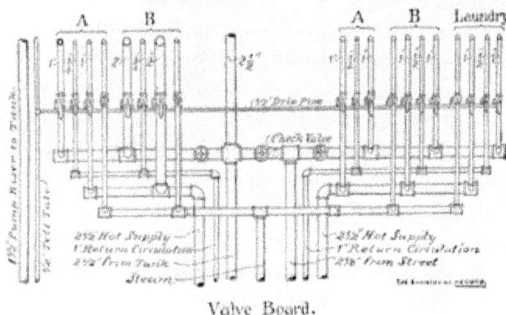

Valve Board.

FIG. 2.

PLUMBING IN A NEW YORK INFIRMARY.

pipe lines, and all valves, bends, cleanouts, traps,
and other pipe fittings, and from these drawings and
the comprehensive specifications the estimates and
bids for contracts were made. The accompanying
illustrations are prepared from these original drawings, which show the principal plans and elevations.
The plans for the plumbing were drawn by A. L.
Webster, of New York City, and the work was done
by W. H. Alexander, of Englewood, N. J.

The water supply from the street is commanded
by a 2½-inch brass gate valve and wheel in a covered
masonry box at the front area wall, whence a 2½-
inch pipe is carried under the basement floor in a
masonry main drain trench, and on the cellar ceiling
to the valve board in the basement. Here branch
lines and control valves are arranged as shown in
Figs. 1 and 2. Before reaching the valve board the

the tank and discharges **with a brass** flap into a telltale sink in the cellar.

Street pressure hot and **cold** water is arranged to
supply the cellar boilers, basement and first-story fixtures, and front and rear yard hydrants. All hot
water for the building is furnished by two galvanized-
iron heavy boilers (200 gallons street pressure and 400
gallons tank pressure), with interior spiral brass tubes,
heated by steam from the steam boiler. Hot tank
and street circulation returns are carried from the
tops of all hot risers to valve board in basement and
returned thence to the hot boilers. All high-pressure
hot risers are extended to and above the house tank
to serve as expansion pipes. All wastes are carried to
telltale sink under the basement ceiling. All water
pipes are heavy galvanized iron, exposed throughout. Traps and back-air connections are lead.

Waste and back-air pipes are of cast iron. All branch connections are Y's and one-eighth bends. Y-branch connections were required to be well turned up. Brass extra-heavy screw cleanouts with ferrules calked into iron pipe are set to command all bends and horizontal sections of drain pipes. All hot and cold-water pipes have 12-inch extensions beyond faucets to prevent water hammer. All pipes exposed to frost are packed with extra-quality mineral wool with painted canvas covers. Hot and cold-water pipes are spaced 2 inches apart, and steam and cold-water pipes 6 inches apart. Unions are placed at frequent intervals to allow the pipe lines to be easily disconnected for changes and repairs. Finished nickel-plated stop and wastes with number tag are put on all branch lines on each floor to shut off individual groups of fixtures. All mains are carried to the valve board in the basement, and all the distributing risers are taken from this point to all fixtures. All branches to the basement fixtures are hung on the basement ceiling and drop to the fixtures. All branches on the upper floors are hung on the ceiling of the room below and rise to fixtures above. All openings in the floors and ceilings for lines of pipe are entirely closed and packed with mineral wool so as to completely seal the opening. All roof joints are flashed with six pounds sheet lead, and all open ends above the roof are protected with wire globe cages. Each vertical column of pipes is solidly and entirely supported on a 12x12-inch brick pier, built before the application of the water tests. All horizontal or inclined lead pipes are supported for the en-

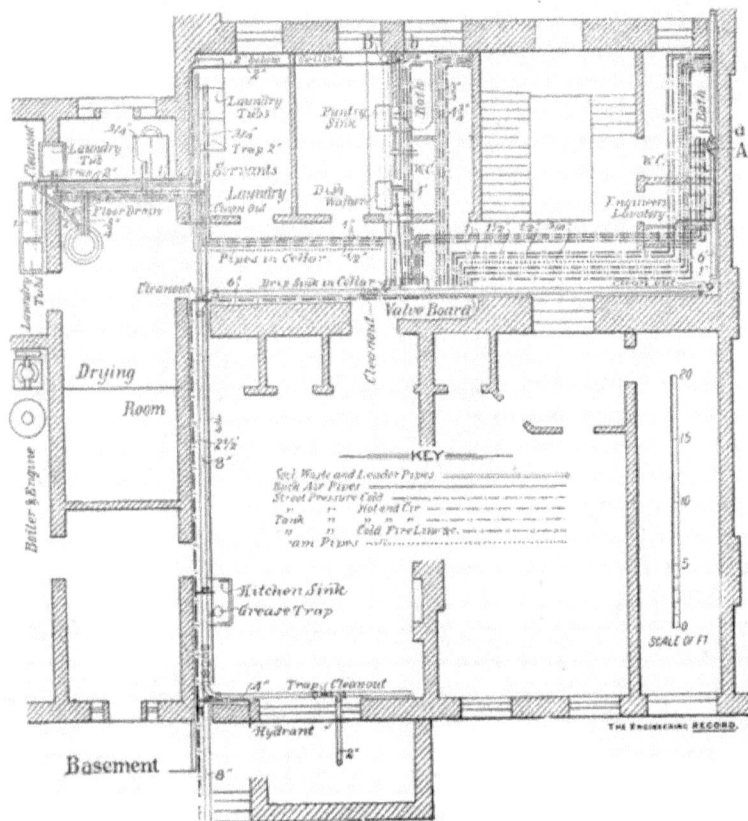

FIG. I.

PLUMBING IN A NEW YORK INFIRMARY.

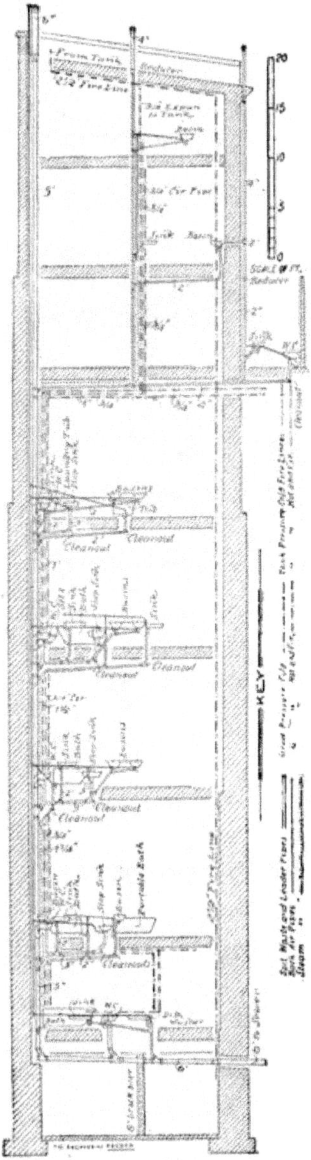

Section-Columns B and C.

FIG. 3.

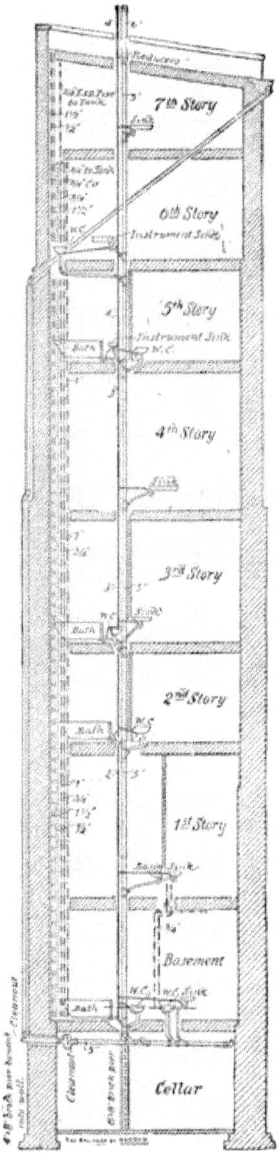

Section-Column A.

FIG. 4.

PLUMBING IN A NEW YORK INFIRMARY.

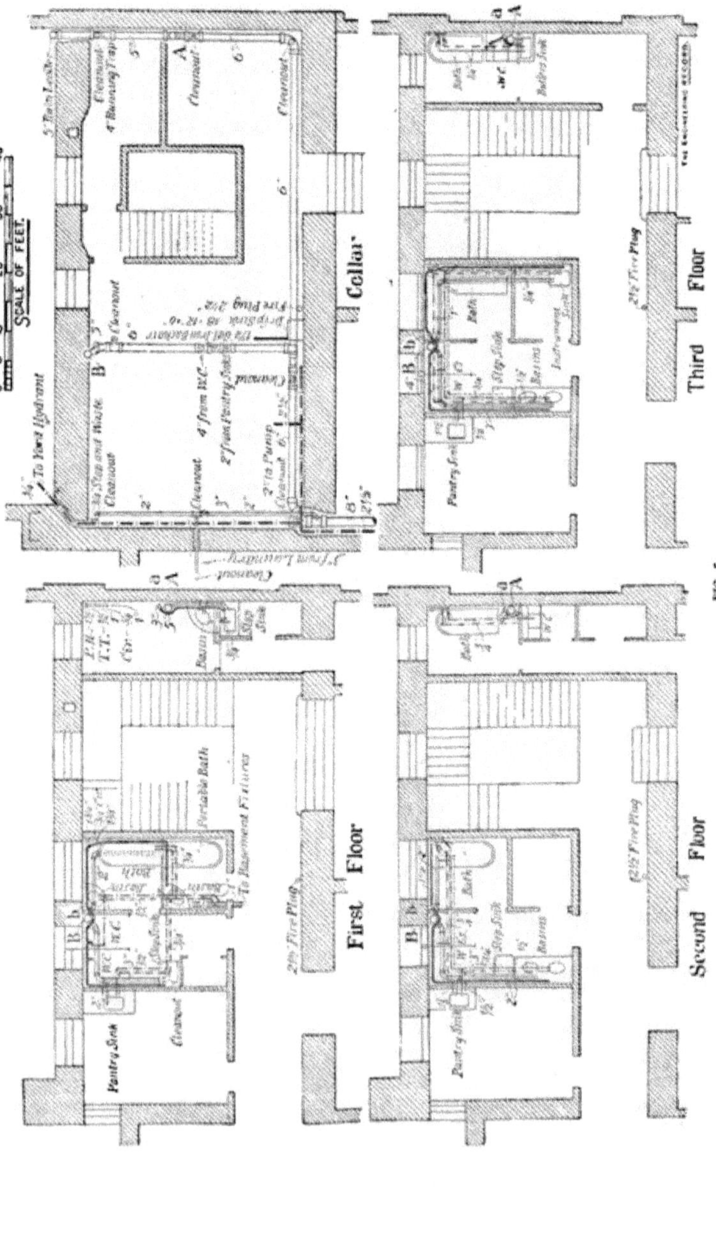

FIG. 5.

PLUMBING IN A NEW YORK INFIRMARY.

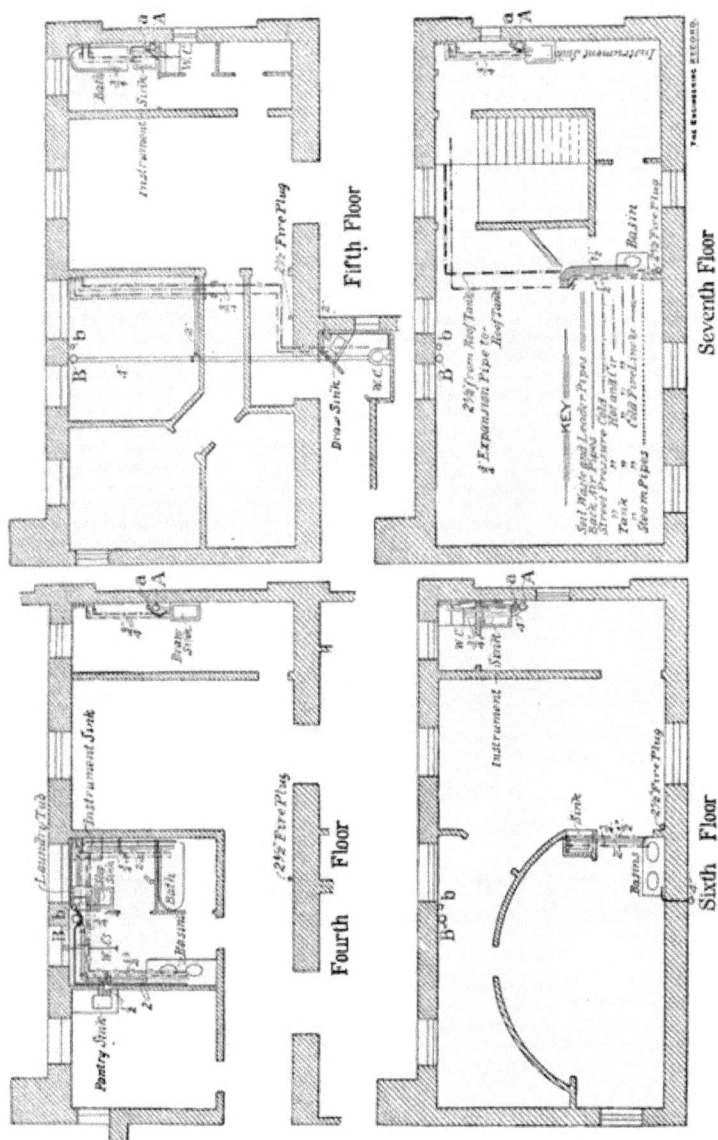

FIG. 6.

PLUMBING IN A NEW YORK INFIRMARY.

tire length on shelves or wooden carrying strips. All pipes are laid to drain completely.

The main drain pipes are laid below the basement floor in a brick drain trench with 16 inches clear width inside. The bottom of the trench is of 3 inches concrete with half-inch Portland cement finish, troweled smooth. The trench begins at the front area wall and extends back with a continuous rising grade of one-fourth inch to the foot. The side walls of the trench are 12 inches thick where more than 2 feet below the surface, elsewhere 8 inches thick. The trench is covered with 2½-inch hammer-dressed bluestone flags flush with the floor, cut with perfect joints, laid in Portland cement, and having full 4-inch bearings on each side wall. There is also a masonry trench under the laundry floor to carry floor drain and a 3-inch and 2-inch line to laundry tubs and laundry machinery. The mortar was mixed 1 part cement to 2 parts sand; concrete 1 of cement, 2 of sand, 3 of 2-inch broken stone. Over the main house trap in the front is set a 16x16-inch iron cover countersunk into the stone and with countersunk lifting ring, and over the rain leader trap a 10x10-inch cast-iron cover countersunk flush with the surface. At the front wall is set an 8-inch running trap with brass screw-cover cleanout and a 6 inch fresh-air inlet to the front curb with brick box and iron grating let into the flag, hinged and with street wash lock and key.

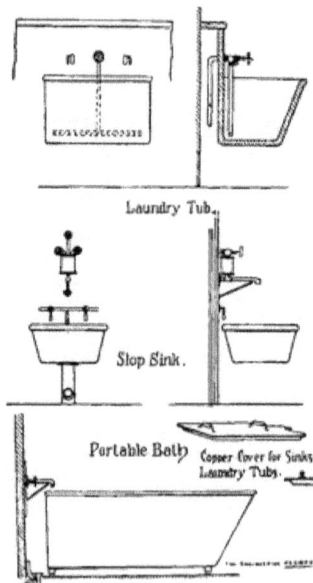

Laundry Tub.

Slop Sink.

Portable Bath

Copper Cover for Sinks Laundry Tubs.

FIG. 7.

THE cast-iron pipes must have the following average weights per lineal foot: Two inches 5½ pounds, 3 inches 9⅝ pounds, 4 inches 13 pounds, 5 inches 17 pounds, 6 inches 20 pounds, 7 inches 27 pounds, 8 inches 33½ pounds. All joints in cast iron pipe are made with picked oakum and pure soft pig lead, well calked home. For each joint in cast-iron pipe 12 ounces of lead was specified to be used for each inch of diameter of pipe in which joint was made,

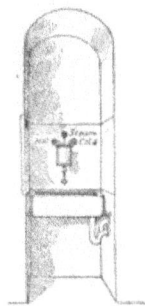

Sixth Floor Alcove Sink

and no joints were allowed to be covered or painted before being tested under water pressure. Afterwards pipes and fittings were painted three coats of lead paint. All wrought-iron pipe, plain galvanized or otherwise treated, is "standard" pipe, factory-tested to 300 pounds per square inch. Pipe to 1¼-inch diameter butt-welded, larger sizes lap-welded. All lead soil, waste, and vent pipe is drawn pipe of the best quality and of the following weights per lineal foot: One-half inch one pound, three-fourths inch 1¼ pounds, 1 inch two pounds, 1¼ inches 3½ pounds, 2 inches 4¾ pounds, 3 inches seven pounds, 4 inches eight pounds. All connections of lead and iron pipe are made by "heavy" brass ferrules of the same size as the lead pipe, threaded and screwed into the hub of the iron pipe. Fixture connections with iron pipe have short lengths of heavy lead pipe where not exposed to view. There are no safes under any of the fixtures. When the work was completed ready to set the fixtures, the soil, waste, and drain pipes were tested by water pressure maintained without leakage, from the level of the main house trap to the top of the highest pipe, and after the entire completion of the work it was subjected to the peppermint test.

The number and location of fixtures is as shown in the accompanying table:

	Cellar.	Basement.	First.	Second.	Third.	Fourth.	Fifth.	Sixth.	Seventh.
Water-closets	..	3	2	3	3	2	2	1	..
Baths	..	..	2	2	2	1	1	..	..
Sitz baths	..	..	1	..	..	..	..	..	..
Sinks	1	3	2	1	1	1	1	2	1
Slopsinks	..	..	1	1	1	1	1	..	..
Basins	..	..	3	3	2	2	2	..	..
New laundry tubs	..	5	..	..	..	..	..	..	..
Old laundry tubs	..	3	..	..	1	..	..	..	..
Laundry mach.nes	..	3	..	..	..	..	..	..	..

The new laundry tubs are of brown glazed earthenware, and each has a ¾-inch nickel-plated crosspipe at the bottom, capped at both ends, perforated with small holes, and supplied with steam through a ¼-inch nickel-plated globe valve with wood wheel handle. All are arranged so that water may be boiled in the tub by injecting steam into it as shown in Fig. 7. All tubs have stiff wire-bound planished copper movable covers with lip turned down inside all around and with two lifting handles on each cover with rubber guard. The roll-rim kitchen sink is of brown glazed earthenware with bronzed iron legs, marble back, cap, and ends, with porcelain-lined iron body grease trap. The scullery sink is of earthenware, with nickel-plated recess standing overflow, set on iron brackets with marble back, caps, and ends along the entire wall at the back and ends of sink space. There are 24'x17'x6' butlers' earthenware sinks on the first and third floors and 23'x16'x6' earthenware tea sinks on the first, second, third, and fourth floors, all with recessed standing overflows On the fourth, fifth, and sixth floors are earthenware draw sinks with 18-inch Italian marble back, sides, and caps. The sixth-floor sink is set in an alcove with Italian marble back and sides 18 inches high, and rests on galvanized-iron pipe through marble so as to be free all round. It has a steam jet turned down into the sink and hot and cold water and steam combination cock with globe valve with wood handle. On the third, fourth, fifth, sixth, and seventh floors are five 30'x20'x7' earthenware instrument sinks with ¼-inch nickel-plated brass steam supply to the bottom of the sink with perforated crosspiece and globe valve with wood wheel handle. Each sink has a stiff wire-bound nickel-plated brass or copper movable cover with two lifting handles with rubber handholes and a lip turned down inside and neatly fitted to make a tight joint. All is finished so that water may be boiled in the sinks by steam jets.

There are nine stationary porcelain-lined bathtubs with nickel-plated double compression faucets, rubber-tube coupling and nickel-plated sprinkler. Also one portable indurated fiber 5 foot tub on wheels with nickel-plated draw-off cock to discharge into a nickel-plated brass funnel connected with lead trap and back-air complete This tub has on the side wall a nickel-plated combination cock with nickel-plated coupling hose and nickel-plated sprinkler. The washbasins are oval, 19x15 inches, and they and the porcelain water-closets are ivory-tinted, with nickel-plated fixtures.

PLUMBING IN THE ISABELLA HOME, NEW YORK.

(PUBLISHED IN 1885.)

THE Isabella Home is an institution founded by Oswald Ottendorfer, Esq., and completed in 1889, in the northern part of Manhattan Island (at Tenth Avenue and One Hundred and Ninetieth Street), as a hospital and home for sick and aged persons.

It is a large building with granite walls, fireproof interior, and natural wood finish, and designed to accommodate 200 inmates, besides Sisters of Mercy,

nurses, attendants, servants, etc. The architects were William Schickel & Co., of New York, and the plumbing and gasfitting was done by Oliver Barratt, of the same city.

The building is four and five stories high above the basement, and consists essentially of a center portion about 60x200 feet flanked by a hospital wing of 60x100 feet at each end, the whole inclosing three sides of a rectangle, whose fourth side is partly occupied by a detached three-story brick stable building, with carriage-house and coachman's apartments. The plumbing system in the main building comprises four lines of 2-inch fire pipe, with hose cocks on each floor, eight main distribution lines of hot and cold water and six main lines of soil pipe, connected with a private sewer to the city system several blocks away. There are, besides, three main soil-pipe lines in the stable.

All water pipe is of galvanized iron, bronzed, and all soil pipes were extra heavy cast iron tested, after fitting, by air pressure of about 10 pounds per square inch (21 inches of mercury). All pipes were entirely exposed and everywhere accessible, and never closer than 1 inch to the finished wall or ceiling; most of them were offset much further than 1 inch, affording ample opportunity for calking joints, setting fixtures, and for painting around them. All horizontal lines were suspended below the ceilings, and the cleaning holes for soil pipes were placed below the ceilings. There are 18 general toilet-rooms, containing 28 water-closets, six urinals, 36 washbowls, 18 bathtubs, and 18 slopsinks.

There are two servants' rooms in the basement, that together contain two double washbowls, four water-closets, one urinal, and one slopsink. There are, besides, two detached washbowls, nine porcelain sinks for butler's pantry use, etc.; three kitchen iron sinks, three basement sinks for receiving the discharge from drip and waste pipes, etc.; two iron sinks for chambermaid's use, one iron sink in the boiler room and one iron sink in the dead-house. In the coachman's apartments are laundry tubs, two kitchen sinks and boilers, and two water-closets.

In the stables is a horse trough, carriage washer, and numerous hose cocks and draw cocks, and there are about a dozen lawn sprinklers; all water is supplied from the city mains through a 3-inch pipe, and all the soil and drain pipes are served by a 12-inch terra-cotta sewer pipe from the outside walls to the city sewer.

Figure 3 shows the receiving tank and boiler in the basement adjacent to boiler-room. A is a wrought-iron tank about 10x6 feet and 5 feet deep.

Water from the city main is received through the 3 in ch pipe B and ball cock C, and is pumped out through the 3-inch suction pipe D.

The 2-inch emptying pipe F is branched above its valve G to afford a supply to the boilers through the 2-inch pipe E independent of the roof tanks or pumps. H is a 3-inch overflow. The boiler I, about 10 feet long and 4 feet in diameter, is supported above A, as shown by iron beams J J, on columns K K, and contains two 100-foot steam coils that are supplied with live or exhaust steam through L and return it through pipes M and M. N is a 2-inch pipe supplying cold water from the roof tank, and O is a check valve to prevent any upward flow. T is a 1-inch supply to lawn sprinklers, and Q is a ¾-inch branch to sink faucet. R is the 2-inch hot-water supply and has a ¾-inch branch S to the sink faucet. T is the ¾-inch emptying pipe. U is the 2-inch (increased above to 2½ inches) force pipe from the adjacent pump to the roof tanks; it can be emptied by ¾-inch drip pipe V. W is a 1½-inch safe waste and X is a 2-inch trap vent pipe. Y is an iron sink receiving the discharge from pipes H, T, and W, and emptying to sewer through trap Z.

Figure 4 shows diagrams of the twin roof tanks A and A. They are built of ¼-inch wrought iron and are each about 12'x8'x6' deep. B is the 2½-inch force pipe from the pump, filling the tanks through their independent valves C and C. D is a 2½-inch delivery for house supply, and can draw from either or both tanks by regulating valves E E. On the opposite side of the tanks the overflow pipe F discharges into the roof gutter, and the tanks may be emptied through the 3-inch pipe G that discharges into a sink on the next floor beneath.

Figure 5 shows a portion of one tank and its tie-rods and the details of support.

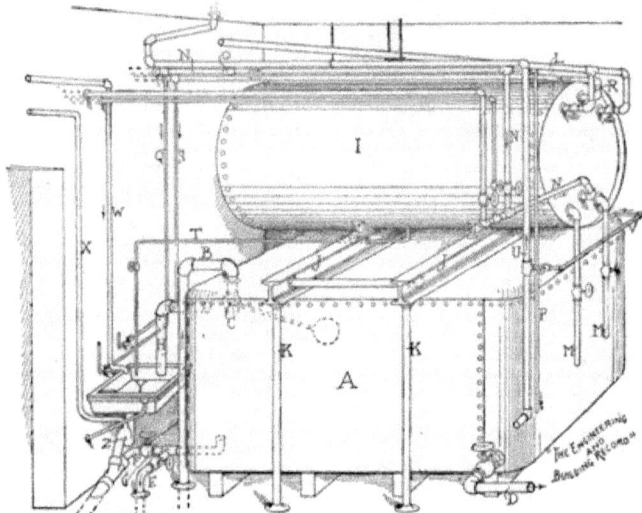

· FIG · 3 · PLUMBING IN THE ISABELLA HOME · N · Y ·

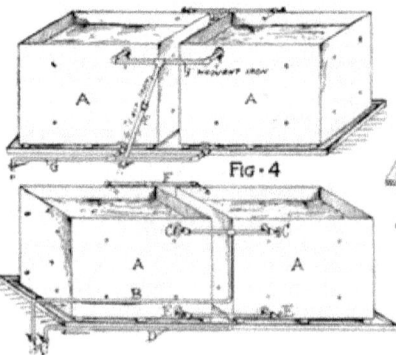

FIG · 4

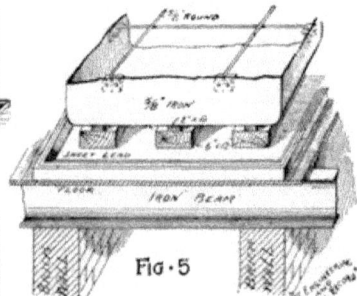

FIG · 5

PLUMBING IN THE
ISABELLA · HOME · N · Y ·

PLUMBING IN THE ISABELLA HOME, NEW YORK CITY.

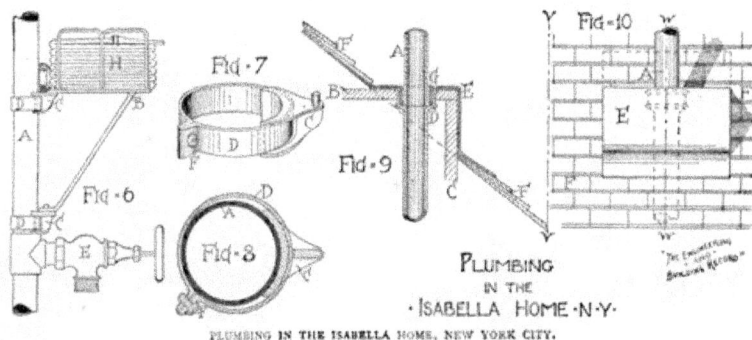

PLUMBING IN THE ISABELLA HOME, NEW YORK CITY.

Figure 6 shows the hose bracket, designed by Mr. Barratt, to fasten on the fire-line pipe instead of against the wall as is customary. A is the fire line and E is the hose cock. H is the hose supported on bracket B that swings on hinge blocks C C. A thin steel band D passes through a slot in C and is sprung over the pipe A and tightly clamped by set screw F.

Figure 7 shows the arrangement of hinge block and band, and Fig. 8 is a section through Fig. 7, showing in solid black the pipe A, that is omitted for clearness, in Fig. 7.

Figure 9 is a section at W W of Fig. 10, and Fig 10 is an elevation from V V, Fig. 9; both show the method adopted for carrying the soil and trap vent pipes, as A, through the roof. B is the top and C the riser of a small board step just at the top of a hub D. A sheet of lead E is flanged into the hub and its upper edge is laid under the slates F, and its lower edge is laid on top of them; the next section of pipe is then joined on and calked at G, as usual.

PLUMBING IN OFFICE BUILDINGS.

PLUMBING IN THE METROPOLITAN BUILDING.

(PUBLISHED IN 1894.)

PART I.—GENERAL DESCRIPTION, PLANS, AND PIPE LINES.

THE Metropolitan Building, at Madison Square and Twenty-third Street, New York City, occupies an area of about 120x140 feet and rises 10 stories above the sidewalk. Several entire floors are occupied by the offices of the Metropolitan Life Insurance Company and the remainder is offered for rental as business and professional offices. The style and proportions of the building are imposing and it is constructed and equipped in accordance with the modern practice for great office buildings. The large interior court and the corner location promote the lighting and ventilation, and the fireproof construction and interior finish, high ceilings, light, abundant tiling, and exposed metalwork give an effective background for the exposed plumbing-work, especially in the airy and commodious subbasement, where the pumps, etc. are most attractively set.

The plumbing was executed by contract by John Toumey & Son, of New York, in conformity to the plans and specification of the architects, N. LeBrun & Sons. The system includes a filter plant, suction tank supply to boilers, toilet-rooms, and washbowls, and elevator system and fire lines, besides the waste, soil, and drip lines and trap ventilation, in all requiring about 10 miles of pipes, including 22 vertical sets of fixtures, most of them with risers about 200 feet in extreme height. The soil pipes from closets and urinals and a few other fixtures are extra-heavy cast iron, as are the trap vent pipes. All other waste and water pipes throughout are brass, tinned inside, and polished where exposed in toilet-rooms, etc. Nearly all closets, washbowls, slopsinks, etc. are fitted up with white marble slabs, white or cream ceramic tiles, oak cabinet-work, and polished brass fixtures and trimmings.

The soil and waste pipes serving each toilet-room are set just below the ceiling of the toilet-room beneath it, and all the angles of the different branches are commanded by cleaning screws at their extremities. All riser lines are set in recesses left for them in the exterior and partition walls, and are in some cases sealed up in plaster and in others are accessible by movable panels. The pump lines, fire lines, and distributing risers ascend through the ventilator exhaust shaft, which also contains the exhaust-steam pipe and elevator pipes. As the vertical lines of vent, soil, and waste pipes were completed, each was successively tested with hydraulic pressure of a max-

imum of 100 pounds by connecting the foot to a temporary steam pump and filling them with water.

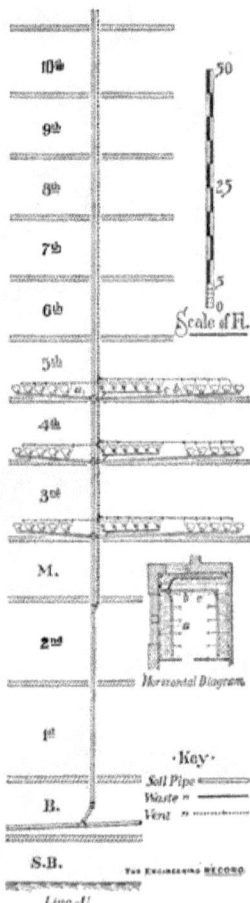

FIG. 2

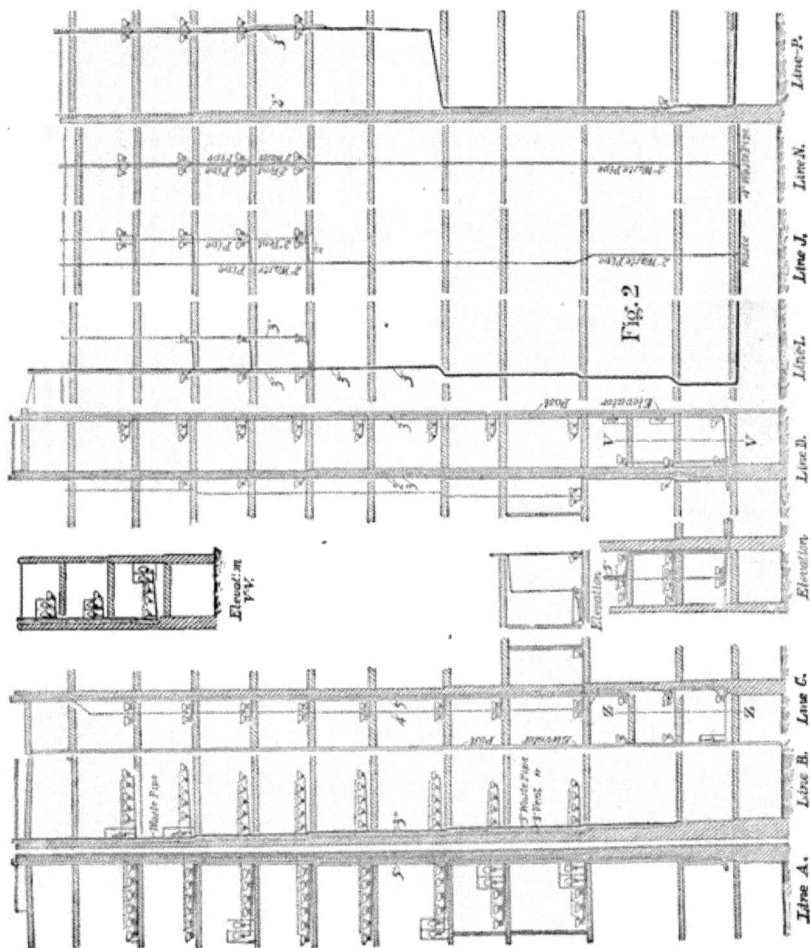

Fig. 2

PLUMBING IN THE METROPOLITAN BUILDING, NEW YORK CITY.

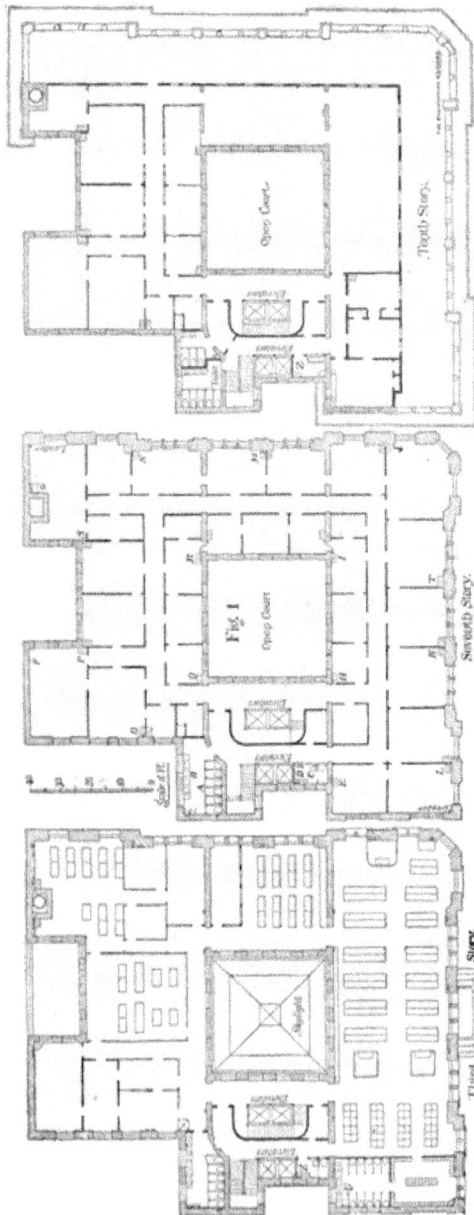

Third Story.

Seventh Story.

Fig. 1
Open Court

Tenth Story.

PLUMBING IN THE METROPOLITAN BUILDING, NEW YORK CITY.

which remained in all instances for several days without receding from the top.

The installation comprises 28 toilet-rooms, containing 110 water-closets, 25 urinals, and 75 washbowls; janitor's toilet rooms, containing four fixtures each; also 12 slopsinks and 175 separate washbowls with cold water only, in the office rooms; one private bathroom in the president's apartments, with a tub, closet, and washbowl; a bathtub and water-closet for the use of the firemen; a 3-inch fire line with a 2½-inch hose valve and a 100-foot reel of 2½-inch hose on every floor; two 3-inch Thomson meters; two No. 5 Loomis filters; a 3,000-gallon suction and a 6,000-gallon roof tank for city water; a 4,000-gallon roof tank for well water, and four pumps; a 400-gallon

hot-water tank and a 250-gallon distributing drum in the subbasement; also a range, washtrays, and separate boiler in the janitor's rooms.

In excavating for the subcellar large quantities of flowing water was encountered in the bedrock, so a well or cistern 8 feet deep and about 20 feet in diameter was made there below the floor, and the concrete floor around it was laid on 2 feet of broken stone, which allowed free subsoil drainage into the well and provides a large quantity of water that is used to diminish the metered supply from the city mains. A 4-foot brick well is built at one side of the cistern, and through its walls the water percolates to the cistern, and above the bottom) of the the foot valve (about 1 foot above the bottom) of the suction pipe of a pump that delivers the water either

to a tank or for street washing, or to discharge into the sewer above the tank, affording a supply for all flush tanks throughout the house. Safe wastes and overflows and leaders from the skylight over the court discharge into this cistern, but the drips from exhausts and the returns from the steam-heating apparatus are automatically pumped back into the boilers.

Figure 1 shows characteristic floor plans and indicates the general arrangement of rooms and location of washbowls, water-closets, etc. As a large number of women are employed by the insurance company, women's toilet-rooms are provided on three floors, similar to that shown in the third floor. Z Z Z is the janitor's room, with a rim-flushing slopsink,

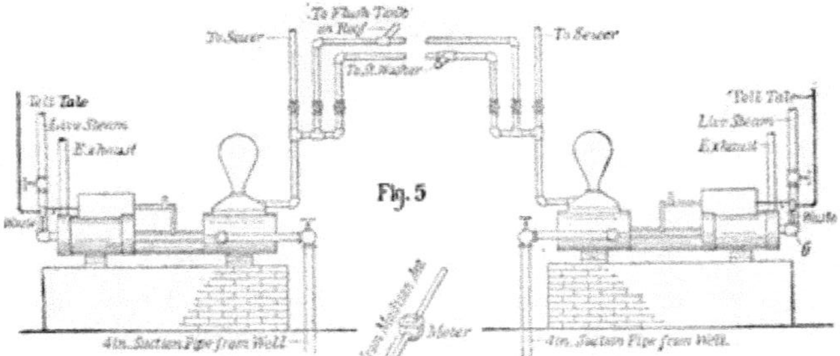

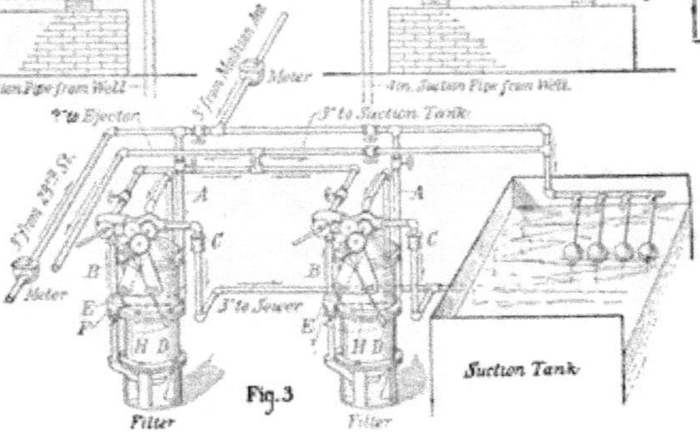

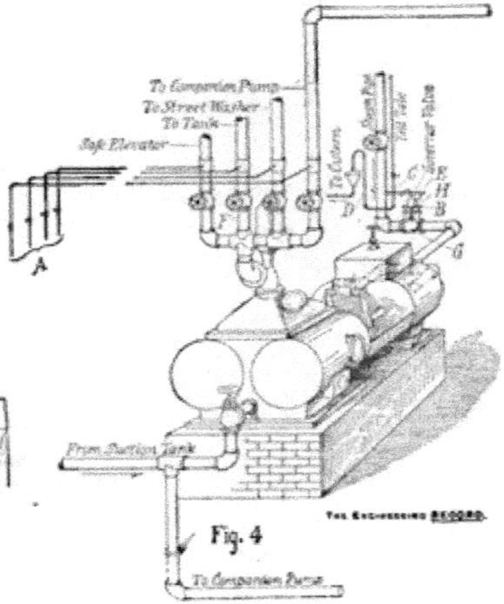

Fig. 4

Fig. 5

Fig. 3

Suction Tank

Filter

PLUMBING IN THE METROPOLITAN BUILDING, NEW YORK CITY.

provided with hot and cold city water and cistern water for flushing. In one room there is a bathtub and in another a butler's sink for washing dishes from the women's lunch hall.

Figure 2 shows conventionally the arrangement of line U, which is an additional one put in after the completion of the original plans, to provide for three special women's toilet-rooms, the fixtures in which are arranged on branches a b c as shown in the horizontal diagram. In the elevation branches b and c are shown as revolved into the plane of branch a, for clearness. In this figure the soil pipes are shown by

double lines, waste by single full heavy lines, and the trap vent pipes by heavy dotted lines.

PART II.—PUMPS, FILTERS, DISTRIBUTION SYSTEM, AND HOUSE TANKS.

CITY water is received through two 3-inch Thomson meters, one on a pipe from the Twenty-third Street main and one on a pipe from the Madison Avenue main. These supplies unite inside their meters to deliver to two No. 5 Loomis filters, which work under pressure and discharge through a 4-inch pipe, and from 1½-inch ball cocks into a 3,000 gallon open iron

suction tank at the pump level in the subbasement about 30 feet below grade. From this tank the water is pumped about 200 feet high into a 6,000-gallon iron roof tank, from which it is delivered back to the basement in two vertical pipes. One 3-inch line supplies the 2½ inch hose cocks for fire service on every floor, and has also direct pump connection duplicated to four separate pumps, and the other 3-inch line leads to a distributing drum in the cellar, whence horizontal branches supply riser lines up to the different groups of fixtures. The pumps are all duplicated, and the three pairs for house tank, cistern, and boiler-feed service are interchangeably connected, and are used alternately so as to keep each one in regular working order.

Figure 2 shows conventionally the general arrangement of the vertical soil and vent lines, the position of the branches and fixtures being sometimes revolved into a different vertical plane to bring them into side view. All soil pipes are shown by double lines, all waste pipes by full heavy lines, and all trap vent pipes by dotted lines. A and B are respectively the lines for the closets and washbowls in the principal toilet-rooms (see Fig. 1, seventh story, p. 210). C and D are lines serving the janitors' closets, which contain slopsinks, basins, and water-closets. I is a line direct for one set of washbowls, and having branches through the floor under the doors to basins in adjacent rooms through which no vertical pipe was carried. J and N are lines to office washbasins in the upper floors, and P is a line with double horizontal offsets to avoid passing through a wide open floor space in the lower stories.

its main overflow, and exerts a pressure through E that drives the piston down in cylinder H and shuts off steam until, the overflow clearing, all the water in the telltale escapes and the valve is opened by its spring.

Figure 5 shows the connection of the pair of alternate pumps that lift the water from the cistern to the flush tank reservoir on the roof or discharge it to the sewer. A third pipe delivers it to the street washing hose, which is usually supplied with Croton water to avoid danger of discoloring the marble front. It has an automatic regulating valve G corresponding to that shown in Fig. 4.

Figure 6 shows the arrangement of house tanks for general purposes and for flushing. The flushing tank is set a few inches higher than the former so that its safe waste can discharge into the pan of the former. Both are of the usual riveted boiler-plate construction with internal tie-rods, and are intended to be always pumped full, the pumps being arranged to work automatically until the water escapes through the telltale, which is the only provision for indicating

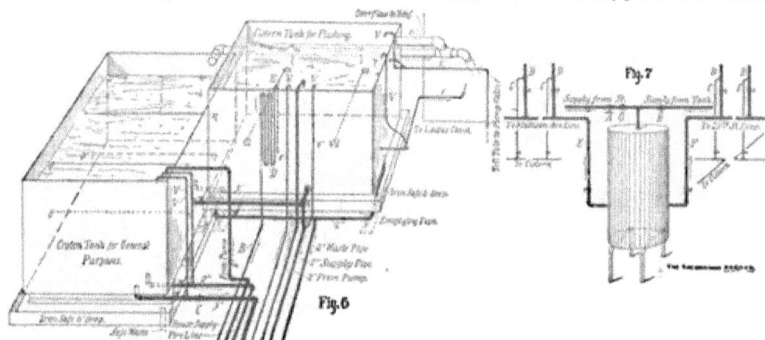

Fig. 6

Fig. 7

Figure 3 shows the arrangement of the filter plant. Each filter has a rated capacity of 1,280 gallons a day, and receives its supply through pipe A and delivers it through pipe B. By reversing lever D the discharge valve is closed, and the filter is washed until the escaping water to the subdrainage cistern seen through glass C becomes clear. E is a small tank to contain alum, which can be introduced into the supply through pipe H, in proportions controlled by the graduated lever F. Figure 4 shows the connection of one of the duplicate Worthington house pumps, the discharge of which is also connected to the boiler-feed and fire-line pipes. A A A, etc. are emptying pipes which are carried down below the concrete floor and discharge into the broken stone surrounding the cistern, where the subsoil waters are all collected. Ordinarily all the discharge valves are closed except F, and the throttle valve is left open so that steam may be freely admitted to valve G, which is automatically controlled by the Ford's pump governor B, which consists essentially of a piston-rod attached to the stem of the gate valve and held open by a spring. When the tank is full it overflows through the 1-inch telltale C just below the level of

the height of water. The two tanks are entirely independent, and the connection between them is usually closed, but may be opened to admit water from either tank into the other. B is an expansion pipe to relieve the hot-water boiler, which can blow off into the tank. V V, etc. are vents to promote the emptying of riser lines, and C is a check valve opening away from the tank, to permit constant tank pressure on the fire line and to close against direct pump pressure from below and permit the fire pump to operate without wasting water through the tank.

Figure 7 shows the large steel drum, about 30x50 inches, in the subbasement, through which all the cold water supply for the building passes and is delivered through branches E and F to horizontal pipes, each of which half encircles the building and distributes the water to the different riser lines D D, etc. These can be independently cut off, and may be drained through pipes C C, etc. Ordinarily valve A is closed and B is open, but in an emergency valve A may be opened and B closed to shut off the tank, and the street water will supply the system as high as its pressure limit, and the check valve will prevent any escape into the street when it fluctuates.

PLUMBING IN THE WAINWRIGHT BUILD-
ING, ST. LOUIS, MO.

(PUBLISHED IN 1894.)

PART I.—GENERAL DESCRIPTION AND PLANS OF PUMPS,
TANKS, BOILER, AND PIPE DETAILS.

THE Wainwright Building is a 10-story fireproof office structure situated at the corner of Seventh and Chestnut Streets, St. Louis, Mo. In general dimensions it is about 117x127 feet and in plan it is U-shaped, being divided into one front and two rear sections by a long open court 30x75 feet in the rear. In the basement are the engines, elevator machinery, plumbing, heating, and power plant, and some rooms available for rental. In the attic are a barber shop and toilet-rooms, storage and pipe chambers, and janitor's apartments, while the rest of the building is devoted to offices and suites of offices, 210 in all. Each office has a washbowl supplied with hot and cold water, and there are slopsinks on every floor, general toilet-rooms in the attic, detached closets and urinals in several stories, and ordinary kitchen and bathroom plumbing in the janitor's apartments. The total list of fixtures comprises 37 water-closets, 15 urinals, 226 washbowls, 11 slopsinks, one common sink and one drip sink, two bathtubs, and three washtrays. The toilet-rooms and washbowls are fitted up with nickel-plated piping, white marble slabs, and polished-oak cabinet-work. Main lines of water pipes are all of galvanized iron, and connections to faucets, traps, and vents are made with lead branches. The soil and vent pipes are subjected to a water-pressure test, and the general features and specifications of the system and workmanship were those usual in standard modern practice—exposed connections, accessible pipe lines, careful trap and local ventilation, and simple direct arrangement and the

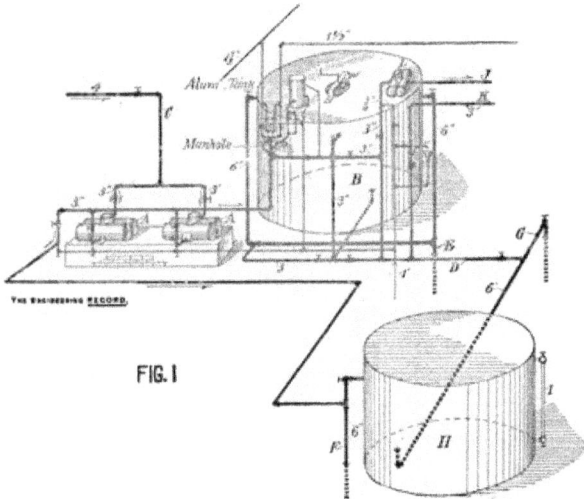

FIG. 1

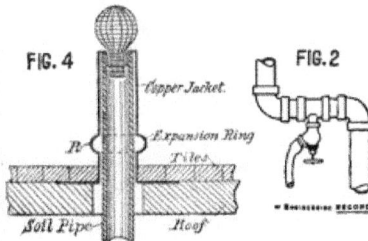

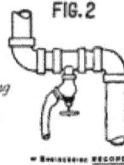

FIG. 3
Open Court

FIG. 4
Copper Jacket.
Expansion Ring
Tiles
Soil Pipe.
Roof

FIG. 2

PLUMBING IN THE WAINWRIGHT OFFICE BUILDING, ST. LOUIS, MO.

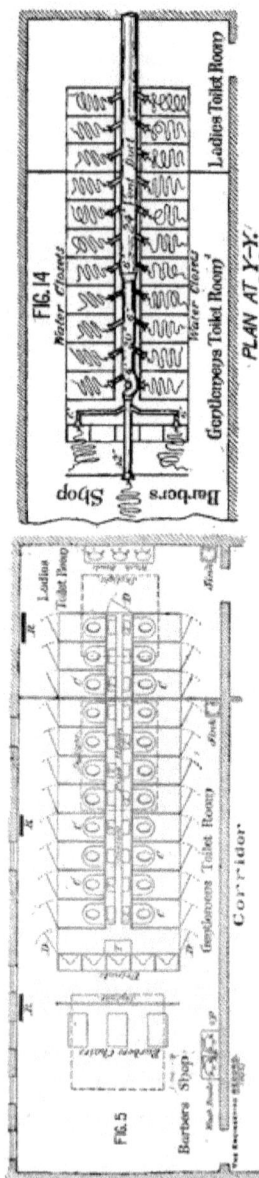

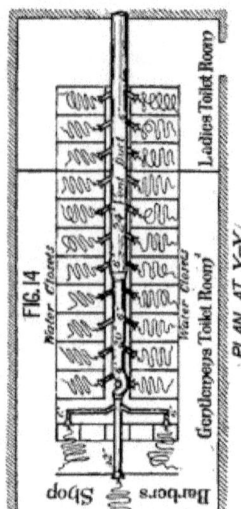

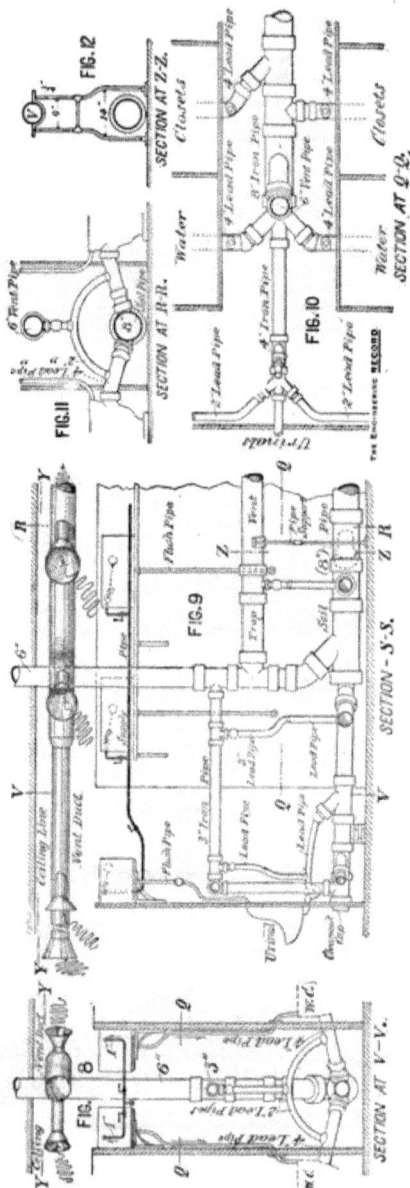

PLUMBING IN THE WAINWRIGHT OFFICE BUILDING, ST. LOUIS, MO.

use of very strong and heavy materials being the principal features of the work, which was executed on a contract price of about $25,000.

Water from the street mains is received through a 4-inch pipe C, Fig. 1, and passing through two 3-inch Worthington meters A A is delivered to the Jewell filter B, which discharges through pipe D to the 8x5-foot suction tank H, which supplies the pump through its suction pipe G. F is an overflow pipe and I is a water glass. A special 2-inch meter L is provided to record the supply to the steam boilers through pipe J, but it can be cut out and the supply taken through pipe K.

Two duplicate Smith-Vaile duplex pumps are connected to pipe G for house and fire service and are cross-connected with two 7"x5"x10" boiler feed pumps, so that each of the four is available for any part of the duty. A 5-inch fire line rises to the roof in each wing of the building, and at every floor has 100 feet of 2 inch hose connected as shown in Fig. 2. These lines are under tank pressure through check valves and there is an automatic governor to start the pump whenever the pressure in tank falls below 70 pounds.

The boiler feed pumps are controlled by automatic regulating valves operated by float attachments, and the house pumps are controlled by Fisher gravity governors, both automatically regulating the supply of steam so that they will begin and stop pumping when the water reaches the required levels. The suction and discharge headers of the tank pumps are so connected as to be available for re-enforcing the elevator pumps when special service, such as lifting safes, etc., is required. The tank pumps normally discharge into a 5,000-gallon iron roof tank, from which a 4-inch distributing pipe is carried along the attic floor and supplies 13 vertical lines of 1½-inch pipe which descends to the different groups of fixtures, and a separate 2-inch pipe that feeds a hot-water boiler 4 feet in diameter and 10 feet long, from which a riser goes to the attic and there distributes downwards to a system of vertical pipes, supplying washbowls, etc., similarly to the above-mentioned cold-water lines, and adjacent to them. These pipes have a circulation connection at their bottoms to the boiler, and are all accessible throughout by panel doors in the walls. The boiler contains two 18-foot coils of 3½-inch brass pipe, so connected as to be supplied at will with either live or exhaust steam to heat the water.

Figure 3 is a diagram plan of one of the office floors, showing the arrangement of rooms and the location of washbowls, which are set in pairs, each

supplied by separate vertical pipes. Local ventilation for all the toilet-rooms is secured through galvanized-iron ducts exhausting into an attic chamber, where a 25 horse-power Eddy electric motor drives two steel-cased 90-inch Buffalo blowers which discharge the foul air out through the roof.

The ventilating fans consist of a duplex Buffalo 100-inch steel-plate machine, which is in reality two 100-inch pulley fans, but both driven by a common driving pulley. These machines are regularly built with 45½-inch round inlets and 37¼-inch square outlets, the diameter of the blast wheel inside of the casing being 71 inches; width, 35½ inches. There is also employed one 50-inch Buffalo upblast left-hand steel plate exhauster. This machine has an inlet 24½ inches in diameter and a square 18½-inch outlet, with blast wheel 36 inches in diameter by 17½ inches wide.

The soil pipes are all enlarged one size at the top and project about 3 feet above the roof, through which they pass with an expansion jointed flashing shown in Fig. 4, which indicates the construction and operation too clearly to need further explanation.

Charles K. Ramsey, of St. Louis, is the architect, Adler & Sullivan, of Chicago, associated. The plumbing was installed by F. Abel & Co., of St. Louis.

PART II.—DETAILS IN THE BARBER SHOP AND TOILET-ROOMS.

THE principal closet and toilet-rooms are located in the front part of the building on the tenth floor as shown in the plan, Fig. 5. Figure 6 is a perspective sketch from P, Fig. 5. The principal feature of the plumbing-work here is that the pipes are arranged in a special chamber between the rows of closets, and while shut off from the public are located conveniently for the original construction or for sub-

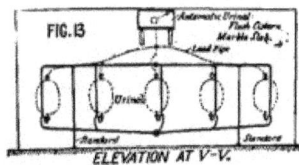

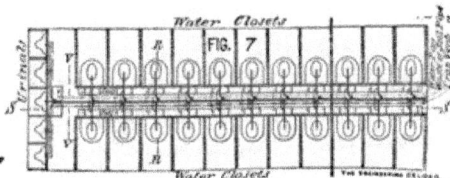

FIG. 6

FIG. 13

ELEVATION AT V-V.

FIG. 7

Water Closets

Water Closets

PLUMBING IN THE WAINWRIGHT OFFICE BUILDING, ST. LOUIS, MO.

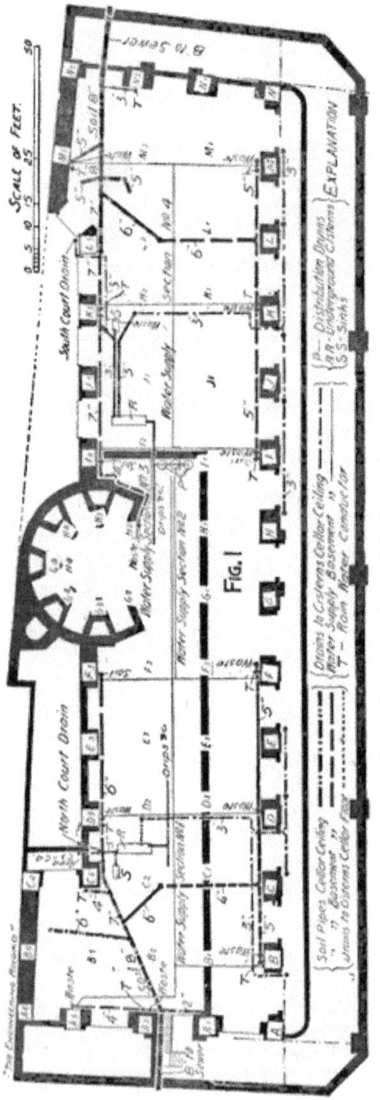

Fig. I

sequent work and cleaning, and are exposed and accessible, besides being directly and systematically arranged. The main soil and vent pipes are of cast iron, and all connections are with lead branches soldered to brass ferrules. Figure 7 is a diagram of the pipe lines, omitting the ventilating pipe or duct, and showing the trap vent pipe a little displaced to separate it from the soil pipe, which is really exactly beneath it. Figure 8 is a vertical transverse section through the pipe chamber at V V, Fig. 7. Figure 9 is a corresponding longitudinal section at S S. Figure 10 is a plan and section at Q Q. Figure 11 is a section at R R to show the connection of the trap vent and soil-pipe connections from the pairs of closets. Figure 12 is a section at Z Z to show the pipe frames supporting the main trap vent pipe V. Figure 13 is an elevation diagram from V V showing the arrangement of urinal pipes on the back of the marble slab N, Fig. 5. The cistern is set to flush automatically every 15 minutes during the daytime, and the pipes are dimensioned so as to give equal supply to each of the urinals. Figure 14 is a plan of the ventilation duct for exhausting the foul air.

PIPE SYSTEMS AND PRESSURE TESTS IN THE HAVEMEYER BUILDING.

(PUBLISHED IN 1892.)

PART I.—GENERAL PLAN AND ELEVATIONS.

THE Havemeyer Building, situated at Cortlandt, Dey, and Church Streets, New York City, for business purposes, is an iron-frame building occupying an area over 200x80 feet, and it is provided with a very extensive water supply, drainage, and trap ventilation system for 16 stories, exclusive of roof drains and subbasement. We precede a comprehensive detailed statement of some interesting features of an unusually severe and prolonged pressure test now being executed by the contractors by the accompanying general drawings of the architect, George B. Post. These are of interest as showing the character of the work, its classification and concentration into distinct groups of fixtures and lines, the proportions adopted for the varying services and the development of the system.

Figure 1 is a diagram plan of the basement, and Figs. 2 and 3 are diagram elevations of the riser lines on the main walls, corresponding to the black circles in Fig. 1, which indicate riser pipes. Three of these pipes are here shown in each group, which actually embraces four or five, the additional parallel galvanized-iron pipes for hot and cold water supply being omitted for sake of clearness.

PART II.—DISTRIBUTION DRUMS AND DETAILS OF PRESSURE TESTS.

FIGURE 4 is a diagram of the distribution tanks at C, Fig. 1. D and E are for cold water under street and tank pressure respectively, and F is for hot water under tank pressure. T is the roof tank.

Figure 5 is a diagram showing the connection by which either or both supplies may be put in service. In construction the rising lines were started at the basement floor about 12 feet above their connections with the sewer under the cellar floor, and a system of horizontal pipes connecting their bottoms, was connected to the house pump temporarily set up for the purpose. As each story in height was successively added to all the lines they were pumped full of water, which was allowed to remain there till another section was set and filled, and so on till when the final height of about 200 feet had been reached the bottom sections had been under a continually increasing pressure for perhaps six weeks. Then the City Inspector was called to inspect the work. This severe test was made in this manner by the contractors, Byrne & Tucker, for their own satisfaction and to facilitate the prompt discovery of any imperfection of material developed by the pressure, and it proved satisfactory to them.

Figure 6 shows the pump P connected up for the pressure test. B is its suction from a reservoir C supplied by independent connections D and E from the city mains on Dey and Cortlandt Streets respectively. The discharge pipe F delivers into systems of pipes supplied by G on the basement ceiling, and H H under the basement floor, which together connect as at K, Fig. 6, with all the vertical rising lines shown in Figs. 2 and 3. System G is formed of the drip and emptying pipes provided for the risers, though they are in some instances moved from their final permanent positions. Pipes H and H are permanent tank-pressure distribution mains, and S S are street-pressure mains. L is an emptying pipe.

Figure 7 shows the connection of test pressure pipe A to the riser lines D D D. A and B are here the

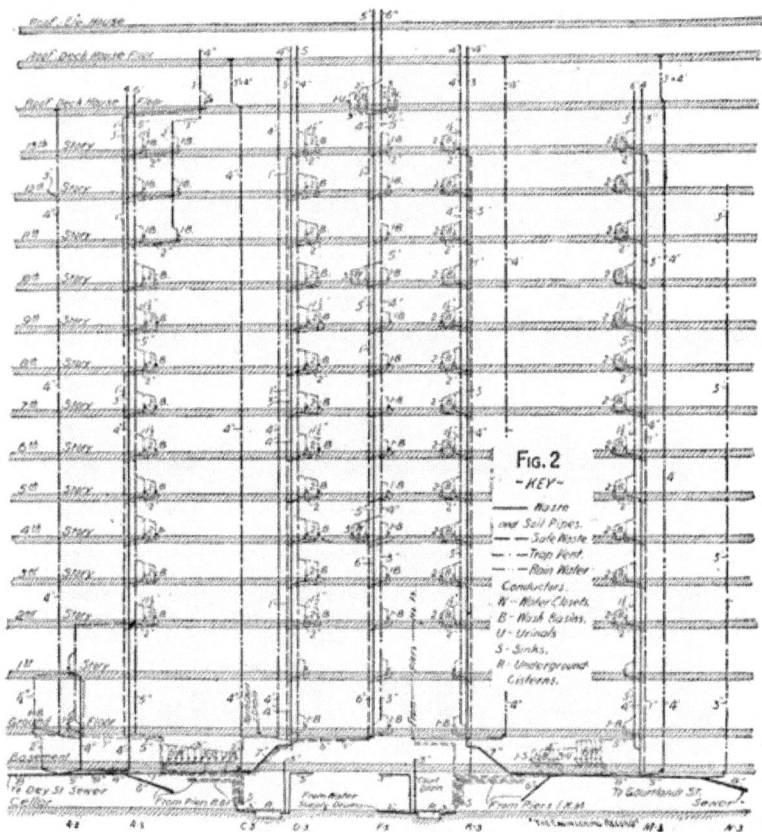

Fig. 2
-KEY-
— Waste and Soil Pipes.
— Soil Waste
- - Trap Vent
- - Rain Water Conductors.
W - Water Closets.
B - Wash Basins.
U - Urinals
S - Sinks.
R - Underground Cisterns.

PIPE SYSTEMS IN THE HAVEMEYER BUILDING, NEW YORK CITY.

Fig. 11

Fig. 12

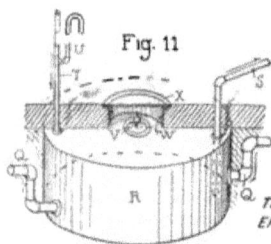

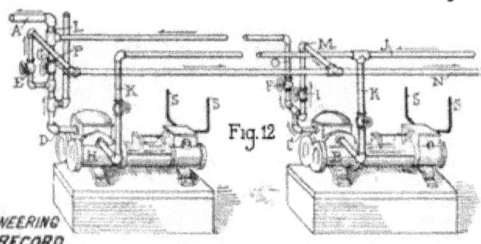

THE
ENGINEERING
RECORD.

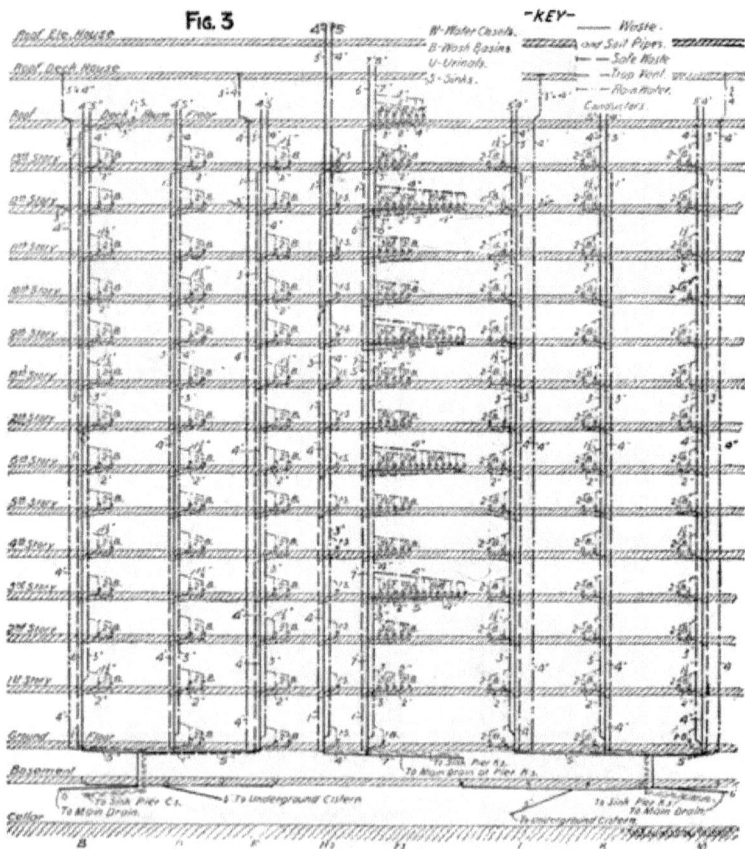

Fig. 3

-KEY-
— Waste.
▨▨▨ and Soil Pipes ▨▨▨
— Safe Waste
— Trap Vent
— Rain Water
Conductors

W'-Water Closets.
B-Wash Basins.
U-Urinals.
S-Sinks.

PIPE SYSTEMS IN THE HAVEMEYER BUILDING, NEW YORK CITY.

permanent supply and riser lines for cold water and remain permanently as shown after pipes F F F are disconnected and the tee E is closed. Eventually lines D D D are continued to the sewer, and separ-

ate tests are to be made of these additional joints and those at the fixtures.

Figure 8 shows the method of connecting pressure test pipe T by screwing it into the cap C, which

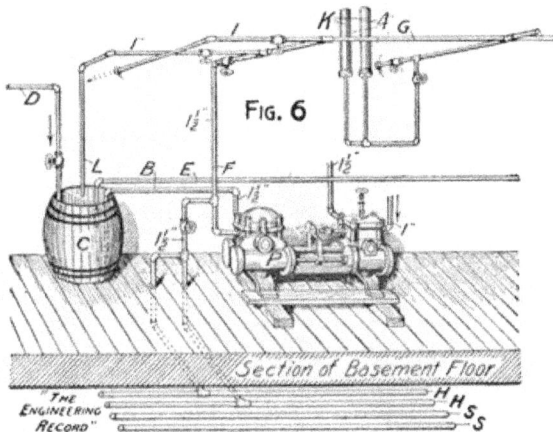

FIG. 6

Section of Basement Floor

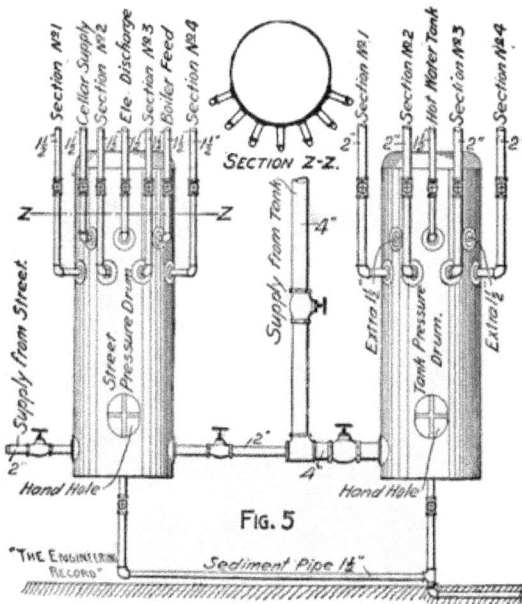

SECTION Z-Z.

FIG. 5

Sediment Pipe 1½"

PIPE SYSTEMS AND PRESSURE TESTS IN THE HAVEMEYER BUILDING.

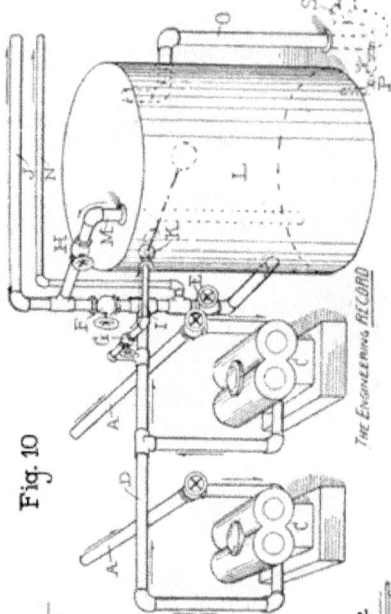

Fig. 10

THE ENGINEERING RECORD

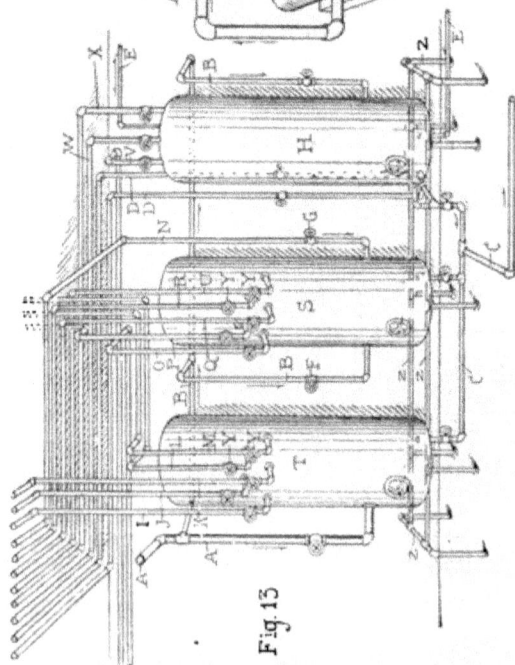

Fig. 15

screws into wrought-iron riser R, and Fig 9 shows the method of connecting test pipe T to .he cast socket S. which is calked on the cast-iron riser C.

PART III.—ARRANGEMENT OF METERS, SUCTION-TANK PUMPS, AND DISTRIBUTION TANK.

THE water supply for the building is received through two 4-inch pipes A A, Fig. 10, one of which is supplied from the regular city distribution main, and the other from a special street fire line. All the water passes through the 4-inch Worthington meters C C, which deliver through the 4-inch pipe D. Ordinarily valves E and F are closed and G and H are open, so that the suction tank L is filled through the 2-inch branch I and ball cock K, and the pumps draw through the 4-inch suction pipe J and branch M, but by closing H and opening F the suction is through J, direct from the meter pipe D. If valve F is closed and E opened the suction tank will be filled direct from the meters without reference to the ball cock. N is a separate 2-inch supply to the cellar distribution tanks, O is a 4-inch overflow, and P is a 1½-inch emptying pipe which, together with O, discharges into one of the 3-inch pipes Q Q which carry the cellar drainage, drips, etc. to the iron tank R, Fig. 11, which is about 7 feet in diameter and 30 inches deep, and is set several feet below the level of the street sewer into which its contents are periodically pumped through the 2½-inch suction pipe S of a steam pump. A 2-inch pipe T is connected with the exhaust head above the roof and brings its condensation water to the tank; its 1½-inch branch U serves as a vent opening at the cellar ceiling. The tank is tightly closed by the manhole cover V, which is accessible through the cast-iron well W, the cover-plate X of which is set flush with the cellar floor.

Figure 12 shows the connection of the 6'x4'x6' Worthington pumps, one of which H is for the house

and roof tank and the other B is for the boiler service. Their supply is through the 4-inch pipe J, as described in Fig. 10, and the suction branches K K and their delivery through pipes C and D. Ordinarily valves E and F are closed and G and I are open so that pump H delivers through branch L to the 4-inch roof-tank pipe A, and pump B delivers through branch M to the boiler feed supply pipe N but by closing valve E and opening G pump H will deliver through branch P to the boilers. and by closing valve I and opening F. pump B will deliver through branch O to the tank. Q is an air chamber. and S S S S are steam pipes.

Figure 13 shows a perspective of the distribution tanks or drums whose connections have been developed from the original preliminary arrangement of two of them shown in Fig 5. The three vertical cylindrical galvanized iron tanks are respectively T, 200 gallons, cold, tank pressure; S. 200 gallons, cold, street pressure; and H, 150 gallons, hot. tank pressure. A is the 4-inch supply pipe from the roof tank with 2-inch branches B B. N is the 2-inch supply from street mains, as shown in Fig. 10. C is the 2-inch emptying pipe. E and E are ¾-inch steam and exhaust pipes to the 4-inch brass steam heating coil inside tank H. D and D are ¾-inch hot-water return-circulation pipes from the systems in the north and south sections of the building respectively. All the other pipes shown are 1½-inch hot and cold water supply pipes to different parts of the distribution system as follows: Cold water, I and J to the basins. flush tanks. etc. in the north section; L and M, the same to the south sections; K and R to the suction tank of the elevator pumps; O to the barber shop and bathtubs; P to the south and Q to the north section, U to the basement toilet-room. The hot-water supplies are: V to barber shop and bathtubs, W to slopsinks, and X to the restaurant. Z Z Z Z,

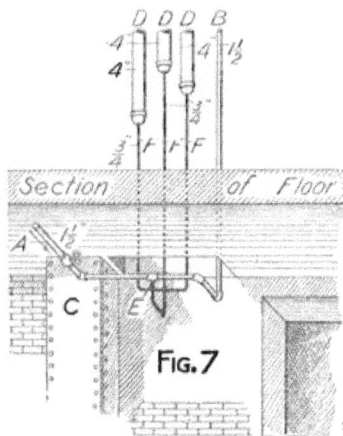

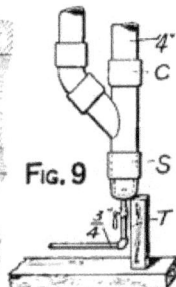

etc. is a pipe frame supporting the tanks. Tanks T and H are always under pressure from the roof tank, and tank S is normally under street pressure, but it also may be placed under tank pressure by closing valve G and opening F. The supply pipes are commanded by valves Y Y, etc., which are connected to the tops of right and left nipples which can be unscrewed so as to disconnect any riser from its horizontal branch into the tank, without moving or disturbing any other connection.

PLUMBING DETAILS IN THE MECHANICS BANK BUILDING.

(PUBLISHED IN 1890.)

THIS building, at 33 Wall Street, New York City, is designed for an office building and to accommodate the Mechanics' Bank, which occupies all of the first floor. The building is about 30 feet front by 70 deep and is nine stories high, exclusive of basement and janitor's apartments on the roof.

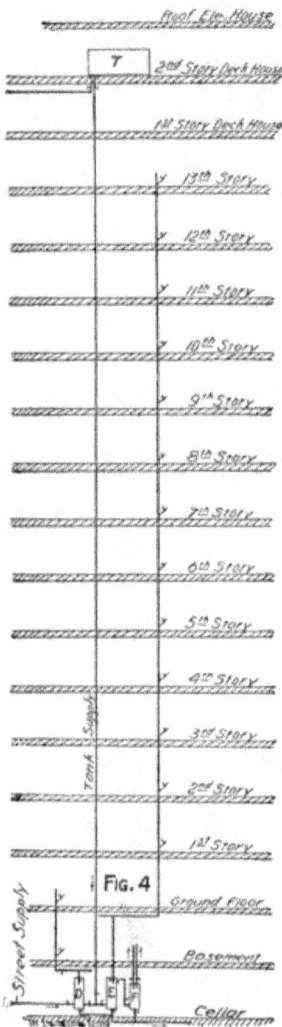

FIG. 3

PLUMBING IN MECHANICS' BANK BUILDING, NEW YORK.

The plumbing includes the following fixtures: One basement toilet-room for the bank employees, containing four water closet sinks, three urinals, and five washbasins. One private toilet-room on the first floor for the bank officers, containing one water-closet, one bathtub, one washbasin, and one urinal. One general toilet-room on the ninth floor that contains eight water-closets, three urinals, and one washbasin. On the floor there is a ladies' toilet-room containing three water-closets and one washbasin.

The eighth floor has a small toilet-room with one washbasin, and there is a siopsink on every floor above the basement, and on each floor from the second to the seventh inclusive there is a toilet-room, one water-closet, one washbasin, and one urinal. The janitor's apartments contain one bathtub, one water-closet, one washbasin, one pantry sink, one kitchen sink, and 13 tray laundry tubs. In the basement engine-room are the hot and cold-water distribution drums, a dripsink, and water-closet.

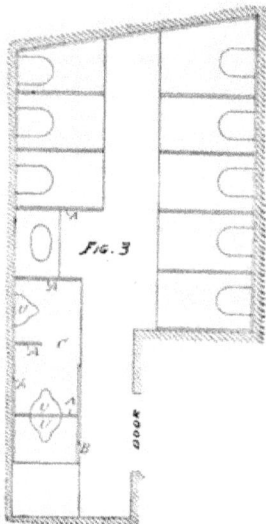

FIG. 4

FIG. 2

FIG. 5

FIG. 4

FIG. 1

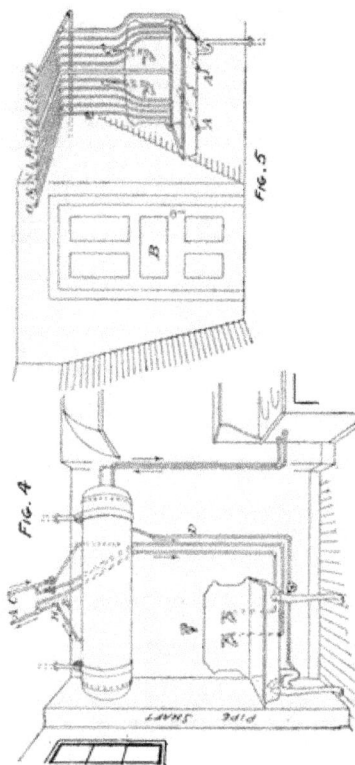

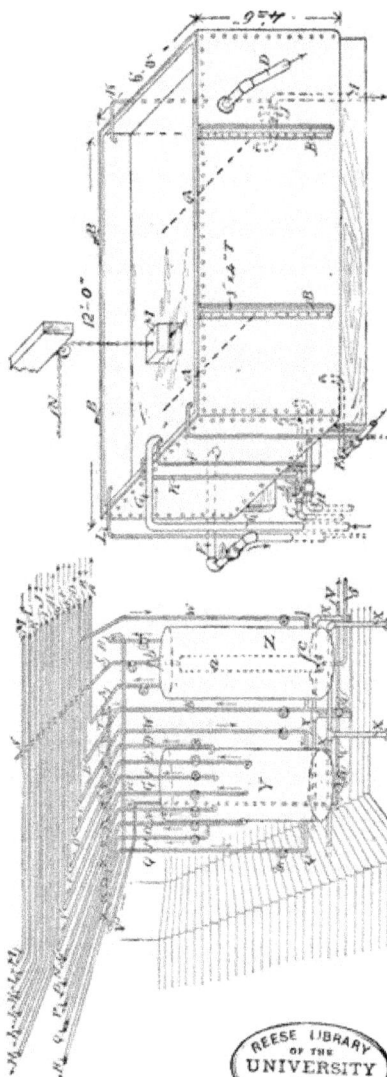

PLUMBING IN MECHANICS' BANK BUILDING NEW YORK.

Figure 1 shows the distribution system of the basement near the pumps. Y is the cold and Z is the hot water drum, each about 2x5 feet, made of galvanized iron and supported by the gas-pipe table X X, etc , set in the footing of the foundation walls. The drum Z contains a 2½-inch pipe *a*, supplied with live or exaust steam through a pipe S.

The arrangement of the pipe *a* is similar to that made by Byrne & Tucker for the Mills and the Duncan Buildings in New York, described respectively on p. 203, Vol. VI., and p. 296, Vol. VIII., of THE ENGINEERING RECORD. The pipe S passes through a stuffing-box *b*, and is screwed into a wrought-iron head plate that is welded into the top of the pipe *a*. The return steam pipe *s* is screwed into a hollow rivet that is cast on the bottom of the cap *c* that is screwed into the bottom of the boiler and tapped to receive the pipe *a*. The stuffing-box permits the pipe S to slide back and forth through it with the expansion movement, which is taken up by the spring of the horizontal branch S S. The large pipe *a* is used instead of the usual steam coil to prevent noise when the full head of steam is turned on, and to heat the water rapidly. Cold water under tank pressure is received through the 2-inch* pipe C and supplied to the distributing drums through branches W and W. Q is a 1½-inch pipe supplying water under street pressure. Ordinarily its valve V is closed, but by opening it the drums Y and Z may be filled directly from the street mains, or the basement fixtures, which are branched from Q, may be supplied from the tank.

Cold water is distributed through the building from the drum Y through the 1½-inch pipe D and the 1-inch pipes E, F, and G to different lines of washbasins, and through the 1-inch pipes N, O, and P to the various flushing cisterns.

Hot water is delivered from the drum Z through the ¾-inch pipe R to the engine-room sink and through the 1¾-inch pipe A to all the slopsinks and to the officers' toilet-room. B is a ¾-inch return-circulation pipe, H is a 2-inch safe waste from the roof tank, and I, J, K, and M are 1-inch safe wastes from the toilet-rooms. L is a 2 inch delivery pipe from the pumps to the roof tank. V is a drip pipe through which all the rising lines may be drained by opening their cocks T T, etc. The drums Y and Z may also be drained into it through their emptying pipes U U.

Figure 2 shows the roof tank, built of ¼-inch iron plate and 2x2-inch angles with ¾-inch tie-rods A A, and tee-bar stiffeners B B, etc. The tank is filled through the 2-inch pump pipe G, overflows directly upon the tiled roof through the 3-inch pipe D and may be emptied also directly upon the roof through the 1½-inch pipe E. F is a 3-inch fire line with hose couplings on every floor. C is the 2-inch supply pipe to the distributing drums Y and Z (Fig. 1), H is a 1½-inch supply pipe direct to the janitor's apartments, and I is a 2-inch supply pipe direct to the ninth-floor toilet-room. K K K are vent pipes to facilitate the discharge of all the water in the supply lines C, H, and I, after their valves J J J are closed. L is a vent

* The sizes given for pipes are correct but the sketch is not drawn to scale.

pipe from the hot-water drum Z, Fig. 1. M is a float whose chain N operates the index of the tank gauge in the pump-room. The float M is made of a butler's pantry copper sink with a sheet of copper tightly soldered over the top.

Figure 3 is a diagram of the ninth-floor toilet-room, which is about 12 feet wide by 22 long, and contains eight water-closet sinks, three urinals, and one washbasin. It has a tiled floor, oil-finished woodwork, and white marble wainscot and panels A A, etc. The irregular shape of the room made it necessary to arrange the urinals as shown, two of them, U U, placed back to back opposite the entrance and screened by a 7-foot marble slab B. C is a marble safe.

Figure 4 shows the suspended horizontal boiler in the janitor's kitchen. Cold water is received through the pipe C and hot water through the pipe H; the branch B supplies the kitchen sink and the branch A supplies the laundry tubs, bathroom, etc. The janitor's bathroom is finished with unusual elegance, having marble washbasin table, royal porcelain bathtub, ivory-finished water-closet, and nickel-plated exposed brass pipes. The floor and wainscot are finished with white and tinted ceramic tiles.

Figure 5 shows the dripsink S in the engine-room. It is supported on pieces of 1-inch gas pipe A A, leaded into the footing of the foundation wall over which it is set so as not to occupy much floor space. An adjacent water-closet at B is also arranged so as to be convenient for the firemen, etc., and not take up unnecessary floor space. Q is the drip pipe from the roof-tank safe, and H, I, J, K, and M are drip pipes from the toilet-room safe wastes. R is the hot and Q the cold supply (Fig. 1). N N are safe wastes, P is a trap vent pipe, and O is the drip pipe from the steam radiators.

C. W. Clinton, of New York, was the architect of this building, and Byrne & Tucker, also of New York, did the plumbing.

PLUMBING IN THE UNION TRUST COMPANY'S BUILDING.

(PUBLISHED IN 1890.)

PART 1.—ROOF, SUCTION, AND DRIP TANKS AND PUMP CONNECTIONS.

THE Union Trust Company's new building on Broadway, New York City, is about 100 feet front, 120 feet deep, and has 13 stories above the basement. It is designed to accommodate a bank on the first floor and the offices of the Union Trust Company on the second floor, while the upper floors are fitted for offices; a restaurant, kitchen, etc., being built on the main roof.

The plumbing comprises, in general, the hot and cold water supply to the toilet-rooms and to basins in all office rooms, and the drainage from all fixtures. The most of the water is pumped through the 4-inch pipe A, Fig. 1, to the iron roof tank T, that is about 14 feet square and 8 feet 6 inches deep, and is closed by iron plates over the top. The tank rests on wooden sills B B B, laid in a sheet-iron lined safe S, which is supported by wooden beams C C C, resting on the floor directly under the sills B B B. The tank overflows through a 4-inch pipe D to the roof gutter, and may be emptied, also to the roof gutter, through a key valve E and a 2-inch pipe F. H is a 2-inch supply pipe to the eighth and ninth floor toilet-rooms, and G is a 4-inch supply pipe to the distribution drums in the cellar. I I are 1-inch vent pipes to the pipes G and H. J is a vent pipe from the hot-water system, and L is the 1½-inch safe waste, discharging on the roof. M is a wire cable from the tank float operating the index of the indicator in the pump-room.

Water under street pressure is delivered through a 2-inch pipe A, Fig. 2, and ball cock B to the closed iron suction tank C, about 5×5 feet, that is in the cellar near the pumps. This tank overflows through a 4-inch pipe D, which has a trap with a 24-inch seal that is preserved by the constant discharge from the steam drip pipe I. E is a valve for emptying the tank F and G are 1¼-inch pipes to empty the pump delivery to the tank and the 4-inch supply pipe from the tank to the distribution drums. H is the 4-inch suction pipe to the pumps.

Figure 3 shows the connections of the special tank pump A and one of the boiler pumps B. H is the 4-inch suction pipe from the tank C, Fig. 2, with a branch E to the pump B. D is the 4-inch delivery pipe to the roof tank with a branch F from the pump B. G is a 2-inch branch connected with the delivery pipe of another steam pump, generally used for the steam boilers. Ordinarily the valves M and N are open, while I is closed, and the tank is filled by the

pump A; but if the valves I, K, and L are open the pump B can also be turned to the tank, and by opening the valve J the third pump may also be used for the tank in case of fire or in any other emergency.

Figure 4 shows the cellar drip tank A, which is simply an iron shell about 6 feet long and 30 inches in diameter, to receive the overflow from the suction tank C, Fig. 1, the waste from emptying the pipes of the distribution drums, discharge from drip-sinks and cellar drainage, all of which is below the

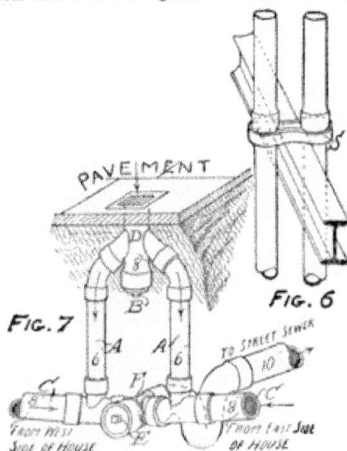

FIG. 7

FIG. 6

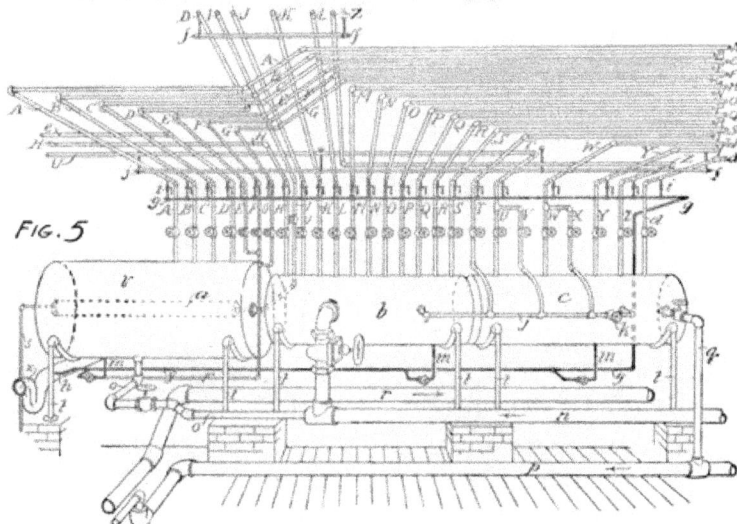

FIG. 5

PLUMBING IN THE UNION TRUST COMPANY'S BUILDING, NEW YORK CITY.

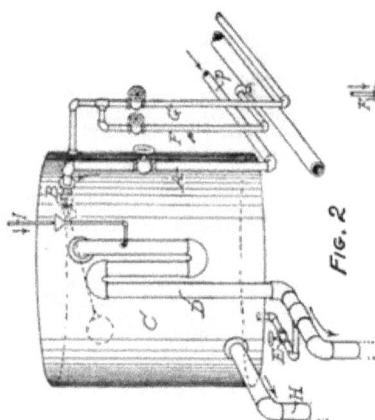

FIG. 1

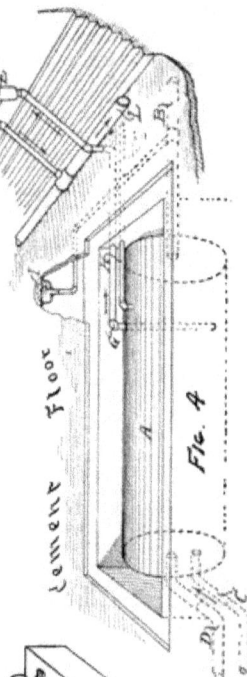

FIG. 2

Cement Floor

FIG. 4

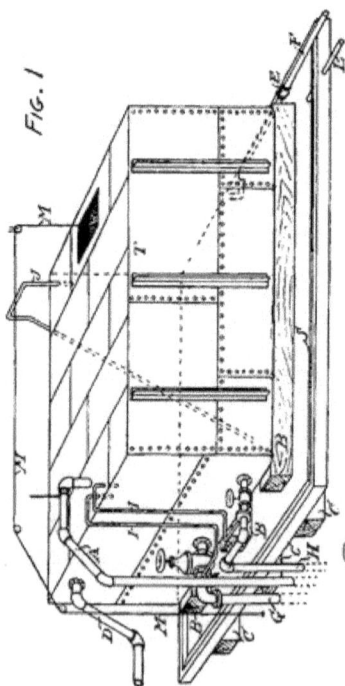

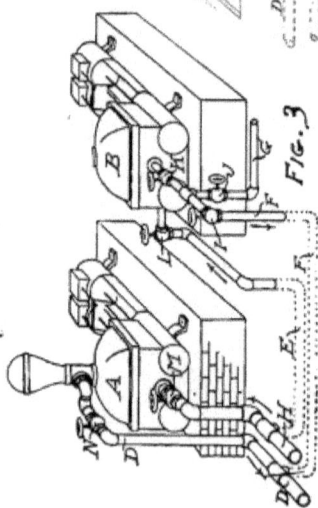

FIG. 3

street sewer level. B is a collector from the waste pipes of the floor strainers I I, etc., and the dripsinks. C is the emptying pipe from the distribution drums, and D is the emptying pipe from the tank C, Fig. 2. E is an aspirator to which steam can be admitted through the valve F to produce suction and draw the contents of the tank A out through pipe G, which discharges through a branch K to the house drain L that empties into the street sewer. H is a vent pipe to maintain atmospheric pressure in the tank A.

PART II.—DISTRIBUTION DRUMS, SEWER VENTILATION, AND PIPE SUPPORTS.

THE supply of hot and cold water to all parts of the building is controlled entirely from the distribution drums in the cellar, shown in Fig. 5. They are of galvanized steel, about 5 feet long; v is for hot water under tank pressure, and b and c for cold water under tank and street pressure respectively. Tank water is received through the 4-inch pipe n and its 2-inch branch o. Water under street pressure is received through a 2-inch branch q from a 4 inch pipe p to the suction tank C, Fig. 2. Live or exhaust steam is delivered through a ¾-inch pipe I to the 2½-inch pipe a, and, after heating the water in the drum v, escapes through a ½-inch pipe s.

The pipe A is a hot-water 1-inch supply to two slopsinks, and B is a 1½-inch supply to all other slopsinks. C and D are 1-inch supply pipes to the president's and directors' rooms. F is a ¾-inch and E and G are ½-inch hot-water return-circulation pipes from the tops of the lines B, C, and D, and connect above its valve with the sediment pipe m of the drum v. The pipes H, J, K, and L are 1-inch tank cold-water supplies to different lines of washbasins. M is a 1½-inch supply pipe to a group of washbasins and urinals, N is a 1½-inch supply pipe to a group of basins, urinals, and slopsinks, and O, P, Q, and R are 1-inch supply pipes to lines of washbasins. S is a 1-inch supply pipe to the directors' room. T is a 1-inch supply pipe to the president's room. U is a 1½-inch cold-water supply pipe under street pressure to a basement toilet-room, and V is a 1-inch supply pipe

to the same. W is a 1½-inch supply to another basement toilet-room. Y is a 1½-inch supply pipe to the elevator tank. Z and d are 1-inch supply pipes to the front and rear cellar sinks; j is a 1½-inch by-pass connecting the drums b and c, the valve (k) of which is usually closed to keep the two systems separate, but may be opened to admit tank water to the drum c if the street water is turned off. The lines of pipe, W and U, are generally supplied by street pressure, the valves w and u being open and the valves x and y closed. By reversing these valves, however, the supply is under tank pressure.

All the rising lines may be emptied through their valves i i, etc. into the drip pipe g g that discharges into the trap of a 5-inch rainwater leader z. The drums v, b, and c may also be emptied by branches m m m, through the pipe g; t t, etc. are pipe legs supporting the drums; f f, etc. are pipe hangers supporting the horizontal lines from the iron floor beams; p is the 4-inch pump delivery pipe to the roof tank.

Figure 6 shows the method of supporting stacks of cast-iron pipes from the iron floor beams by the welded iron strap S, carefully fitted just under the hubs.

Figure 7 shows the special fresh-air inlet at the main sewer trap. Each of the main house sewers C C has a branch A to a double Y connection D at the sidewalk grating, and any dirt or other obstruction can be very readily removed through its cleaning hole at B and another at E in the double Y connection F with the main sewer.

The plumbing in this building was done by Byrne & Tucker, of New York City.

PLUMBING IN THE PRUDENTIAL BUILD-ING, NEWARK, N. J.

(PUBLISHED IN 1894.)

THE Prudential Building at Newark, N. J., is located upon one of the principal streets of that city, is perhaps the most prominent business structure in the place, and is a large and thoroughly equipped

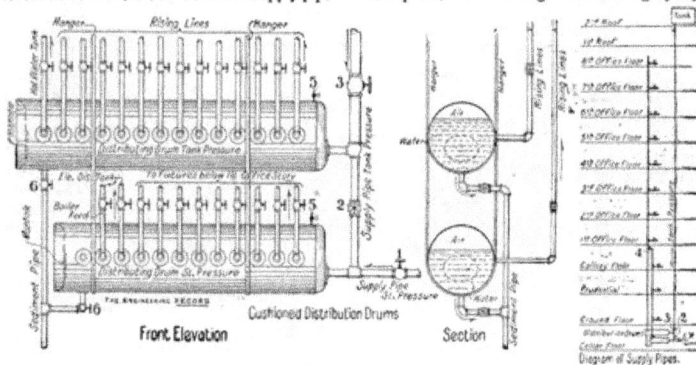

Front Elevation   Cushioned Distribution Drums   Section

Diagram of Supply Pipes.

PLUMBING IN THE PRUDENTIAL BUILDING, NEWARK, N. J.

modern office building. It is owned by the Prudential Life Insurance Company, and is occupied by its
offices and by tenants. In this building a large
number of fixtures are supplied with water at a considerable pressure, and to prevent danger of injurious
water hammer when one or more faucets may be
rapidly opened and closed simultaneously, a special
provision for air cushioning was made by Mr. George
B. Post, of New York City, the architect, which is
illustrated in the accompanying cut, and has been
successfully applied also in substantially similar cases
which have arisen in his practice. The street and
tank pressure supplies are delivered to small tanks
or drums in the cellar, from which the pipes to the
different riser lines diverge. These pipes are valved
for control at one point. The drums are similar to
those often used for distributing from one or two
mains to several lines, but are of larger capacity and
are intended to be always partly filled with compressed air confined in the space above the level of
the outlets to the distributing pipes. When the system was installed the sediment cocks 6,6 and petcocks 5,5, were closed, thus confining the air to the
drums. Being unable to escape, the air was compressed by the inflowing water until it occupied the
upper parts of the drums as shown in the section.
As it exerted a uniform pressure on the water, it
acted as an elastic cushion for the water and absorbed the impact of shocks. Ordinarily valves
2, 4, 5,5, 6, and 6 are closed and all others are open,
but if it should become desirable to cut off street
pressure, tank water may be supplied throughout by
closing valve 1 and opening valves 2 and 4. Similarly
street pressure can be supplied throughout by closing
valve 3 and opening all the others. Should the air
in the drums become absorbed, valves 5,5 and 6,6 are

opened and all others shown are closed, when the
water in the drums is replaced by air. Then if
valves 5,5 and 6,6 are closed and 1 and 3 are opened,
the air is forced into the upper parts of the drums
and the system operates as before.

Any sediment deposited in the drums can be
cleaned out from time to time by opening valves 5,5
and 6,6, and closing all the others, when the manhole covers may be removed and the drums thoroughly cleaned and scrubbed out, valves 1 and 3
being opened sufficiently to admit water for rinsing
them, which escapes through the sediment pipe.

CONTROL OF HOT AND COLD-WATER DISTRIBUTION IN A MILWAUKEE OFFICE BUILDING.

(PUBLISHED IN 1893.)

THE Mathews Building, at Third and Grand Ave-
nues, Milwaukee, Wis., has been erected for general
business and office purposes at a cost of about $200,-
000. It is a modern six-story building of iron and
brick fireproof construction about 100x110 feet in size,
and 92 feet high above the basement. It is provided
with five toilet and water-closet rooms, has slopsinks
on every floor, fire-hose cocks in each main corridor
and a washbasin with hot and cold water supply in
each of 50 suites of rooms. The plumbing, which
was in accordance with the requirements of architects
Ferry & Clas, cost about $14,000, and was executed
by Halsey Brothers, of Milwaukee. The work conforms to present standard practice, and is designed
to secure special thoroughness and simplicity
throughout.

The head in the city mains supplies sufficient pressure for the required service, but an attic tank was
provided to store a supply for possible interruption in the street mains, for fire
purposes and to insure abundant supply
at those times of the day when the maximum draft is made on the mains. This
tank is filled through ball cock M, and
delivered through valve N connected to
the separate rising pipes P and Q. It was
desirable to have a simple and direct connection between this pipe P and the hot-
water heater F and the distribution risers
I I, etc., which furnish cold water to the
different floors, and the accompanying
sketch shows the arrangement designed
to distribute the water among all, whatever their respective drafts may be. Cold
water from the city main is received
through a check valve in the 3-inch pipe
C and delivered to the 3-inch header B,
which distributes it to the 1-inch risers I I,
etc., supplying different floors. The 2½-
inch pipe D furnishes cold water to the
hot-water boiler F. The tank riser P connects directly with header B and the
end of boiler feed. D. G is a 2½-inch
header for the distribution of hot water,
which leaves the boiler through pipe H and

WATER DISTRIBUTION IN THE MATHEWS OFFICE BUILDING, MILWAUKEE, WIS.

returns through circulation pipes K K, etc., which are taken from the highest points of the risers T T, etc., that supply the different floors. L and N are drip and waste pipes, V V, etc. are valves to empty the rising lines, U is a safety valve, W is a catch basin, and X X are steam connections to the coil heating the boiler. All the supplies are individually controlled by valves R R, etc. Usually valve O is closed, and all others except V V V, etc. are open; but they are arranged so as to cut out any part of the system for repairs, extensions, etc. without affecting any other part. The rising lines all converge to ascend through a vent shaft S, in which they are accessible and, like all the rest of the plumbing-work here, exposed.

The figure is prepared from a sketch made to illustrate the arrangement and operation, and is not drawn to exact scale or position.

PLUMBING IN MANHATTAN LIFE INSURANCE BUILDING.

(PUBLISHED IN 1894.)

PART I.—GENERAL CONDITIONS, REQUIREMENTS, AND FEATURES OF THE SYSTEM, ARRANGEMENT AND OPERATION. DESCRIPTION OF WATER, DRAINAGE, AND TRAP VENTILATION PLANT. ENUMERATION OF PRINCIPAL APPARATUS AND OUTLINE OF ITS OPERATION, METHOD OF DISTRIBUTION, PLANS OF PIPE CELLAR, MAIN SEWER LINES, CHARACTERISTIC OFFICE-FLOOR INSTALLATION AND DETAIL OF PIPE SHAFT.

No PERSON who has been in New York City during the winter of 1893–94 needs a labored explanation to aid in identifying the Manhattan Life Insurance Building as the lofty structure which towers above the finial of Trinity Church steeple, just across Broadway, and the building, which is remarkable even in an era of tall structures, has been made familiar through the medium of illustrations to all who are interested in architectural development. The pneumatic caissons and other details of the foundation were described in THE ENGINEERING RECORD of January 20, 1894. The building is about 120x67 feet in size, has a height of over 300 feet from the cellar floor to the main roof, and has 19 stories devoted to the uses of the Insurance Company, to tenants'

offices, and to the operating plant and equipment. It has a complete system of plumbing and drainage conforming to the requirements of Messrs. Kimball & Thompson, the architects of the building, and Mr. William Paul Gerhard, consulting engineer for sanitary work, which was installed by J. W. Knight & Son, contractors, under the supervision of Mr. Gerhard.

The system comprises a supply of Croton or artesian-well water in every toilet room throughout the building and for fire service, steam-heated hot water to slop-sinks, and the drainage and ventilation of all water, soil, and drip lines. Water through the city mains is received through a 4-inch pipe, meter, and gate valve and discharged in the cellar through two 2-inch

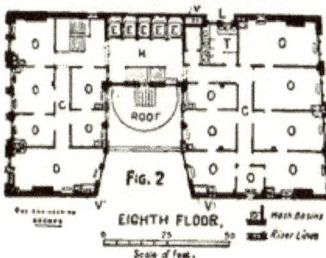

FIG. 2

EIGHTH FLOOR.

Scale of feet.

ball cocks into a 2,000-gallon open iron suction tank 12'9"x7'4", set on brick foundations and provided with hinged wooden cover. This tank has emptying and overflow pipes and a supply to the suction pipes of all the pumps, three of which are for house and fire service, two for the elevators, and two for the boiler feed water, the house and feed pumps being interchangeably connected. The house pumps lift the water about 300 feet into a boiler-iron 5,000 gallon house tank 11'6"x8'x7'6". From this tank 1½-inch riser lines supply the distribution branches to the washbowls on each floor above the seventh, and a 2-inch pipe supplies the steam hot-water heater in the cellar. A 4-inch line also supplies an auxiliary 2,000-gallon tank on the eighth floor, which is filled through ball cocks and connected with a drum in the cellar, from which risers lead to all the fixtures on the

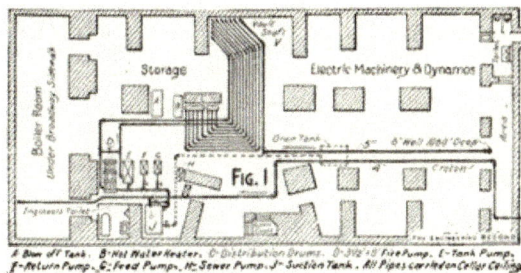

FIG. I

A Blow off Tank. B Hot Water Heater, C Distribution Drums. D-3½x4½ Fire Pump, E-Tank Pump, F-Return Pump, G-Feed Pump, H-Sewer Pump, J-Suction Tank. All Pipes carried on Cellar Ceiling.

PLUMBING IN THE MANHATTAN LIFE INSURANCE BUILDING, NEW YORK CITY.

second, third, fourth, fifth, sixth, seventh, and mezzanine floors. These are controlled from one central point and are subjected to a pressure much less than would be imposed by direct communication with the roof tank. Adjacent to this distributing drum in the cellar is another one which is connected directly with the street pressure and delivers it to riser lines supplying all fixtures on and below the first floor.

A 6-inch artesian well 2,000 feet deep has been drilled in the cellar, and water from it is delivered by a separate pump to a roof tank, the counterpart of the one described for the Croton water, and supplies all the water-closet and urinal flushing cisterns direct to the eighth floor, and from an auxiliary tank below the eighth floor, thus effecting a considerable economy of metered water purchased. The roof tanks and the eighth-floor tanks for street and artesian water and the distributing drums in the cellar for street and tank pressure are respectively cross-connected so that either or both may be supplied from either source. All riser lines are valved at their bottoms, and each horizontal branch is separately valved. All the hot water is under full roof-tank pressure.

A separate 3-inch riser extends from the cellar to the roof tank and has valves and hose on every floor for fire protection. It is under constant tank and pump pressure and has check valves to prevent the escape of water at increased pressure. The two Worthington duplex steam house pumps each has 6-inch steam cylinder, 5-inch water cylinder, 12-inch stroke, and a capacity of 100 gallons per minute, and are connected to draw from the suction tank or from the Croton main direct and to deliver either into the roof tank or to boiler feed pipes. The boiler feed pumps are similarly connected. The house pumps are fitted with sight feed lubricators and Fisher's automatic regulating attachment to start them whenever the tank water falls below a certain level. The height of water in the tanks is also indicated in the engine-room by an electric alarm operated by high and low water floats.

The fire pump is a Worthington new pattern "Underwriters' fire pump," with steam cylinder 14 inches

in diameter, water cylinder 7 inches in diameter, length of stroke 12 inches, and a nominal capacity of 500 gallons per minute, equal to two 1¼-inch fire streams at 250 gallons per minute each. This pump is built strictly and entirely in accordance with the specifications for underwriters' pumps, as adopted by the Boston Manufacturers' Mutual Fire Insurance Company, of Boston, and is fitted up with all the attachments, fittings, etc., therein described, and improved polished-brass sight-feed lubricators. This pump is connected with the 3-inch fire stand-pipe, and has an automatic regulating device for starting

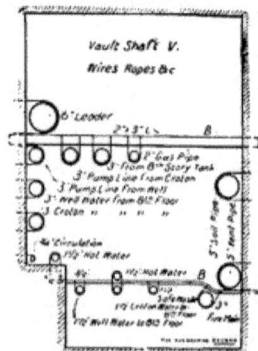

the fire pump automatically in case a fire valve shall be opened on any of the floors. The fire line connects with the house tanks, with 3-inch swing check valve opening downward, and also with the house pump discharge pipe, so that in case the house pump should need repairs it may do service temporarily in filling the house tanks.

The general system of water pipes includes the direct street-pressure pipes from the 4-inch main as follows: 3-inch to the elevator tanks, 4-inch to the suction tank, 2-inch to the boiler feed pump, 4-inch to the house and fire pumps, and 2½-inch to the dis-

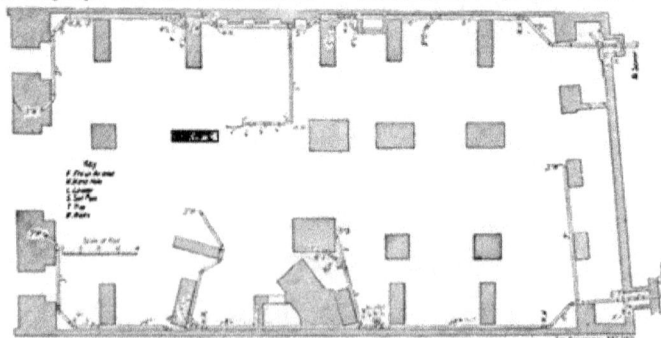

MAIN DRAINS IN CELLAR, MANHATTAN LIFE INSURANCE BUILDING, NEW YORK CITY.

tributing drum; a 3-inch discharge pipe from the house pumps and another from the artesian well pump to the roof tanks; a 3-inch fire line from pump to roof tank, a 4 inch pipe from roof tank to the eighth-floor tank, a 4-inch pipe from the eighth-floor tank to the cellar distributing drum, and a 2-inch pipe from the roof tank to the hot-water heater in the cellar. All of these main pipes are supplied with shut off gate valves, and have no branches taken from them. There is a separate falling main from the artesian roof tank to the general toilet-rooms, with branches on each floor for water-closets and urinal flushing cisterns only. It is 3 inches in diameter above the second floor, where it is reduced to 2 inches and continued to the pipe cellar, where it runs horizontally around the ceiling and supplies the various private toilet rooms with separate vertical risers.

From the Croton pressure distributing drum there are run separate 1½-inch lines to each of the toilet-rooms in the basement, subbasement, and first floor. From the tank-pressure distributing drum there are run separate risers to each vertical line of office washstands 1½ inches in diameter up to the eighth floor. For each vertical line of private toilet-rooms there are run 1½ inch risers up to the eighth floor, and similar 1½-inch lines are run down from the roof tanks to the eighth floor. For the large group of general toilet-rooms the risers are 3 inch pipes from the cellar and from the roof to the eighth floor. From these vertical rising supply lines the branches to the fixtures are as follows—viz., for supplying each washbasin, one-half inch; for supplying each water-closet and urinal cistern, one-half inch, for supplying each slopsink three-fourths inch; for supplying each bathtub, kitchen or pantry sink, three-fourths inch.

Wherever a branch line supplies more than a single of the fixtures named it is proportionally increased in sectional area. The branches for supplying each general toilet-room begin at 1½ inches in diameter, and those supplying private toilet-rooms at 1 inch in diameter, and are reduced in size as the various branches to fixtures are taken off. Each horizontal branch from a rising line supplying a group of fixtures is provided with separate shut-off valve, in order to control each toilet-room separately. Branches are provided in the rising lines for each story, and where there are no fixtures tees for possible future use are left tightly plugged. Separate full size shut-off valves are provided at each fixture (both on the cold and hot water supply) at each water-closet and each urinal flushing cistern. The rising hot-water lines for slopsinks in toilet-rooms are 1½ inches in diameter, of tinned and annealed brass. They are carried up to the highest fixture without any reduction in size, and a ¾-inch circulation pipe is taken from the highest fixture back into the hot-water tank or boiler. The hot-water riser is extended upwards and turned over the top of the roof tanks. All supplies to fixtures (except flushing cisterns) are provided with large-size air chambers. All horizontal lines are arranged neatly and symmetrically so that they do not unnecessarily cross each other have no depressions or sags, nor are bent up in such a manner as to become air-bound.

The main vertical pipes are run in the ventilation shafts (V V, Fig. 2), or in wall recesses where they are accessibly inclosed by movable wooden panels, screwed on. Hot and cold water pipes do not touch each other, and are usually separated 3 inches in the clear. All supply pipes throughout the building are so graded and valved that they may be readily and completely emptied.

Figure 1 is a plan of the cellar showing the location of the tanks, pumps, distributing lines, etc. there. Figure 2 is a typical floor plan showing the arrangement of washstands, water-closets, etc. in the eighth floor, to which the other rented floors are similar. At each riser line are four pipes, a trap vent, a soil, a cold wa'er supply, and a safe waste, all run in wall recesses except at the two ventilation shafts V' and V. which are accessibly located in the foul-air flues. In V' there are six pipes and in V there are 15 pipes, including the fire line or stand-pipe, all mains to and from the roof and intermediate tanks, pump risers, rainwater leaders, gas mains, etc.

PART II.—SIZES AND WEIGHTS OF PIPES, CONNECTIONS, FITTINGS, ARRANGEMENT OF SUPPLIES, ETC., ROOF DRAINAGE, BACK AIR, DRAWING AND DESCRIPTION OF ROOF TANKS.

ALL lines of cold water supply pipes, except the branches under the fixtures, are lap-welded extra-heavy (not the standard) galvanized-iron pipes, warranted to be tested by hydraulic pressure of 500 pounds per square inch. Concealed branch supply pipes at all fixtures are heavy drawn AAA lead pipes, required to weigh as follows:

½ inch pipe to weigh 3 pounds per foot.
⅝ " " " " 3½ " " "
¾ " " " " 5 " " "
1 " " " " 6 " " "

In all public and private toilet-rooms, except employees', all exposed supply pipes at fixtures (not the rising supply mains) are nickel plated brass pipes, with nickel plated brass hangers, hold-fasts, escutcheons, etc. Numbered polished-brass tags are attached to all shut-off valves in the building, except at the shut-off valves directly at the fixtures, the use of which is obvious, and corresponding printed lists are prepared, giving number, location, and description of every shut-off valve in the building so as to avoid any possibility of mistake or confusion in their operation.

There are in this building, exclusive of the fire valves, etc., the following apparatus and fixtures—viz.: One drip tank, about 63 water-closets, 15 slopsinks, about 52 urinals, 20 washbasins in toilet-rooms, about 166 office washstands, one bathtub, one kitchen gas range, one kitchen and two pot sinks, two engineer's sinks and one pump sink, two grease traps, one kitchen boiler, heated by Vulcan gas-burning attachment, one hot-water tank, two roof tanks, two intermediate tanks, one suction tank, two distributing drums, three pumps, one water meter; total about 343 fixtures. The location of these fixtures is: 15 in pipe cellar, 16 in basement, eight in first floor, 17 in second floor, 19 in third floor, 19 in fourth floor, 19 in fifth floor, 19 in sixth floor, four in seventh floor, 24 in mezzanine floor, between seventh and eighth,

20 in eighth floor, 19 in ninth floor, 19 in tenth floor, 19 in eleventh floor, 19 in twelfth floor, 19 in thirteenth floor, 18 in fourteenth floor, 17 in fifteenth floor 17 in sixteenth floor, 27 in seventeenth floor and on roof.

The plumbing in general is " exposed work "—i. e., all fixtures are arranged in an open manner, and all pipes and plumbing-work kept exposed to view. The finish at fixtures is of nickel-plated brass in all general and private toilet-rooms and of all office washstands, and of lead and galvanized wrought-iron silver bronzed for all employees' fixtures. All toilet-rooms are fitted up complete with marble partitions and have no woodwork, except the doors to closets, which are short flap doors, paneled, trimmed, with brass hardware, the latch operated from the inside, and with rubber striking tips and single-action

of the hot-water faucet, so that in case hot water shall be wanted in the future a hot-water faucet can be fitted up. The standard size for bowls is 15x19 inches for all 22x33 inch square slabs, and 14x17 inches for all corner slabs and for the smaller square slabs. Over each urinal or set of urinals is placed a patent automatic flushing cistern, copper-lined, and encased in marble of design to correspond to the water-closet cisterns. The capacity of each tank is such that each urinal receives a one-gallon flush. With each of the 19 fire valves (one in each story) there is connected 75 feet of 2½-inch three-ply heavy unlined linen fire hose, guaranteed to stand a pressure of 300 pounds per square inch.

All main drain, soil, waste, and leader lines are of heavy asphalted wrought-iron pipes of standard make. The pipes and their fittings are of a uniform

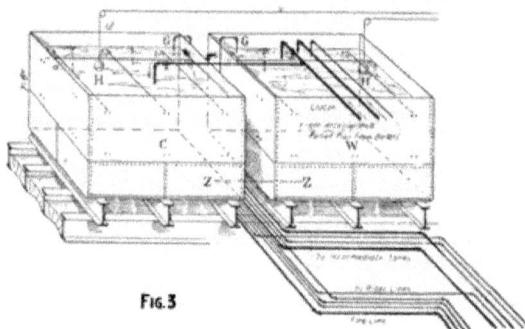

FIG. 3

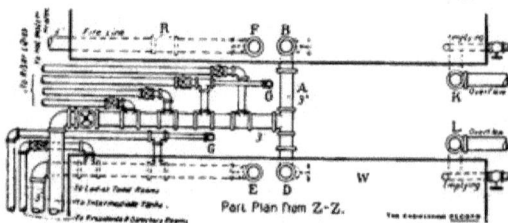

PLUMBING IN THE MANHATTAN LIFE INSURANCE BUILDING, NEW YORK CITY.

butts, bolted to marble jambs with brass spherical-headed bolts. Marble platforms 1½ inches thick are placed under all water closets, slopsinks, urinals, and washbasins, and under kitchen and wash sinks, and bathtub.

The washstands and all urinals and water-closets are supplied with cold water only. All slopsinks, kitchen sinks, pantry sinks, engineer's sink, and bathtubs have both hot and cold water supplies. All washstands are supplied with cold water only, but in the engineer's bathroom there is both hot and cold water. Where basins have cold water only a raised marble button is provided in place

thickness of not less than one-fourth inch. All pipes and fittings were tested by hydrostatic pressure and by hammer test at the pipe mills, and a written guarantee was filed that such test had been applied. All vent pipes and vent fittings are galvanized. The weight of heavy wrought-iron pipes is as follows:

6-inch pipe,		18¾ pounds per foot.		
5	"	"	14½	"	"
4	"	"	10⅞	"	"
3	"	"	7⅝	"	"
2	"	"	3½	"	"

All joints in iron pipe are screw joints, made absolutely tight by a mixture of red and white lead.

The 1½-inch hot-water riser has several swivel joints to provide for the expansion and contraction due to temperature changes. Underground pipes are laid in a well-graded trench, and after inspection were bedded and covered with cement, leaving exposed only the handholes for access, which are surrounded with brick manholes and furnished with cast-iron frames and covers. All connections between lead and iron pipes are made by means of extra-heavy brass screw ferrules, or nipples, screwed into the iron fittings, and connected with the lead pipes by wiped joints, and all ferrule joints were included in the test of the soil pipes. All iron pipe fittings for branches are Y or half-Y branches, except on upright lines, where curved or sanitary T Y's are used, and on vent pipes, where common T branches are used. No quarter-bends nor ordinary offset fittings are used on soil or waste pipe stacks. One-sixth, one-eighth, and one-seventh bends, or the new pattern (45 degrees) offset fittings are used. All outlets not used for connections are properly closed with brass trap screws and kept accessible. All connections between vertical stacks and horizontal pipes are made with Y branches and eighth bends. All junctions of branches with the main house sewer are by Y branches. All vertical lines of soil and waste pipes are provided with Y branches or T Y's for outlets in each story. In cases where such outlets are not used they are securely closed with brass trap screws. On horizontal lines T Y branches are not used, but they are used on upright lines.

Back-air pipes run along each upright line of soil or waste pipe, and for office washstands are 4 inches in diameter (except 5 inches where there are water-closets on the line). Vent pipes for urinals, basins, and slopsinks in toilet-rooms are 4 inches in diameter, and along soil pipes are 5 inches in diameter. All back-air lines run through the roof independently, and are dripped at the bottom of the line, and all T branches for vents from fixtures are set above the overflow point of the fixture so that the vent line cannot act as a waste pipe in case of stoppages.

Branch waste and vent pipes and flush pipes for fixtures are drawn " D " lead pipes, weighing as follows:

1¼-inch, 2½ pounds per foot				
1½	"	3½	"	"
2	"	5	"	"
3	"	6	"	"
4	"	8	"	"

All water-closet lead bends are of eight-pound lead. In all toilet-rooms, whether public or private, except those for employees and janitors, all exposed waste, vent, and flush pipes are of heavy seamless drawn brass heavily nickel-plated. All back-air pipes for traps, as far as exposed to view, including couplings, etc., are of heavy seamless brass, nickel-plated. All safe waste lines are of galvanized wrought iron, 1¼ inches inside diameter, all put together with screw joints, and with 1½x1-inch branches for outlets in safes on each floor. All drip-pipe lines are carried independently at the ceiling of the cellar and made to discharge through finished hinged brass

flap valves separately and openly into a trapped and water-supplied sink in the cellar. Along each vertical line of soil pipe and waste pipe is a line of 1½-inch drip pipe from the marble safes or platforms under the fixtures. These drip pipes are carried down to the engineer's sink in the pipe cellar, each line separately, and the mouth of each is protected by a stamped brass-hinged flap valve. Waste-pipe lines for office washstands are 3 inches in diameter (except 5 inches where there are water-closets on the line). Waste pipes for urinals, basins, and slopsinks in toilet-rooms are 4 inches in diameter. All soil pipes are 5 inches in diameter.

The roof drainage is carried down through four 4 and one 6-inch extra-heavy asphalted wrought-iron inside leaders, each trapped at the foot of the stack by a full-size extra-heavy cast-iron trap, and connected by Y branches and eighth bends with the two sewer lines. All outside leaders for domes, skylights, etc. are connected with brass or copper ferrules and calked joints to the rainwater inlets in the wrought-iron leaders.

Figure 3 shows the arrangement and connections of the two 5,000-gallon roof tanks, which are located in a separate tank-house built on the main roof. Each tank rests on three rolled-steel girders, which distribute its weight over four lines of floor beams, and is made of boiler-iron, lap-riveted and stayed by 10 intersecting internal tie-rods. The tanks are 7 feet 6 inches deep, and are intended to hold 7 feet of water. The bottoms of the tanks are connected by a 3-inch flanged cast-iron pipe A with vertical ends, having gate valves B and D, through which the main house supply is drawn. Ordinarily it is intended that the tank W shall be filled with well water for use in flushing cisterns only, and tank C should be filled with city water for other general purposes. When this is the case valve D is closed and valve B is open, supplying the intermediate tanks, boiler, and riser lines through pipe A and its connections. Valve E is open to supply the special cistern system, and valve F is always open to insure a constant tank-pressure fire service. R is a check valve closing toward the tank with the pressure from the fire pump. By reversing valves B and D the whole supply may be taken from the well tank W, and by opening both of them the water in both will be mixed and equalized. G G are vent pipes to promote the emptying of the riser lines, and H H are copper floats operating the electric alarms indicating in the engine-room. The vertical outlet pipes at B and D rise 6 inches above the bottom of the tanks.

PART III.—CELLAR DRAINAGE AND SEWAGE PUMP, TEST-ING OF PIPE LINES, HOT-WATER HEATER, DETAIL DRAWINGS AND DESCRIPTION OF INTERMEDIATE PRESSURE TANKS AND BASEMENT DISTRIBUTION DRUMS.

THE drips, tanks, wastes, etc., in pipe cellar, being below the level of the street sewer, are drained into a mason's cesspool. The contents of this tank are emptied periodically by means of a pump, and discharged through a proper waste pipe into an elevated, trapped, and water-supplied sink in the basement, with waste to sewer. The tank is a round, open well, of 500 gallons capacity, 6 feet in diameter, and 3 feet deep. There are two sewer connections with the street sewer in New Street, each 6 inches in diameter, trapped by 6-inch traps and fitted with 5-inch fresh-air inlet pipes.

After the pipe lines were completed all openings of soil, waste, drain, and vent pipes and the ends of horizontal drains were securely closed by means of soil-pipe plugs, the lead bends of branches properly soldered up, and braced where required to withstand the pressure, and the whole system of piping was filled with water to the top of the building. The water remained at the original level for 12 to 48 hours without signs of leaking. This test was made in two parts, sections of the pipes being first filled up to the thirteenth story, as they were put in place from the bottom up.

Figure 4 shows the arrangement and connections of the intermediate tanks designed to relieve the ex-

cessive pressure on the lower sections of the pipes. No special provisions having been made for their reception it was necessary to limit them in size and place them in two tiers in a small closet on the eighth floor, where they are filled through the overflow pipes K and L from the roof by Croton and well water tanks respectively, the roof tanks having their electric indicators arranged to show when the pipes K and L are full and overflow begins. Provision for a separate pump line to the eighth floor has been contemplated, so as to avoid raising the water unnecessarily high for the lower floors, but, as the expense of pumping is only slightly increased by lifting the water an increased distance after it is once started, it was thought best, for convenience and simplicity, to arrange it as shown. In the four corners of the room were set special riveted steel columns C C C C, which support at convenient heights 10-inch rolled I beams, two of which carry each tank, so as to leave it well exposed and accessible for connections, inspection, and the manipulation of its

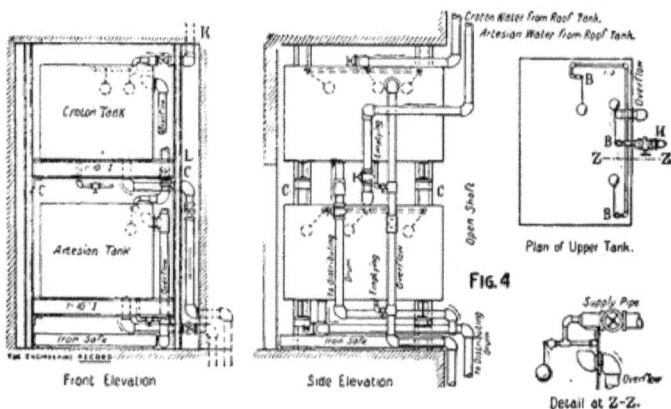

Front Elevation   Side Elevation   FIG. 4   Plan of Upper Tank.

Detail at Z-Z.

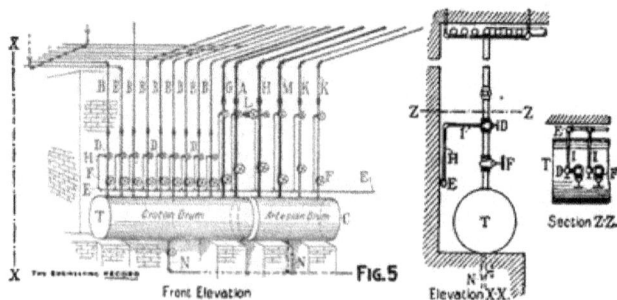

Front Elevation   FIG. 5   Elevation XX.   Section Z·Z.

PLUMBING IN MANHATTAN LIFE INSURANCE BUILDING, NEW YORK CITY.

valves, and to allow the convenient arrangement of pipes and valves with economy of space. Croton water from the roof tank is delivered through the 3-inch falling main K and a gate valve to the three ball cocks B B B, arranged as shown in plan on a horizontal branch pipe, which is fastened to the upper edge of the tank, and having check chains to limit the motion of the floats. These tanks are entirely independent of and unconnected with each other, and all their pipes are carried up and down 'hrough the adjacent large ventilation shafts at the rear.

Figure 5 shows the connections to the distributing drums in the cellar. They are made of boiler-iron and tested by a hydrostatic pressure of 500 pounds per square inch. Drum T for supplying the basement to the seventh floor inclusive with Croton water is 8½ feet long, 2 feet in diameter, and holds 200 gallons. It is supplied from the eighth-floor intermediate tank C by the 3-inch pipe A, and distributes water to the different lines of fixtures through the independent 1½-inch risers B B, etc., each with a separate gate valve F and emptying valve D, by which its contents may be discharged into drip pipe E when D is open and F closed. G is a separate 3-inch main from the main roof tank to supply the drum directly if necessary, but its valve is ordinarily kept closed. Drum C, of about 100 gallons capacity, is intended to receive the artesian well water supply through 3-inch pipe H from the intermediate tank shown in Fig. 4, and distribute it to all flushing cisterns below the ninth floor through the 1½-inch pipes K K, but by opening valve L, which is usually closed, it may be supplied from the roof or intermediate tanks. M is a 3-inch main from the artesian roof tank W, Fig. 3, and its valve is usually closed.

It will be noticed that the pipes and valves are arranged with symmetry and connected with unions which allow the disconnection of any one without disturbing the rest. The valves allow each line to be separately cut out and emptied for new connections, alterations, etc. N N are 1½-inch emptying pipes. Adjacent to these distributing drums, as shown in Fig. 1, is set on wrought-iron standards or supports a boiler-iron closed hot-water tank of 250 gallons capacity, 2½ feet diameter by 6½ feet long. This tank is warranted to have been tested by a hydrostatic pressure of 500 pounds per square inch. It is supplied from roof-tank pressure and provided with manhole, emptying pipe, and proper supply connections from the falling main. Inside of the tank is a 1-inch tinned brass steam coil, with connections to live and exhaust and return steam pipes, and a Powers automatic heat regulating attachment to shut off the steam supply when the temperature of the water rises to 150° Fahr. The tank has one 2-inch delivery pipe supplying the hot-water fixtures and a 1-inch return circulation pipe entering the emptying pipe just above its valve. There is a 2-inch cold-water roof-tank supply pipe, and the tank has a non-conducting covering and a manhole opening.

PLUMBING IN THE BANK OF AMERICA BUILDING, NEW YORK.

(PUBLISHED IN 1883.)

PART I.—TANKS, BOILERS, ETC.

IN the Bank of America's new 10-story building in Wall Street, New York, the first floor is occupied by the bank and its offices and the other stories are arranged as single offices and suites, each of which is supplied with hot and cold water. There are also public and private water-closets, urinals, etc.

The architect of the building was Charles W. Clinton, New York, and the plumbing, some details of which we shall describe, was done by John Tourney, New York, whose foreman in charge of the job was John B. Donovan.

To prevent possible interruption of the supply and to insure its sufficiency, water is taken from two separate branches of the city mains—viz., from Wall Street, through the 2-inch pipe B, Fig. 1, and from William Street through the 1½-inch pipe A. Both pipes deliver through ball cock C to the receiving tank T, which is on the cellar floor adjacent to the heating and distributing system. By opening valve D and closing valve E water is drawn from Wall Street only; by reversing the valves, from William Street only, and by opening both valves, from both streets. F is the 2-inch suction pipe to boiler, tank, and fire pumps, and G is the sediment pipe for emptying the tank. H is the overflow pipe.

Figure 2 shows the heating and distributing system, in which portions of tank T and pipe A, B, H, and F of Fig. 1 reappear. C is a boiler containing a coil that receives live or exhaust steam from pipe G and discharges it through pipe I and trap E. J is the 2-inch pipe from the roof tank and its 1½-inch branch K supplies cold water to the boiler C. N is the hot-water pipe from boiler to the different floors. O is the hot-water return-circulation pipe. M is a branch from tank pipe J and supplies distributor D, from which the 1-inch pipes P P, etc., supply the various upper floors. Q is a direct supply from street pressure to the bank offices. S is a pipe discharging all safe waste drips into a sink in boiler-room. U is a branch from Wall Street main direct to the suction pipe of the tank pump. V is a drip pipe draining the rising lines and discharging into tank T. R is a sediment pipe draining D. A sediment pipe hidden behind it empties the boiler C.

Figure 3 shows the arrangement of tank pump Z in the cellar near the receiving tank. A is its suction pipe that is supplied directly from the Wall Street main, through pipe U, or from receiving tank T, Fig. 1, through pipe F. B is the delivery pipe that is connected to the roof tank pipe E by branch G, and to the fire line pipe by branch J. Ordinarily valves C and P are closed and valve D is open, and the pump delivers to the tank, but in case of fire valve D may be closed and C opened, and the pump worked on the fire line. I I are branches from the fire and boiler pumps that connect by pipe H with the pipe E and provide for the filling of the tank if pump Z should be disabled. K is a drip pipe emptying the tank pipe. O is the hot-water return-circulation pipe (see Fig. 2 also). M and L are cold-water pipes supplying engine-room sink and water-closet. N is an extra supply for boilers to furnish water when pumps are not working. P is a valve to which hose may be attached for washing the floor, etc. Q is a check valve.

Figure 4 shows the engine-room sink that is supported by the polished-brass pipe frame and standard A, has hot and cold taps, H and C, and receives the drip and safe waste pipes and the overflow from tank T, Fig. 1.

PART II.—ROOF TANK, PIPE LINES, AND DETAILS.

FIGURE 5 is a sketch of the roof tank A that is about 6x12 feet and 7 feet deep, built of ⅜-inch iron, with 2½x2½-inch corner angles and stiffening bars of 4x4 T iron and 2x3 angles. It rests on 5x6-inch yellow pine sticks B B, that raise it from

the iron safe C that is provided, notwithstanding that it is above the iron roof surface D, from which all water drains directly to the gutters. E is the 4-inch fire line. F is the 2 inch house supply. G is the 4 inch overflow that discharges directly on the roof, and has a 10-inch copper flaring top H, intended to increase the rapidity of discharge. I is a 2-inch sediment pipe emptying the tank. It also discharges on the roof. J is a 2-inch pump pipe, and K is a 1-inch relief pipe from the boiler C, Fig. 2. L is a 2-inch safe waste discharging in the sink, Fig. 4. M is an 18-inch copper float operating the gauge index in the engine-room.

This building contains about 40 water-closets, 20 urinals, 50 washbowls, and six slopsinks, exclusive of the janitor's apartments, which have a kitchen sink and washtubs with marble panels and safes, one toilet-room containing urinal and water-closet, another

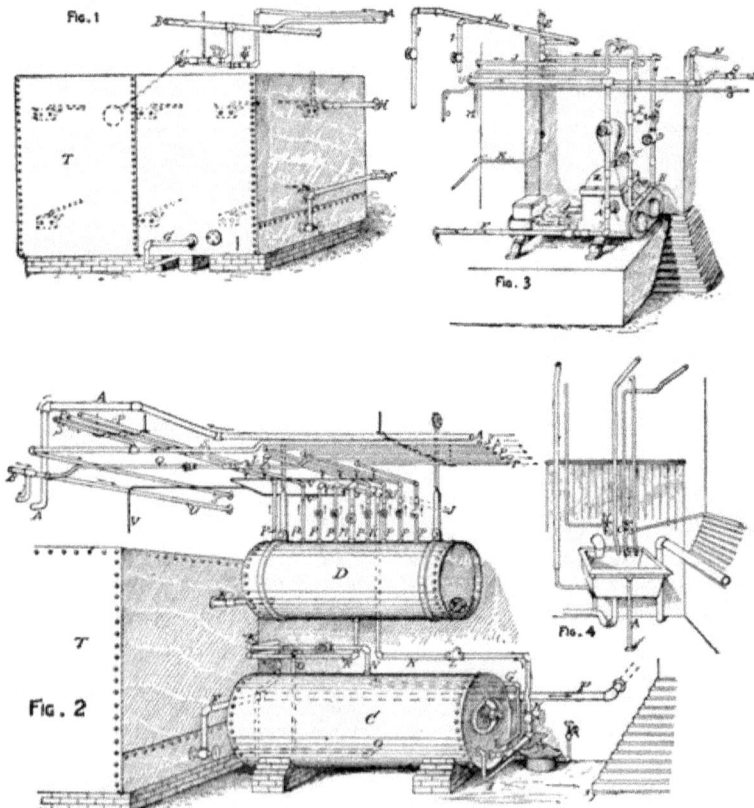

PLUMBING IN THE BANK OF AMERICA BUILDING, NEW YORK CITY.

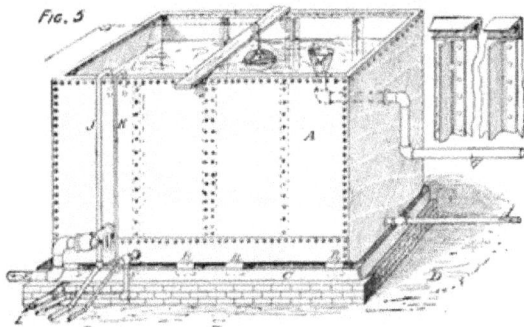

FIG. 5

containing bathtub and water-closet, another with a sink only and two washbasins. All the water pipes are brass, and where exposed are polished and lacquered.

There are 20 lines of soil, waste, and vent pipes from 3 to 6 inches in diameter, and rising to about 150 feet above their lowest points. These lines were all filled to the top with water, and it is stated that

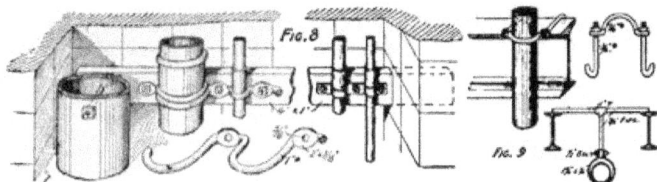

FIG. 8

FIG. 9

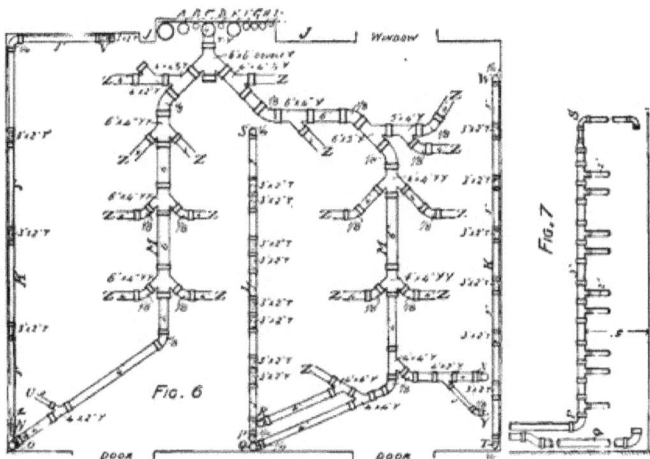

FIG. 6

FIG. 7

PLUMBING IN THE BANK OF AMERICA BUILDING, NEW YORK CITY.

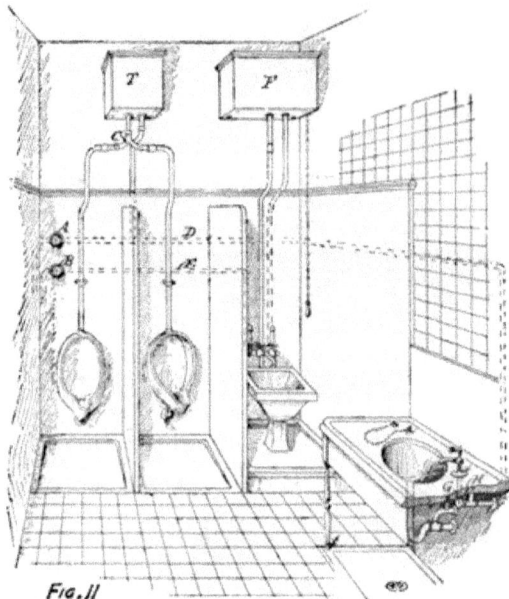

FIG. 11

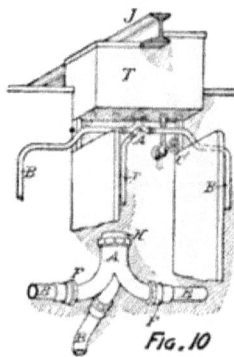

FIG. 12

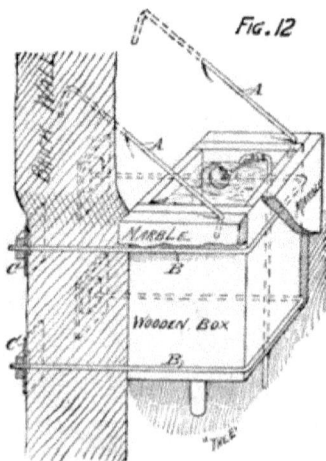

FIG. 10

PLUMBING IN THE BANK OF AMERICA BUILDING, NEW YORK CITY.

no leaks were found, and the water did not settle perceptibly in any of them during the five or ten hours' test. In the eighth-floor toilet-room, 16x20 feet, it was necessary to crowd in 16 water-closets, four urinals, two washbasins, and a slopsink. The complicated arrangement of soil and vent pipes used here is shown in diagrams, Figs. 6 and 7. J J is a pipe shaft containing all the rising lines, A is a 4-inch vent pipe from lower floors, B is the gas main, C the main soil pipe, D is a 1½-inch hot-water pipe from the basement boiler, and is continued to open above the roof tank, E is a 6 inch rainwater leader, F is the 2-inch pump pipe to tank, G is the 2-inch supply pipe from tank, H is the 2-inch drip pipe from tank safe and lower safes, I is the 1-inch hot-water return-circulation pipe.

The soil pipe C extends above the roof, and its main branches, M and M, connect at O and Q with vertical pipes extending above the roof. K, L, and K are the trap vent pipes and rise above the roof from points N, P, and T. At each tee and at S, V, and W a 2-inch lead branch connects with the trap of the corresponding fixture. At Z Z, etc. are water-closets, at X a slopsink, at Y two washbowls, and at U four urinals. The lines M and M are laid beneath the floor tiles. The vent lines, K and K, are behind the wainscot, and L is concealed in a fireproof partition that crosses the room. The walls and ceiling of this room are plastered and tinted, the floor tiles, wainscot, and other panels are of Italian marble, and the cabinet-work (which is all elevated from the floor) is of oil-finished oak. The shaft J J is faced inside with glazed brick, and to avoid defacing these two special iron supports, shown in Fig. 8, were built in the first or bank story, and one in each of the other stories, to carry the rising lines of pipes.

Figure 9 shows different methods of supporting the pipes from the iron floor-beams.

Figure 7 is an elevation of vent pipe L in Fig. 6.

PART III.—WATER-CLOSETS AND DETAILS.

FIGURE 10 shows the automatic tank T and flush pipes B B B that serve three urinals in the basement. The tank has a capacity of 10 gallons, and is usually set to flush every 15 minutes; it is a wooden box lined with copper, faced with marble panels and neatly fitted around iron floor joist J. The supply is through a ball cock, and is controlled by valve C. The discharge is through the special brass three-way branch A with ground couplings F F F connecting it to the flush pipes B B B that are smoothly and symmetrically curved.

Figure 11 is a general view of a room on the seventh floor. The walls and floor have white ceramic tiling and the paneling is of white marble. The supply pipes D and E are controlled by valves A and B, and are easily accessible behind the marble panels. The automatic flush tank T is similar to that shown in Fig. 10, except that its branch C is two-way. The flush tanks, shown in Figs. 10 and 11, and all others in the building are ingeniously supported as shown in Fig. 12, where the ¼-inch brass rods A A are leaded into the marble at the lower end, and into the brick at the upper ends, and the ½-inch rods B B are

tightly screwed up against iron washer plates C C on the back of the wall. A furring strip (omitted for clearness in the illustration) separates the marble from the wooden box far enough to permit the rods B B to be placed as shown.

PLUMBING IN THE CONSTABLE BUILDING.

(PUBLISHED IN 1895.)

PART I.—WASTE AND VENT AND HOT AND COLD WATER PIPE LINES, FITTINGS, VALVES, AND CONNECTIONS, ARRANGEMENT OF PIPES IN BASEMENT, CONNECTIONS OF RISERS TO DISTRIBUTION MAINS, AND SECTIONS SHOWING HOT AND COLD WATER RISER SYSTEMS AND SOIL-PIPE LINES.

AN office and store building has just been erected by Marc Eidlitz & Son, builders on the northeast corner of Fifth Avenue and Eighteenth Street, New York City, for the estate of Henrietta Constable according to the plans of William Shickel & Co., architects. The plumbing has been executed by T. J. Byrne, with Arthur H. Napier, C. E., as consulting engineer. The building has a frontage of 100 feet on Fifth Avenue and a frontage of 200 feet on Eighteenth Street, and consists of a basement and 12 full stories, with bulkheads, tank-house, toilet-room, and janitor's apartments, forming a thirteenth story over portions of the building. The first and second stories are arranged for large stores and the remainder of the building for offices singly and en suite.

The plumbing comprises a supply of cold water to washbasins throughout the building, hot water to the slopsinks on every floor and to the main toilet-room; filtered water to each corridor, water under tank and pump pressure for fire service, public and private toilet-rooms, and complete independent domestic apparatus in the janitor's apartments. There is also the necessary system of tanks, pumps, filters, and meters, beside the main and distribution supply pipes, soil, vent, and sewer pipes. Some of the essential and characteristic features of this installation are here described and illustrated from original sketches and the working drawings and specifications. All exposed drains in the basement, and all soil, waste, and vent pipes and branches, and all leaders are made of standard welded steel tubing. All such pipe and fittings (except where exposed in basement) are thoroughly coated inside and outside with a good asphalt varnish. Exposed pipe in the basement is tar-coated on the inside only and painted on the outside. All fittings, traps, etc. for steel pipe are special extra heavy, recessed threaded, cast-iron drainage fittings. Branch fittings and ells have threads tapped at a grade. Reducing fittings are used instead of bushings, and no steam fittings or cast bushed fittings were used.

All cold supply piping, mains, pump, tank, heater, and filter connections, cold risers, fire lines, and safe wastes (except ½-inch pipe) are of galvanized wrought-iron pipe, with heavy galvanized malleable and galvanized cast-iron fittings. All lead traps are of six-pound and eight-pound lead, with brass trap screws; water-closet bends are of eight-pound lead.

All hot supplies, all exposed fixture branches, and all ⅜-inch supply pipes are of brass. Brass waste and vent branches at traps are of "iron pipe size," and all brass traps were specially made to permit the use of this heavy pipe. All supply and fire lines have finished brass gate and globe valves, with iron or brass wheel handles, and are plated or polished to correspond with pipe. Valves 2 inches and over are gate, others are globe valves, and 4 inch valves have an iron body. Connections between cast-iron and steel drain pipes are made with special calking fittings. Joints between lead and cast-iron, wrought-iron, or steel pipe are made with brass ferrules and

soldering nipples the size of the fitting, with the lead run through the ferrule. The joints between pipe stacks and the roof are made by special roof fittings and sleeves of 20-ounce cold-rolled copper, extending 18 inches on all sides of the pipe fittings under the tile and its chambered sleeve above. All changes in direction, and fixture connections are made with Y branches and 45-degree elbows, or on upright lines only, with the special long 90-degree Y's. All branch fittings and bends on waste and soil lines, etc. have threads accurately tapped to give a uniform grade of one-fourth inch or one-half inch per foot. All sewers and drains and horizontal

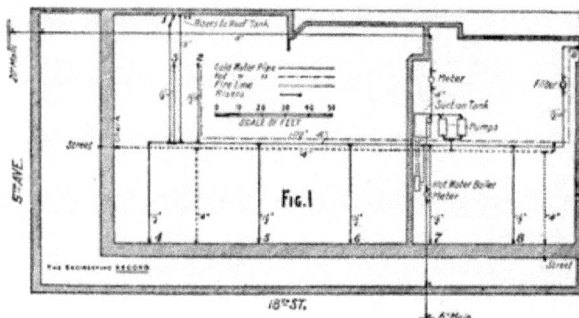

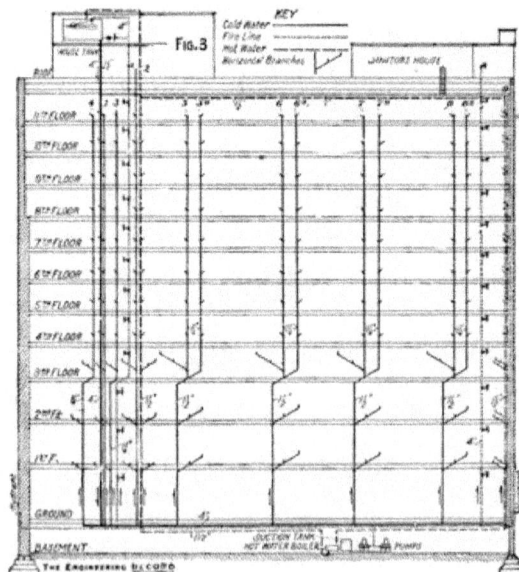

PLUMBING IN THE CONSTABLE OFFICE BUILDING, NEW YORK CITY.

wastes, or their branches have a uniform fall of at least one-fourth inch per foot, and as much more as may be practicable. The rainwater leaders are of 4, 5, and 6-inch steel pipe, with special running traps. There are three 5-inch and two 4-inch lines of soil pipe and 11 lines of 3-inch waste pipe, the latter all increased to 4 inches just above the highest connection.

All branches are in general of steel. At the ends of soil and waste pipe branches, at angles, and on all iron traps there are set Y-branch or trap hub cleanouts of the same size as the pipe, closed with special heavy cast-brass plugs. The water-supply lines in the basement are run exposed on the ceiling, but throughout the building they are covered in with the waste lines. All lines and branches are graded so

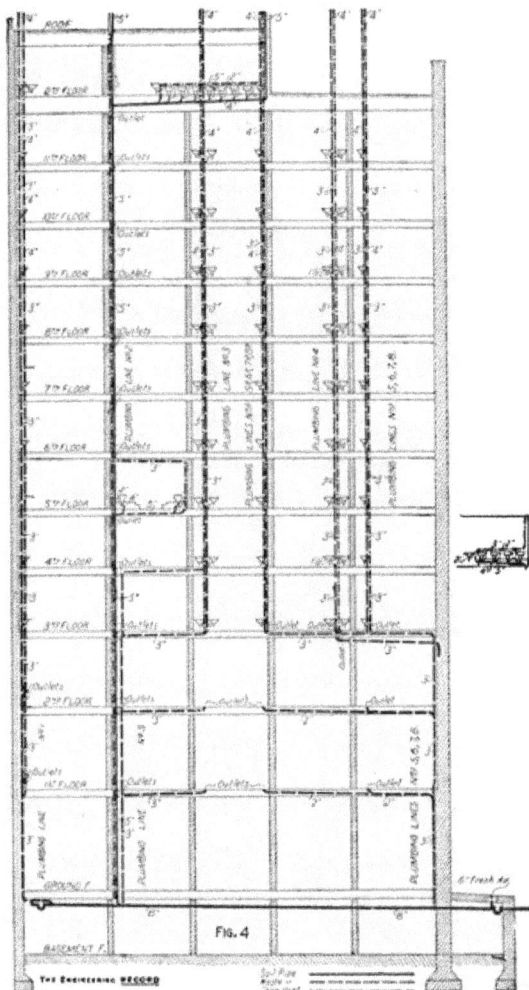

FIG. 4

PLUMBING IN THE CONSTABLE OFFICE BUILDING, NEW YORK CITY.

as to completely empty at the lowest point in the basement. Drip pipes with globe valves are provided for all mains, risers, returns, etc., and are run to the sink or receiving tank in the basement, so that any or all parts of the supply system may be emptied for repairs.

Figure 1 is a diagram of the basement showing the arrangement of the pumps, etc., and the approximate location of the hot and cold water distributing mains and riser branches on the ceiling.

Figure 2 shows the arrangement and support of the group of pipes in one set supported from the basement ceiling.

Figure 3 is a longitudinal section of the building showing the arrangement of the hot and cold water risers on one side of the building. Horizontal

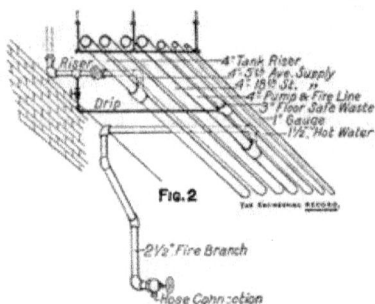

FIG. 2

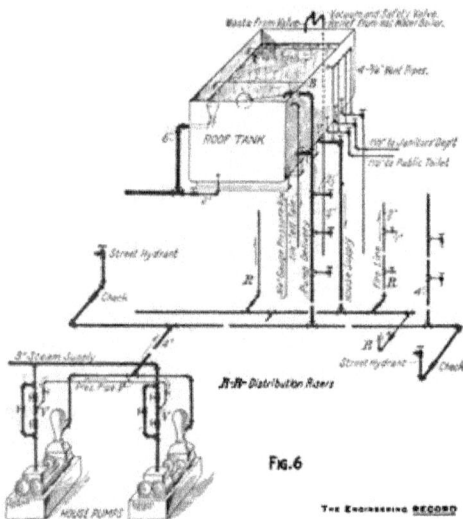

FIG. 6

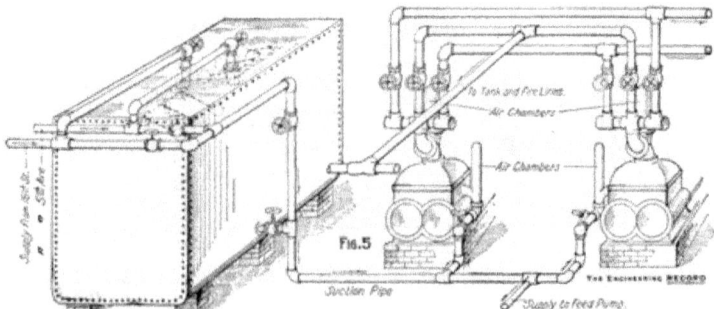

FIG. 5

PLUMBING IN THE CONSTABLE OFFICE BUILDING, NEW YORK CITY.

branches are taken off from the vertical pipes up to the fourth floor, above which level the lines are direct to the rows of superimposed fixtures.

Figure 4 is a tranverse section showing the principal lines of soil waste, and vent pipes and the arrangement of risers to avoid obstructing the stores on the lower floors and to connect with one of the three separate lines of sewer pipe which cross the building tranversely.

PART II.—PUMPS, SECTION TANK AND ROOF TANK.

FIGURE 5 shows the arrangement of suction tank, pumps and main pipe connections. Water from both the Fifth Avenue and Eighteenth Street mains is taken in through 4-inch branches each of which has a 4-inch meter with a check valve on the outlet side of the meter set so as to prevent water from one main escaping through both meters and into the other main if it should be broken or emptied, or to prevent a crossflow when the pressure in the two mains varies. Each main is separately controlled by a valve and can independently fill the suction tank through four 2½-inch ball cocks. The suction tank is about 12'x7'x8' deep, made of ⅜-inch steel plates, with tee-bar and angle-iron stiffeners and cross tie-rods, and holds about 5,000 gallons. It is considered an open tank in that it is provided with an overflow, and the water it contains is not under pressure, but it is tightly closed with an iron-plate cover to keep out dust, and has a hinged manhole door to give access to the interior. All its pipe connections are made with inside and outside flanges and the galvanized-iron 4 inch overflow and 2-inch emptying pipes (not shown here) are run to the receiving tank, which is set below the basement floor. It is made of boiler-iron and is about 14'x9'x5' deep, with riveted iron cover with two manholes with heavy cast-iron covers flush with the floor. This tank receives surface drainage from the light court engine-room floor, and overflows from all basement tanks, etc., and is pumped out to the sewer. The pumps are connected as shown, to be supplied either from the street main direct or usually through the suction tank, and are connected together for regular house and fire service and cross-connected to re-enforce the boiler feed pumps if necessary. There are in all five Blake duplex pumps. The two 10'x4½'x12' shown are for fire and house service, two 6'x4'x6' for boiler feed, and one 6'x4'x6' for receiving tank, drip tanks, etc.

Figure 6 shows the connection of pipes to the roof tank and the connections to the pumps. When the tank is full its rising main is closed by the ball cock B and is immediately subjected to pressure by the action of the pump. This pressure is instantly transmitted through the small pipe P to the Kieley automatic regulating valves V V, which shut off the steam and the pumps stop. As soon as the water in the tank is lowered, the ball cock opens, the pressure in the pump discharge pipe, and consequently in pipe P, is relieved, valves V V are opened by springs, and the pumps immediately start up. Of course the automatic valves V V are provided with by passes, and the pumps can be governed by hand or arranged to

work up to a heavy fire pressure much in excess of the pressure of the tank head.

The 6 500-gallon house tank is about 16'x7'6"x8' deep, constructed of tank steel plate, with all seams riveted and calked. It is stiffened and braced by T bars set vertically about 4 feet apart round the tank, and with two sets of tie-rods. Heavy 3x3-inch angle iron is riveted round the top, and the tank is covered with ¼-inch plates (not here shown), with two large manholes with frames and covers. All pipe connections to tank are made with inside and outside riveted

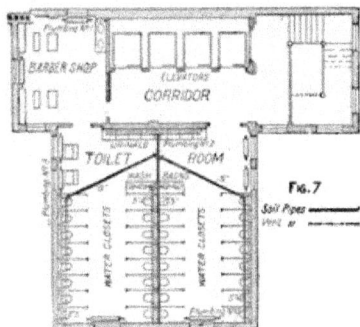

FIG. 7
Soil Pipes ———
Vent ——

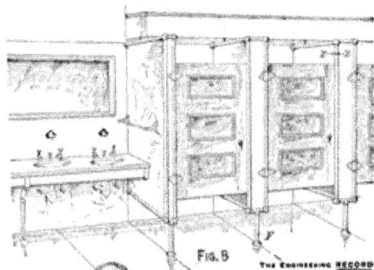

FIG. 8
THE ENGINEERING RECORD

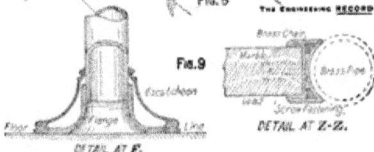

FIG. 9

DETAIL AT F.

DETAIL AT Z-Z.

flange joints. The galvanized standing overflow pipe opens with a large copper funnel about 6 inches below the top of the tank and runs down to the main roof, with a metal flap over the opening. The tank-room floor is protected by an iron pan, which wastes through a 3-inch galvanized iron waste pipe into the tank overflow. A separate ¼-inch galvanized pipe is connected with a Schmidt's best pattern hydraulic indicator gauge, set in the boiler-room to show the depth of water in the tank in feet and inches.

PART III.—TOILET-ROOMS AND DETAILS OF BRASSWORK.

FIGURE 7 is a plan of the main toilet-room in the twelfth story, and shows the arrangement of fixtures and location of vent and drain pipes, the latter being run between the floor and the ceiling of the eleventh story. Between the two rows of closets in the center of the room there is a high double wainscot, the marble walls of which inclose a narrow chamber, which separates them and leaves room for the waste branches and for the trap vent and water supply. The flush cisterns rest on top of these walls. On each of the lower office floors there is a small toilet-room containing two urinals and one slopsink, and on the sixth floor there is a toilet-room for women with eight water-closets and four washbasins. All these toilet-rooms and the twelfth-floor one are handsomely finished with marble floor, tiled walls, and heavy white marble slabs for partitions and the 7-foot wainscot. The partitions are fitted with special heavy nickel-plated brass trimmings, as shown in Fig. 8, which is a sketch of part of the sixth-floor toilet-room. All the exposed edges of the marble are trimmed with 1½-inch brass pipe, polished and plated, and connected at corners and right angles by spherical couplings, so that where the four pieces around the edges of a slab are screwed tightly together they form a secure frame to hold it in position and attach it to the walls, doors, etc. These frames are locked on by being fitted closely into a concave

brass bedplate or chair, which is flanged over the edges of the marble.

Figure 9 is a detail cross-section of the brass chair, which was rolled down from a tube to approximately the required form, and then accurately shaped by being drawn through dieplates. The small interior areas shown are small brass tubes, put in to fill up open space and to re-enforce the walls of the large tube.

Figure 12 is a plan of the fourth floor and is typical of the office stories. It shows the arrangement of washbowls and the location of riser lines and the crossing of waste and vent branches under the floor to the wall risers. The washbasins that are located in pairs have, on the upper floors, vertical flues between them (not here shown) in the thickness of the partition wall or in its enlargement, through which the riser lines are carried as shown in Fig. 13, which illustrates the general arrangement and the use of a

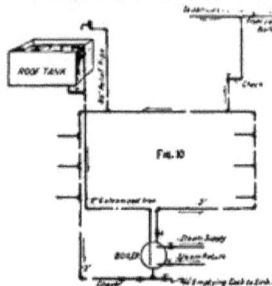

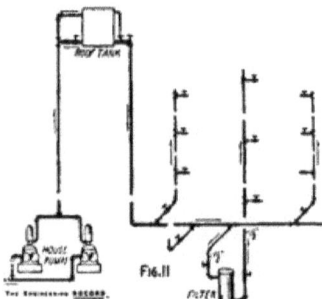

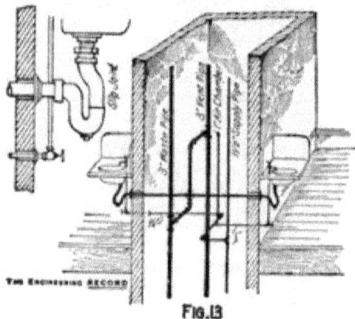

PLUMBING IN THE CONSTABLE OFFICE BUILDING, NEW YORK CITY.

special connection B uniting the wastes and trap vents from both basins. Each office basin wastes through a 1¼-inch nickel-plated trap with 1½-inch waste and vent branches. Where two office basins come together a double waste fitting is used, with 2-inch waste and vent branches. These basin branches are generally of steel, and the waste fitting is screwed directly into the iron-pipe fittings. In the sketch, which is a conventional one, not drawn to accurate position or scale, but merely intended to show clearly the general arrangement of pipes, the size of the shaft or hollow-wall space is exaggerated, and the pipes are separated in the drawing much more than is actually the case. This is done to avoid confusion and to show the connections distinctly. Actually the distance from the trap to the vent pipe is about 6 inches. All the basin branches were capped and tested under pressure. Each office basin is supplied with cold water through ½-inch nickel-plated supply, with ¼-inch nickel plated angle valve and self-closing cock. Each branch has a 12-inch air chamber of 1-inch galvanized pipe behind the casing. The basins in the toilet-rooms are the same as in the offices, except that they have also a hot-water supply.

PART IV.—HOT-WATER AND FILTERED-WATER SYSTEMS, REQUIREMENTS FOR FIXTURES AND CONNECTIONS, LOCAL VENTS, SLOPSINKS, URINALS, AND WASHBASINS.

FIGURES 10 and 11 are conventional diagrams which are not drawn to scale nor position, but are simplified to show principles and operation. Figure 10 illustrates the hot-water supply and circulation system. Water is delivered under tank pressure to the 400-gallon boiler in the basement, and is there heated by an interior coil of 2-inch brass pipe about 30 feet long that is supplied with either live or exhaust steam. The heated water rises from the top of the boiler through a main that runs along the basement ceiling across the center of the building and over to the side wall, thence rises to the twelfth story, and crossing horizontally to the opposite side descends and returns full sized to the under side of the boiler, thus forming a complete circulation system of itself. Distribution branches are taken at intervals from this main to supply toilet-rooms and slopsinks, and a check valve at the bottom, opening towards the boiler, prevents a reverse draft of water and permits free circulation. A connection is made with the janitor's kitchen boiler so as to permit him to draw from the main system through a check valve that closes downwards to prevent the escape of water from his boiler if the pressure there should be the greater. A relief pipe terminates in a Bryant safety and vacuum valve, wasting into the roof tank. All the hot-water pipes are of tin-lined brass.

Figure 11 shows the relation of pump and tank service to the filtered water supply. A branch from the tank-pressure house-supply main in the basement connects with a Loomis filter, which delivers through a special riser line to faucets in one vertical set of slopsinks in the corridors, where drinking-water can be secured on each floor. Regular emptying and washing valves, etc. for the filter are of course provided, but are not here shown.

Figure 14 shows a slopsink supplied with hot, cold, and filtered water controlled by the three upper valves, and also commanded by three lower valves at L, behind the sink, to cut off the supply for repacking or repairing the valves. Delivery is through a special long and heavy spout, designed for this work, and re-enforced by a brass knee brace underneath.

Figure 15 shows the connections of the urinals, where access is had for cleaning out the trap through a 1¼-inch brass ferrule that extends through the slab and is capped in front under the urinal. The flues in general are rectangular and so proportioned that each urinal has a branch 2 inches in diameter, and the main vents an area equal to all the branches. From the twelfth-floor toilet-room urinals a 6-inch vent is run well above the roof, with 8-inch jacket and globe ventilator. From the basement toilet-room a 12x18-inch heavy galvanized sheet-iron vent duct, with white enamel register face, is run on the ceiling to the space in the chimney around the inclosed boiler flue. All local vent urinal flues and branches are of 20 ounce cold-rolled copper with soldered joints. All fixtures are set open without wood casings, and all

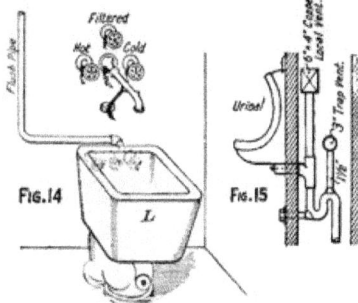

FIG.14   FIG.15

traps, wastes, vents, supplies, and fittings about all fixtures are also exposed and generally nickel-plated. Each water-closet has a large-sized plain pine copper-lined syphon cistern, which is cased in marble. The flush-tank cistern is on top of the marble wainscoting and overhangs, so that the position of the flush pipe is on the face of the marble, and it is made perfectly straight, without bend, curve, or offset. Each water-closet is connected to the soil pipe by a 4 inch lead bend with heavy brass floorplate. The vent from the lead bend is a 2-inch lead and steel branch. Each cistern is supplied through a ½-inch nickel-plated branch with separate globe valve and hush pipe on the ball cock. All urinals waste through 2-inch lead trap and waste, with 1½-inch nickel-plated trap screw brought through the face of the marble back, and with 1½-inch vent branch. Each urinal, or set of two to four urinals, is flushed through a 1¼-inch nickel-plated flush pipe from an automatic syphon cistern set and fitted like the water-closet cisterns. The main copper local vent runs behind the marble as high up as practicable, and a 2 inch branch drops down to each urinal waste. These are of lead, wiped to the brass waste fitting and carried up to the copper vent. Besides the house tank, suction tank, receiving tank, filter, and two water meters, there are 46 water-closets, 29 urinals, 203 washbasins, nine slopsinks, one sink, and in the janitor's apartments one sink, one bathtub, three washtrays, one boiler, and one water-closet. The entire plumbing and drainage system was tested by the plumber with a force pump and mercury gauge under an air pressure equal to 20 inches of mercury. and the gauge column did not show any appreciable loss of pressure in five minutes.

AUTOMATIC SUBSEWER DRAINAGE IN LARGE BUILDINGS.

(PUBLISHED IN 1895.)

The cellar and basement floors of large city buildings are often at or below the level of the adjacent city sewer, and when this is the case it prevents a gravity drainage of water, sewage, condensations, washings, etc., into it. This is the more likely to be the case as the buildings increase in height, and have proportionately deeper and heavier foundations, the excavations for which it is naturally desirable to utilize for such purposes as underground space may be convenient for. These uses under modern conditions of construction are many and diverse; cellars and subcellars furnish an inconspicuous position for machinery and steam plant, storage, etc ; steam boilers

in cellar vaults are more easily reached for repair or renewal, especially if under the sidewalk, than if in the center of the building. In hotels these parts of the structure are used for supplies, stores, wine bins, etc , and in newspaper offices the large presses are installed in the basement, and sometimes the typesetting and stereotyping departments as well are below the street level.

When the foundations are on piles it is a desideratum that the timber should not extend above permanent ground-water level; indeed in some cases it has been planned to artificially irrigate the subsoil for the benefit of the piles, so that the cellar floor and walls are likely to be exposed not only to certain moisture, but to possible hydraulic pressure, which may readily penetrate ordinary masonry or water-proofing Beside this, the lowest parts of the steam and sewerage systems are likely to fall below the flow line of the street sewer, and thus several causes may contribute to deposit waste liquid in the cellar that must be mechanically removed and should be constantly disposed of, automatically and with positive certainty. Formerly it has been customary to collect all drainage in a cesspool or tank below the lowest floor, and periodically pump its contents out into the sewer. Objections have been made to this method, and another system has been developed, which consists essentially in collecting the subsewer drainage in air-tight iron vessels into which, as soon as they are filled, air

AUTOMATIC SUBSEWER DRAINAGE IN LARGE BUILDINGS.

pressure is automatically **admitted on top of the** liquid, which operates to **force it out through a sealed** outlet, up and into the **sewer above. This method is** called the Shone ejector **system and was explained** fully on page 358 of Volume **XXVII.** of **THE ENGI-** NEERING RECORD, where the details and operation of the ejectors and the service in several installations were described. It was adopted for the sewage collection at the Columbian Exposition in Chicago, and has been provided in recent large buildings in Chicago, plans of two of which have been sent to us as typical of improved plant for metropolitan buildings by Urban H. Broughton, Assoc. M. Inst. C. E., Engineer and Manager of the Shone Company, Chicago. From these the following description has been prepared:

Figure 1 is a basement plan of **the** Chicago *Daily News* Building, and shows the arrangement of drainage pipes from areas, floor strainers, engines, elevators, and other machinery, etc. (but not including any sewage), emptying into a depressed brick-walled catch-basin, whence it flows to a pair of Shone ejectors, which deliver it to the city sewer at a higher level.

Figure 2 is a plan of the east end of the basement of Marshall Field & Co.'s large new mercantile building, where the sewage from two large sets of water-closets and surface drainage is piped through back-pressure valves to catch-basins whose contents flow to two ejectors that have a capacity of delivery to the sewer of 50 gallons per minute each, and are inclosed

in a covered circular, brick-walled, water-tight chamber, about 9 feet in internal diameter and 7 feet deep.

Figure 3 is a conventional vertical sectional diagram at Z Z Z Z, Fig. 1. showing the relative vertical positions and features of arrangement of air compressors, receiver, ejectors, and drainage connections to the ejectors and to the sewer.

Figure 4 is a plan and elevation of the two ejectors set in the standard manner, but with their chamber not covered.

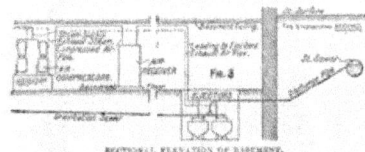

SECTIONAL ELEVATION OF BASEMENT.

Figure 5 is a diagram of an ejector in section to illustrate its operation. The sewage gravitates through the inlet pipe A and flap valve G into the ejector and gradually rises therein until it reaches the under side of the bell D. The air is thus confined at atmospheric pressure inside this bell, and the sewage continuing to rise around it lifts it, together with the spindle, etc., which opens the compressed-air admission valve E. The compressed air thus automatically admitted into the ejector presses on the surface of the sewage, driving it through the bell-

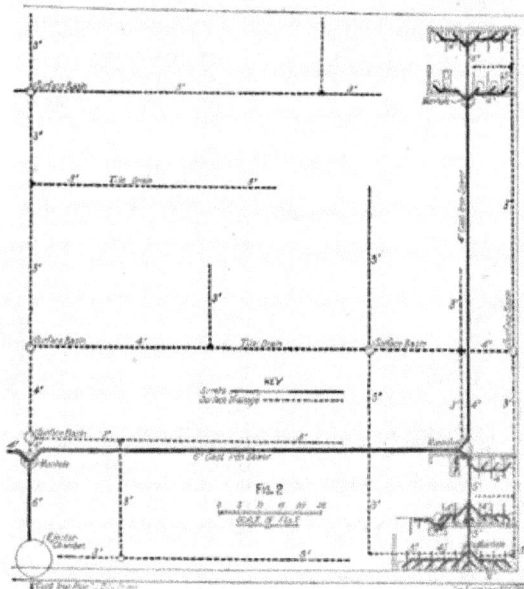

AUTOMATIC SUBSEWER DRAINAGE IN LARGE BUILDINGS.

mouthed opening in the bottom, and through flap valve H and outlet pipe B, into the iron sewage discharge pipe. When the air pressure is admitted upon the surface of the sewage, the valve G on the inlet pipe A falls on its seat and prevents the fluid escaping in that direction. The sewage passes out of the ejector until its level falls to such a point that the weight of the sewage retained in the cup C, which

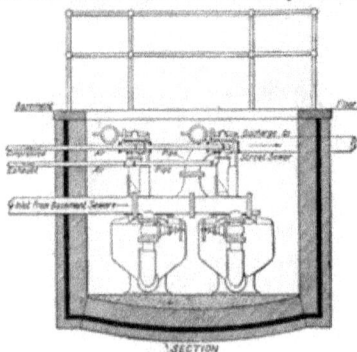

Fig. 4

SECTION

PLAN

SCALE OF FEET

THE ENGINEERING RECORD.

Fig. 5

SECTION OF EJECTOR

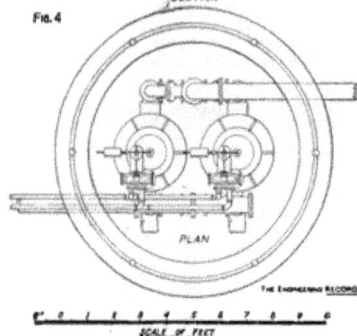

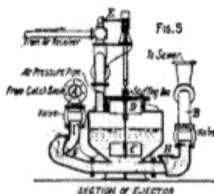

is no longer supported, is sufficient to pull down the bell and spindle, thereby reversing the valve E, which first cuts off the supply of compressed air to the ejector, and then allows the air within the ejector to exhaust down to atmospheric pressure. The outlet valve H then falls on its seat, preventing back flow from the discharge pipe, and the sewage again flows through the inlet commencing to fill the ejector

once more, and so on. The position of the cup and bell is so adjusted that the compressed air is not admitted to the ejector until it is full of sewage, and the air is not allowed to exhaust until the ejector is emptied down to the discharge level. Thus the ejector discharges a specific quantity each time it operates.

In this mechanism the working parts are few and simple and not liable to injury by the sewage. There is no piston friction, the valves do not obstruct the pipes, bottom discharge promotes complete removal of solids and sediment, and the periodic evacuation may constitute a desirable flush. To work the ejector a small air compressor is employed which can be bolted to a wall in a convenient place in the engine-room of the building any distance from the ejector. The compressor delivers air into a tank or receiver and from this the air is conveyed to the ejector by means of a wrought-iron pipe of a small diameter. When steam is turned on to the compressor the whole apparatus is automatic. When the pressure requisite to discharge the sewage is attained the compressor stops, automatically starting up again when the pressure of the air is reduced in the receiver by reason of the discharge of the ejector. The only attendance required is for occasional oiling.

Mr. Broughton writes that "the ejector chamber is generally built of brick, with an asphalt course. The sewers for these building are ordinarily cast-iron sewer pipes, and the subsoil drains are tile pipes laid in the ordinary manner. In nearly all the buildings we put two ejectors of 50 gallons capacity per minute each, although in a few we have put two of 100 gallons capacity each."

PLUMBING IN THE AMERICAN SURETY BUILDING.

(PUBLISHED IN 1896.)

PART I.—WATER-SUPPLY METERS, FILTER SUCTION TANK, PUMP, UPPER AND INTERMEDIATE TANKS.

THE special conditions and requirements of the plumbing in a modern tall office building involve so many points of difficulty and require such a degree of skill and experience and such good construction in the design and installation that the arrangement and execution of the work, while conforming to the same general principles in different cases, still present diverse features and exhibit special details in each different building, showing how similar requirements have been met and like difficulties overcome by modified plans and varied details. As an example of typical requirements and complete system of plumbing service the work in the American Surety Building, New York, illustrates the general characteristics of a sanitary installation for one of the loftiest commercial structures yet erected, and also shows the special details of construction and arrangement adopted to conform it to the exactions of position and severe conditions of high-pressure extended lines, elaborate service, and efficient operation that obtained for this particular building.

This building, on the corner of Broadway and Pine Streets, is about 90x85 feet and 21 stories, or nearly

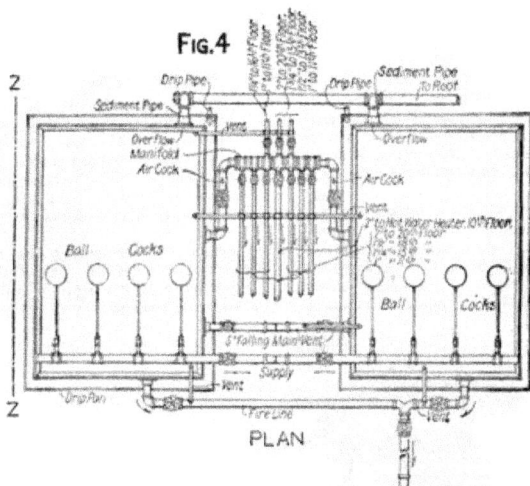

FIG. 4

PLAN

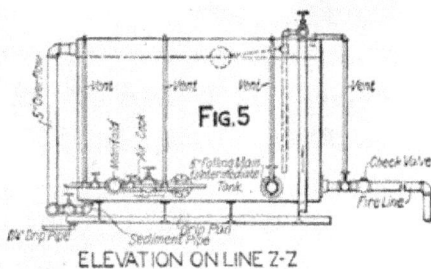

FIG. 5

ELEVATION ON LINE Z-Z

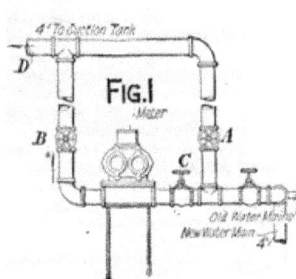

FIG. 1

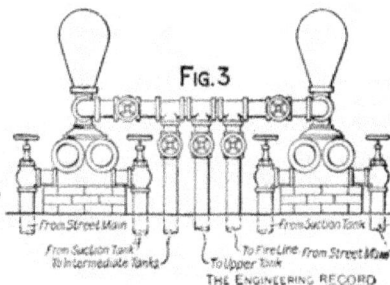

FIG. 3

THE ENGINEERING RECORD

PLUMBING IN THE AMERICAN SURETY BUILDING, NEW YORK CITY

311 feet in extreme height above the sidewalk. It is intended for about 125 suites of offices, exclusive of all the rooms on four floors, which are occupied by the American Surety Company, and contains a large mechanical plant for the different branches of power, heating and lighting service, etc. The general construction is of the modern fireproof steel-cage system, and the equipment throughout is intended to be of the most improved and complete nature, as designed and approved by the architect, Bruce Price, of New York City.

Mr. E. A. Rogers was the architect's assistant in charge of construction throughout the entire time of building. The plumbing contract was let to James Armstrong for about $45,000, and was executed by him to conform to the general plans and details and comprehensive specifications upon which the esti-

pipe D is connected to two No. 8 Continental filters,[*] which are provided with the necessary washout and waste and gate valves arranged to have the water pass through the filters or go around them through a by-pass, and so that either or both the filters may be thrown in or out of service at will. The 4-inch delivery pipe E from the filters supplies the 3,000-gallon suction tank through four 2-inch ball cocks operated by copper floats so as to automatically shut off the supply when the tank, which is essentially an open one, in that it is not designed to receive any pressure, is full. This tank is in a pit under the machine-room floor and it is built of ¼-inch boiler-iron, 3x3-inch angles and tie-rods, furnished with an iron cover and lock and set in a safe pan with 3-inch waste for drainage. There is a 3-inch valved emptying pipe and a 5-inch galvanized-iron overflow

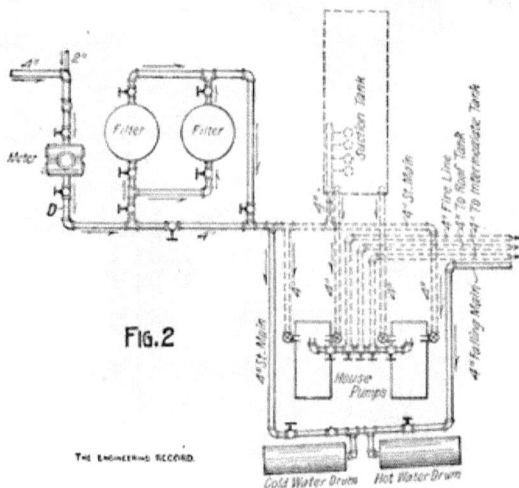

FIG. 2

THE ENGINEERING RECORD.

PLUMBING IN THE AMERICAN SURETY BUILDING, NEW YORK CITY.

mates and bids were based. The plan, arrangement, and details of operation of the system were designed by Mr. Rogers, and its installation was superintended by him, and from his original data the following description of the characteristic features and operation has been chiefly prepared.

The water supply is taken from the city mains through one 4-inch and two 2-inch pipes which are connected by 4-inch brass unions and gate valve with a 5-inch Westinghouse meter provided with a by-pass to enable it to be cut out if necessary without shutting off the supply to the building, as shown in Fig. 1, where valve A in the by-pass is usually kept closed and all the other valves open. By closing valves B and C and opening A the meter can be thrown out of service without interrupting the supply. Beyond the meter the 4-inch delivery

pipe, not shown in Fig. 2, and the pumps are supplied by 5-inch suction pipes laid in iron-covered boxes. All pipes connected to the tank are screwed into riveted iron flanges. The two 14"x7"x10" Worthington steam pumps are supplied, as shown in Fig. 2, both from the suction tank and directly from the street main E, and they are cross-connected and discharge into the tank or fire systems as shown in Fig. 3. Each pump is fitted with a Fisher governor set so as to automatically close the steam valve when the water in the house tanks reaches the upper level required, and to open it as soon as the water falls below that maximum level.

[*] The arrangement and connections of these filters are conventionally indicated in Fig. 2 to show the operation of the system, but they are not drawn to exact scale or position.

AMERICAN PLUMBING PRACTICE.

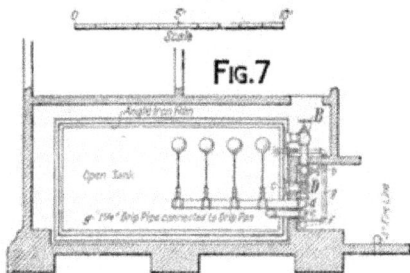

FIG. 7

PLANS

FIG. 8

FIG. 6

SIDE ELEVATION

END ELEVATION

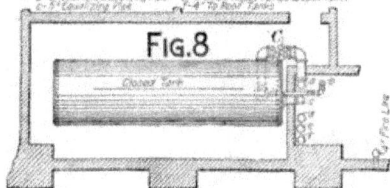

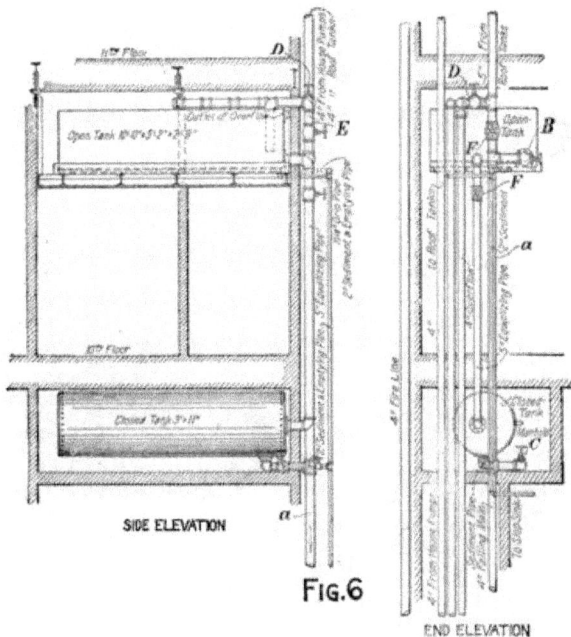

PLUMBING IN THE AMERICAN SURETY BUILDING, NEW YORK CITY.

The normal service of these pumps will be to deliver all the water used in the building (except that taken directly from the mains for basement lines) to the house tanks on the main roof. These tanks supply the intermediate tanks on the eleventh story, and having a connection with the fire line, which would utilize their whole contents for a gravity head for immediate use before the operation of the fire pump. Either pump, when in high-pressure service, must be cut off from the other or house pump and then operates against a check valve that cuts off the roof tank from an upward flow. The pump connection to the intermediate tank is usually closed, but may be opened if it is desired to pump directly into that tank, when the service must be controlled by the electric high and low water gauges, the indexes of which show the heights of water in all tanks on dials in the engine-room.

Figure 4 is a plan of the main house tanks, which are situated in a room on the twentieth floor about 350 feet above the level of the pumps, and have a combined capacity of 10,000 gallons, which affords storage estimated to be sufficient, with that of the suction tank, for 24 hours' supply.

Figure 5 is an elevation at Z Z, Fig. 4, showing the vertical pipes, drip pan, special supporting girders, etc. The tanks are filled through eight 2-inch ball cocks that automatically close when the tank is full. Then the continued action of the pumps against the closed valve produces an increased pressure that operates the regulating valve, and shutting off steam stops the pumps until the drawing of water from the tanks opens the ball cocks, relieves the pressure, and steam is again admitted to the pumps. The ball cocks discharge through hush pipes, extending nearly to the bottom of the tanks, and all the delivery pipes from the tanks have controlling valves near the tank, and just below them vent pipes extending up to above the tank so as to facilitate the emptying of the pipe by admitting atmospheric pressure on top of the water inside when the upper valve is closed. All the toilet-rooms, washbasins, and slopsinks above the tenth floor are supplied directly from the upper tanks by the lines taken from the header shown in Figs. 4 and 5, and each of these separate small falling mains has a drip cock at the bottom to empty it into a sink if necessary. As before explained, the 4-inch fire line is direct from the pump to the tanks, and is arranged to operate both under tank pressure and direct high pump pressure. At every story a 3-inch fire valve is set on it, and fitted with 100 feet of 2½-inch three-ply heavy unlined linen fire hose, tested and guaranteed to a pressure of 300 pounds per square inch, and wound on a swinging bracket reel. The overflow and emptying pipes discharge on the roof, where their contents can be received in the rainwater leaders and their open ends are protected by brass flap valves.

Since the pressure due to the head of the supply from the upper tanks, which reaches 140 pounds maximum, would be excessive for the fixtures in the lower part of the building, all supplies below the eleventh floor are taken from intermediate tanks placed in the tenth and eleventh stories as shown in the elevations, Fig. 6. These tanks are in effect two sections of one tank, and are designed to operate as one, but were constructed separately to enable them to utilize the limited portions of space that could best be assigned to them without obstructing the floors or infringing on rentable room. Portions of upper parts of the lavatories in the tenth and eleventh stories were provided with double ceilings, and in these spaces suspended by iron straps from the steel floor beams above were placed the tanks constructed of the dimensions required to fit their given position. The upper tank is a rectangular open one supplemented by a closed cylindrical one about 13 feet below it, with which it freely connects with a 5-inch equalizing pipe. The normal supply to the tanks is through four 2-inch ball cocks attached to a 4-inch vertical main opening into the bottom of the upper tanks. There is also a 4-inch rising main direct from the steam pumps through which the tanks can be independently filled by opening a valve that is usually kept closed. The supply to the basement drums, whence distribution is made for the lower stories, is through the 4-inch pipe a, valves B, C, and D being open, and valve E being closed. By opening valve E and closing valves B and F, the lower system can be supplied directly from the roof tanks. An overflow is provided for the open tanks, but none is of course needed for the lower closed tank. Each tank has a separate valved connection to the waste pipe, through which its contents may be independently emptied into a basement sink.

Figure 7 is a plan of the open and Fig. 8 is a plan of the closed or auxiliary tank.

PART II.—HOT-WATER AND COLD-WATER DISTRIBUTION.

As before stated, all the water used in the building, except for some purposes in the basement and cellar, is ordinarily first pumped up to the twenty-first story, and is either distributed from there by separate lines that run from those tanks to all fixtures above the tenth floor or is drawn (through an overflow pipe) to an open eleventh-floor tank that supplies all lower stories, thus reducing the maximum pressure to about 70 pounds, or one-half of what would be due to the extreme height from the pumps to the upper tank. The eleventh-floor tank serves merely to regulate the head and store a small supply, hardly more than enough to insure abundant provision for sudden severe draft. It is not conveniently accessible for constant examination and regulation, and is so arranged as never to require any attention except in case of accident or periodical inspection and cleaning unless some unusual necessity occasions it to be filled direct from the pumps or to be cut out of service, when its valves would have to be reversed. Its discharge pipe, the falling main shown in Figs. 6, 7, 8, is connected in the basement machine-room with the hot and cold water drums shown in Fig. 9, and also outlined in Fig. 2. These drums also have a direct connection to the street-pressure supply by a 4-inch pipe, in which a check valve is set, opening towards the drums so that tank water could not escape into the street through it if its valve should accidentally be left open. From these drums separate rising mains are carried as required for all the water supply, for plumbing fixtures, up to the eleventh story. The cold-water drum is of steel tested and guaranteed to 600 pounds pressure per square inch; it is 24 inches in diameter by 74 inches long, with manhole and cover, and is supported on iron standards.

The rising mains vary in size according to their required service, and in some instance diminish in size upwards, starting, for example, as for line B, at 1½ inches up to the sixth floor, and thence running 1¼ inches to the ninth floor, the section being thus proportioned to the fixtures beyond it. The construction of drums and connection of manifold for the convenient connection of pipes, the arrangement and valv-

manded by a Powers automatic regulating attachment adjusted to shut off steam when the temperature of the water exceeds 200 degrees. This drum is supplied like the cold-water one, from the intermediate eleventh-floor tank, and distributes hot water through similar risers connected to its upper manifold to all the lavatories up to the eleventh floor, and to all the isolated washbasins up to the seventh floor. From the top of each of these rising lines a return circulation pipe one size smaller than the smallest or uppermost section of the hot water pipe is brought directly down to the lower manifold that communicates with the drum as shown by the two 3-inch branches similarly to the flow connections above. A vent and relief pipe is carried from the top of one of the risers and inverted, open, above the house tank.

PLUMBING IN THE AMERICAN SURETY BUILDING, NEW YORK CITY.

ing of risers, and provision of drip pipes for cutting out and emptying any line, are clearly shown in Fig. 9. An air vent and pressure relief is secured by extending a 1-inch pipe up from the top of one of the risers and turning it over open above the top of the house tank. Besides the services mentioned distribution is made for boiler-feed pumps, injectors, elevator pit pump, drips, and blow-off tanks, etc., by means of a receiving tank in the subcellar, from which the water is automatically discharged into the sewer by means of a float-valve connected pump. All pipes to pumps and hot and cold water drums are connected up with ground brass flanged unions, so that repairs can be made without disturbing the runs of pipes.

Adjacent to the cold-water drum is a hot-water drum, 24x76 inches, tested and guaranteed to 600 pounds pressure per square inch, and provided with a manhole and cover. Inside this drum is a single loop of large, heavy brass pipe connected with both live and exhaust steam mains and drips to a steam trap and floor drain. The steam supply pipe has a hand valve and is also com-

All flow, circulation, and drip pipes are symmetrically arranged and valved as shown so as to permit the independent operation and emptying, and the drum is cased with non-conducting covering A hot-water supply for the upper part of the building is secured without increasing extreme pressure on the basement heater or lower parts of the pipes by the unusual expedient of making an intermediate heating system and distributing hot water for the upper service from an elevated heater. To this end a No. 3 Berryman feed-water heater is set on steel crossbeams built into the walls of the ventilating shaft in the ninth story and connected up as shown in Fig. 10. Just below the back-pressure valve in the exhaust pipe, which is set at five pounds, a branch is taken out and steam supplied through a pressure regulating valve set at three pounds. Steam circulates from the heater into the main exhaust pipe above its back-pressure valve and escapes freely into the atmosphere above the roof. Both steam pipes have a 2 inch drip connection, and the heater may be emptied through a 2 inch mud pipe, another 2-inch pipe is provided for a surface blow-off and the cold water is introduced in the power part of the heater opposite to the entrance of the two return circulation pipes.

The cold-water supply pipe is taken directly from the manifold at the twenty-first story tank as shown in Fig. 4. The two hot-water distribution mains supply all the main toilet-rooms from the tenth

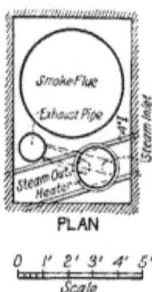

PLAN

0   1'   2'   3'   4'   5'
Scale

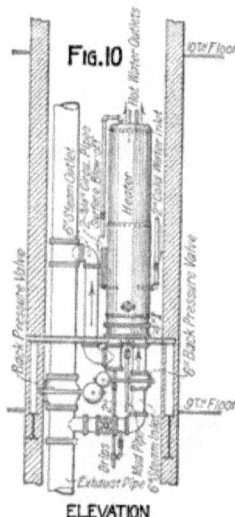

FIG. 10

ELEVATION

to the twentieth stories inclusive, and together with the return circulation pipes are run horizontally in a wood box lined with four-pound sheet lead, with a water-tight cover of 18 ounce copper with the edges turned over and hermetically sealed. Under the heater is placed a heavy sheet-iron drip pan, connected with the above-mentioned box, which has a 1½-inch waste pipe emptying into a slopsink on the eleventh floor. The horizontal portions of the pipes are placed on rollers and firmly secured at the center with the ends left free to act. Each of the supply and circulation pipes is provided with valves so that

they may be turned off independently. The pipes and heater drain into the nearest waste line. The pressure regulating valve in the branch supplying steam to the heater has a three-pound counterweight opposed to its regulating weight so that it is about balanced, and it is connected to a thermostat inside the heater, set so as to shut off steam when the temperature of the water rises to 200 degrees.

PART III — MAIN LINES OF PIPES, DRAINAGE, CESSPOOL TANK AND TESTS.

BESIDES the mains to and from the tanks, there are eight sets of vertical water pipes supplying groups of fixtures, and these lines are run in all cases adjacent to the columns of the main framework of the building. All lines of cold-water supply pipes (except where nickel-plated brass pipe is used) are lap-welded standard galvanized iron pipes, warranted to have been tested to withstand a pressure of 500 pounds per square inch. All hot-water supply and circulation pipes are heavy tinned and annealed brass pipes, warranted to have been tested to stand a pressure of 600 pounds per square inch. All hot and cold water supply branches for fixtures are taken from the rising lines above mentioned, and where exposed they are nickel-plated brass. Branches are of the following sizes: Single basins, one-half inch; two or more basins, three-fourths inch; slopsink and engineer's sink, three-fourths inch; all flush tanks, one-half inch or three-fourths inch, according to size. All supply pipes for hot and cold water are increased one size, from the source of supply to the vertical or falling point. This is in addition to the sizes specified, and shown on the drawings. All rising lines are provided with metal-faced flanged couplers at every other floor, in such a manner that the pipe may be disconnected and sections cut without disturbing section above or below. All pipes of every description are laid to drain completely. All supply pipes were tested by the contractor with a pressure pump and high-pressure spring gauge to a pressure of 125 pounds per square inch. Each column of fixtures, A, B, BB, C, D, E, F, G, H, and I, with the exception of isolated washbasins in offices from the eighth to the twentieth story, each inclusive, has independent risers for hot, cold, and circulation, with stop valve, drainout valve, and drip at the base of each column. The circulation pipe is one size smaller than the hot-water pipe, and connects with the hot-water pipe at the highest point of the fixture, with a branch taken above to supply a fixture to relieve the circulating head from the accumulation of air.

On each of the hot-water risers there are expansion loops on every third floor, beginning at the bottom. All pipes are supported between the loops, allowing the pipe to expand from the first support back to the lower loop from the next support above and so on. Circulating pipes are also provided with loops top and bottom, and in intermediate places where they are connected directly with the hot-water main and branches. Loops are provided on branches, thus avoiding any direct connections or short connections with the main. In these instances care was taken to

allow the branches connecting with mains to expand
with the main. All tank service, hot and circulation
risers have at least three expansion loops in the vertical run. The pipes are clamped and hung midway
between the loops and the ends of the risers. The
clamps are firmly secured to the beams. All water
pipes have frequent heavy flanged ground brass
unions to admit of easy alteration and repair.

All joints and connections between lead and iron
pipes are made by means of extra heavy, carefully
inspected brass screw ferrules or nipples. The ferrules are screwed into the iron fittings and connected
with the lead pipes by means of solder wiped joints.
All main lines of drain, soil, waste, leader, and vent
pipes are of heavy wrought-iron pipes, of the following weights per running foot: 8-inch pipe, 24¾
pounds per foot; 6-inch pipe, 14½ pounds per foot;
4-inch pipe, 10⅗ pounds per foot; 3-inch pipe, 7½
pounds per foot; 2-inch pipe, 3½ pounds per foot. All
pipes and fittings were required to be tested by hydrostatic pressure of 300 pounds per square inch,
and by the hammer test, before leaving the pipe
mills, and the contractor filed a written guarantee that
tests had been so applied. All fittings are of special
wrought-iron, flush, and the pipes when screwed to-

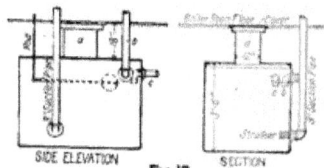

SIDE ELEVATION SECTION

Fig. 12

a-Manhole
b-3" Pipe to 6½ x 3ft Tank
c-3" Drain Pipe

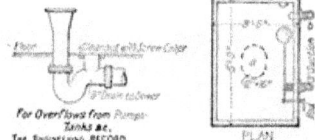

For Overflows from Pumps-
Tanks &c.
THE ENGINEERING RECORD

PLAN

gether have a smooth interior surface. All joints of
iron pipes are screw joints, made with a mixture of
red and white lead and oil. Rising lines of soil, vent,
and waste pipes connect with the main drains and rise
to the roof in columns marked AB, BB, C, D, E, F,

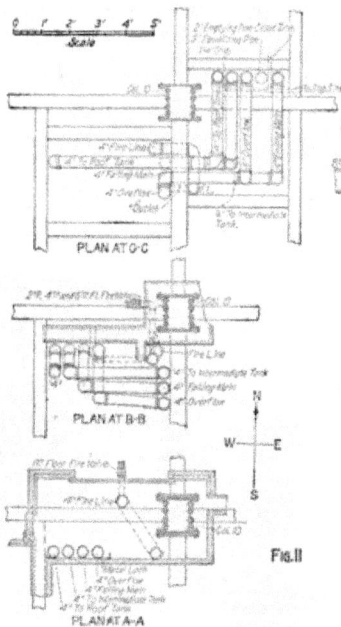

PLAN AT C-C

PLAN AT B-B

PLAN AT A-A

Fig. 11

THE ENGINEERING RECORD

ELEVATION LOOKING SOUTH ELEVATION LOOKING EAST

PLUMBING IN THE AMERICAN SURETY BUILDING, NEW YORK CITY.

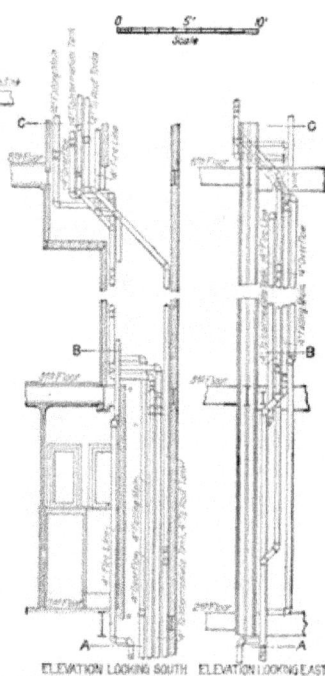

G, H, and I, as shown on the plans. All vent lines increase 2 inches in diameter, before passing through the roof. Where the pipes pass through the roof they are made water-tight by means of 24-ounce copper sleeves fitting around the pipe and extending down on the roof at least 6 inches all round. After the roof was completed, the flashing was overflashed with 24-ounce copper as above. All back-air lines are dripped at the bottom of the line and all the branches for vents from fixtures are set high and above the overflow point of the fixture, so that the vent line cannot act as a waste pipe in case of stoppage.

There are three leader lines of 5-inch extra heavy wrought-iron pipe, marked, carried full size to the roof, and connected there with 12'x12'x12' copper boxes by extra heavy brass soldering nipples. These connections are made with the 5-inch leaders by means of 5-inch pipe connecting with the gutter of the cornice at the twentieth story. These connections

On the house side of the 8-inch main sewer trap there is a 6-inch fresh-air inlet extending up to the sidewalk near the curb, and there provided with a galvanized-iron box and frame 2 feet long, 2 feet deep, and 14 inches wide. Over this box is a brass grating leaded in the sidewalk. All branch connections are Y-T's or Y's and ¼-inch bends. The Y-branch connections are well turned up. Only "long" bends are used. With the exception of the valves directly at the fixtures, all valves on supply lines at tanks, at pumps, etc., are provided with polished cast-brass tags properly numbered to correspond with printed lists giving the location, description, and use of every such valve in the building.

After all the drain, soil, waste, and vent pipes had been run in the building, with the lead or iron bends or branches from fixtures set and connected with the upright stacks, and before any fixtures were set and connected, and the drain permanently connected with the sewers, the tightness of all joints and sound-

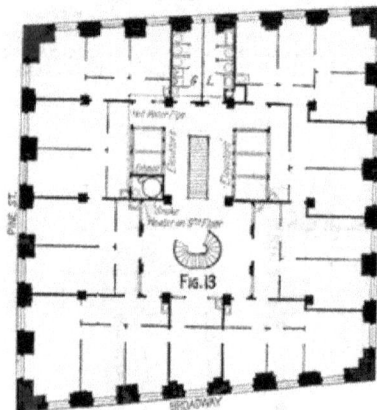

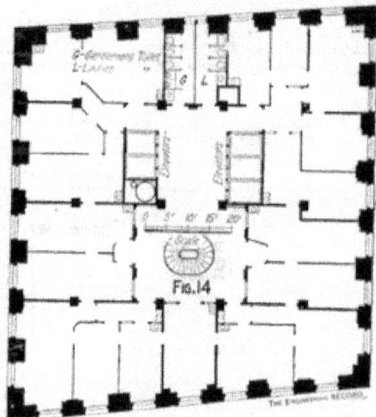

Fig. 13   Fig. 14

PLUMBING IN THE AMERICAN SURETY BUILDING, NEW YORK CITY.

between leader and gutter are made by means of copper boxes, and 24-ounce copper tubes, wiped to extra heavy brass soldering nipples, with heavy soldered joints. Over the mouth of each leader is a heavy brass basket.

These rainwater leaders have 5-inch extra heavy cast-iron traps, provided with two 4-inch brass screw caps. All vent pipes are graded so as to discharge the water collected by condensation, and connected at the bottom with the drain, soil, or waste pipes, in such manner as to avoid obstruction from accumulated rust. The bottom of all vent pipes receives the wash from some fixture. All horizontal pipes are carried with a continuous descent, in no case less than one-fourth inch per foot. Where pipes run along brick walls they are supported on the walls by heavy special wrought-iron semicircular straps, with holes for expansion bolts.

ness of iron pipes, was tested by the contractor, who closed all openings of soil, waste, drain, and vent pipes and ends of horizontal drains. The lead bends or branches were soldered and braced up where required to withstand the pressure, and the whole system of piping was filled with water to the top of the building, and remained for five hours without showing perceptible leakage.

Figure 11 shows the arrangement and manner of carrying the tank riser pipes from the basement to the eighth floor, and the offsets made. Above this point they are run straight in an air shaft, accessible for inspection, etc. Figure 12 shows a wrought-iron cesspool 3 feet deep, 4 feet 6 inches square, strongly riveted and provided with two manholes, set flush with the floor. The tank is made with two compartments, one to receive the sewage, and the other to receive a Kieley governor, which automatically controls a blow-off pump, by which the contents of the tank are automatically pumped out and discharged through a waste pipe terminating in a brass flap valve over a sink that is trapped into the sewer. This arrangement is necessitated because the boiler-room floor is below the level of the street sewer.

PART IV.—FLOOR PLANS AND DETAILS AT FIXTURES.

THE American Surety Company uses the fourth, fifth, sixth, and seventh stories for its own purposes, the first, second, and third stories are devoted to banks and special commercial purposes, and all have washbowls, closets etc., arranged to conform to their special requirements. Above the seventh floor the building is intended to be rented in suites of private

offices, and the arrangement of rooms and plumbing fixtures throughout is similar to that shown in Fig. 13, which is a plan of the eighth floor and typical of all floors above it except the eleventh, shown in Fig. 14, and the thirteenth, which is very much like it.

Figure 15 shows the arrangement and connection for two interior washbasins supplied from one set of risers that are carried in the partition.

Figure 16 similarly shows two basins, the risers to which are run along an interior column.

Figure 17 shows the arrangement and connections when only a single basin is served from an interior column line.

Figure 18 is where two basins are set adjacent to an exterior column, and Fig. 19 is where a single basin is set against an interior wall.

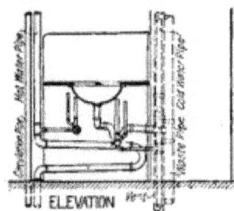

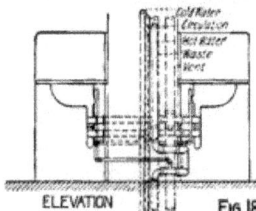

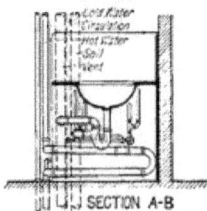

FIG. 17

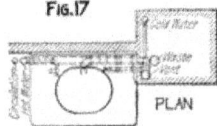

PLAN

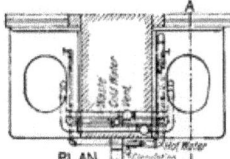

FIG. 18

SECTION A-B

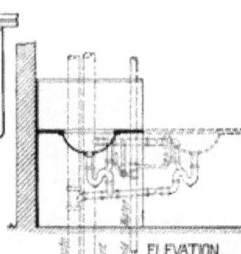

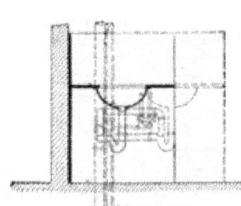

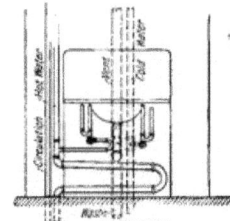

FIG. 15

ELEVATION

FIG. 16

FIG. 19

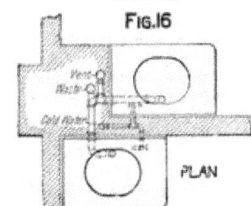

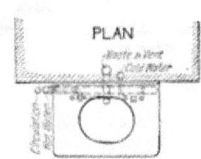

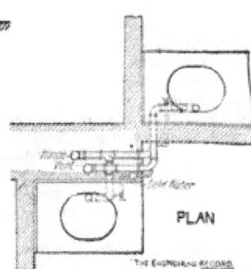

PLUMBING IN THE AMERICAN SURETY BUILDING, NEW YORK CITY.

Figures 17, 18, and 19 show the expansion loop introduced in the hot-water riser to compensate for temperature elongations, and in all these figures the hot and cold water faucets above the basins and waste valve are omitted to avoid confusion.

In all lavatories and at all isolated basins, all exposed pipes at fixtures (not the rising supply mains) are nickel-plated brass pipes. There are separate shut-offs for the supplies to each and every fixture, including shut-offs at each urinal and each water-closet flushing cistern. All water-closets have glazed earthenware traps molded in one piece with the water-closet bowl. All connections between floor flanges of water-closets and iron pipe are made with short lengths of D lead pipe to avoid rigid connection. Besides iron sinks and some fixtures in the subbasement, the list of bowls, etc. throughout the building is as follows :

Story.	Water-closets.	Urinals.	Washbowls.	Slopsinks.
Basement....................	5	5	4	2
First	7	3	3	1
Second	5	2	2	1
Third	5	6	2	1
Fourth	7	4	6	2
Fifth	7	4	3	1
Sixth	8	7	7	1
Seventh	3	7	4	1
Eighth..................	3	7	11	1
Ninth to Twentieth inclusive	80	14	130	14
Total	**135**	**51**	**172**	**91**

PLUMBING IN THE PRESBYTERIAN BUILDING.

(PUBLISHED IN 1895.)

PART I —GENERAL DESCRIPTION, CLASSIFICATION OF WORK, CHARACTER OF WATER, SEWER, AND VENT PIPES AND CONNECTIONS, BASEMENT CESSPOOLS AND DIAGRAM OF WATER-PRESSURE SYSTEMS.

THE Presbyterian Building, at Twentieth Street and Fifth Avenue, New York City, is a large 12-story edifice planned, constructed, and equipped essentially upon the lines of modern metropolitan office buildings, but having some additional features, such as a large first-floor chapel or auditorium, committee rooms, etc. The main part of the building is however devoted to stores and upper-story offices, and by the requirements of such uses the features of the plumbing are mainly determined. Mr. J. B. Baker, C. E., is the architect and engineer, and the plumbing was executed in conformity with his plans and detailed specifications by Messrs. Byrne & Murphy, of New York City.

The plumbing comprises the cold-water street supply, which is received in a suction tank and pumped to storage tanks in the roof house, and thence brought down to distribution drums in the basement, where the separate risers to different lines of fixtures are controlled; three fire lines with hose connections on every floor; hot-water service to sinks, toilet-rooms, and janitor's apartments; wash-basins in offices and elsewhere, public urinals and water-closets, slopsinks, basement sinks, and fixtures in janitor's apartments; waste and soil pipe and trap-vent system; local vents from all water-closets and urinals, roof drainage, and cellar-floor drainage. The fixtures, connections, and piping, etc. in offices and toilet-rooms, etc., present a neat and attractive appearance and workmanlike finish, while the principal features of design and arrangement are developed in the systems of receiving, storing, elevating, distributing, and controlling the water supply, and are chiefly connected with the storage tanks and the pump-room apparatus, both of which are unusually well provided for in convenient and attractive locations.

The special tanks are built in a large room with finished floor and good lights, just beneath the steep-pitched roof, and the room is warmed by a steam radiator. The pumps, hot-water boiler, distribution drum, and valves controlling the separate supply lines are compactly and systematically arranged in one part of the engineer's room, at all times accessible and convenient for his inspection and control. An ample floor space in the basement is devoted to the engine-room, which contains the engines, pumps, dynamos, feed-water heaters, electric switchboard, much of the plumbing apparatus, etc., and is a high, light, and well-ventilated apartment, tastefully finished with glazed tiling, all carefully designed and adapted to the purposes of the large mechanical and power plant of the building. It forms an attractive hall, the apparatus being well displayed and the whole arrangement conducive to sanitary conditions and the proper operation of the different installations there.

The three main house sewers from street sewers to the inside of the house traps are extra-heavy cast iron tested and tarred. All pipes in the soil, vent, waste, and leader system inside of the house traps are of best lap-welded extra-heavy standard wrought-iron steam pipe, and of these all the vent or back-air pipes throughout are galvanized, and all the others are thoroughly coated with asphalt. The connections and fitting of soil, waste, and leader pipes are in accordance with the "Durham System of House Drainage," with screw joints packed with red lead and giving perfectly smooth inside passage at all joints. Vent pipes have ordinary steamfitter's joints. All fittings are extra heavy, of uniform thickness, and those of vent pipes are galvanized. All connections with lead pipes are made by means of soldering nipples and all connections with iron pipes are screw joints packed with red lead.

Horizontal lines of pipes are provided with hand-hole openings and brass screw caps located not more than 20 feet apart. All iron pipes of the water service are of best quality heavy wrought-iron tubing, galvanized, with extra-heavy galvanized-iron fittings and screw joints packed with red lead supported on galvanized-iron hangers and holdfasts. Lead waste and soil pipe where used for short connections is of the following weights: 1½-inch pipe, 3½ pounds per

running foot; 2   pipe, four pounds per running
foot; 3-inch pipe, six pounds per running foot. Lead
water-service pipes are of the following weights:
½-inch pipe, three pounds per running foot; ⅝-inch
pipe, 2½ pounds per running foot; ¾-inch pipe, 4¾
pounds per running foot. Other sizes are extra-heavy
drawn, of uniform thickness "AAA." All are put
up with hard metal tacks and screws, not over 30 inches
apart. All lead pipe in contact with concrete or
deafening is painted two coats of metallic paint.
All brass supply pipes are tinned. All exposed brass
pipes are polished. Hot and cold water pipes are
kept apart everywhere that it is possible; when they
cannot be separated isolated packing is used. Where
water pipes are exposed to frost they are wrapped in
boiler felting and the whole protected by a galvanized-
iron sleeve pipe of proper size.

There are three cesspools under the basement floor,
one in the boiler-room, one in the western storeroom,
and one in the eastern part. At various points cast-
iron catch-basins are built in, with perforated covers,
set flush with the graded floors as follows: One in

the pump-room, one in the dynamo-room, two in the
boiler room, and two in the storeroom. These catch-
basins discharge through 2-inch trapped pipes into
the cesspools. A 3 inch pipe is run from each of the
cesspools to the cesspool in the boiler-room, connecting with the latter 4 inches below its cover. Inside
of this cesspool extending through its wall, are two
pipes with screw caps and 2-inch stop valves, so that
possible ground-water pressure under the main building can be relieved. The boiler-room cesspool is on
a lower level than the others, and is provided with
an automatic steam syphon having an auxiliary water
connection so that the house-tank water pressure can
operate the syphon instead of steam, if necessary.

The general features and arrangment of the main
pipe lines and operation of the water-pressure system
were developed by the architect, and a conventional
diagram of it was made and attached to the contract
specifications. The principal points of it were carried
out in construction, subject of course to modifications
of convenient position and relative location of pumps,
tanks, etc., which will be shown in detail in the

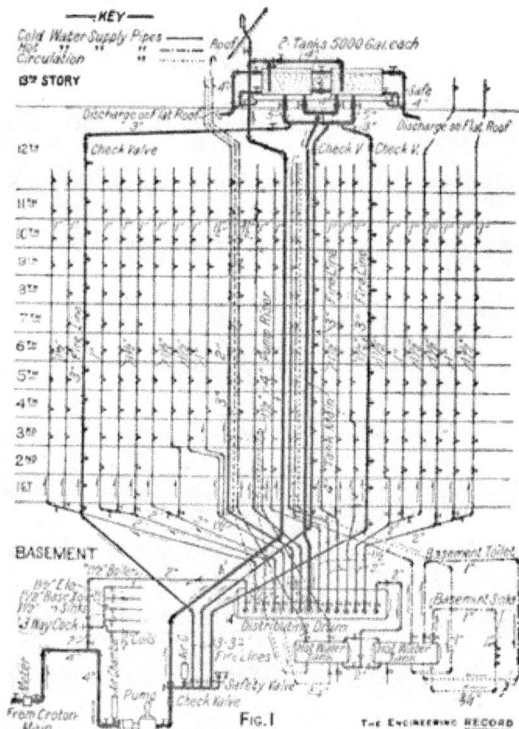

FIG. I

following part. The specification diagram, Fig. 1, needs no additional explanation and forms a desirable supplement to the written specifications, from which some of the data of this article have been prepared.

PART II.—ENUMERATION AND DESCRIPTION OF FIXTURES, LOCAL VENT SYSTEM, DRAWINGS AND DESCRIPTION OF STORAGE TANKS AND CONNECTIONS, AND OF PUMPS, DISTRIBUTION DRUM, HOT-WATER BOILER, SUCTION TANK, RISER LINES, AND OPERATING VALVES IN BASEMENT.

A LIST of the principal fixtures includes 79 water-closets, 24 urinals, seven earthenware hopper slop-sinks, five cast-iron slopsinks, one cast-iron sink in engineer's basement toilet-room, and three cast-iron drip sinks, etc. in the basement. There are 291 washbasins, mostly single, 12½x15 inches, and one porcelain-lined roll-rim bathtub, besides two school sinks temporarily connected up in the basement for use as workmen's water-closets. One end of the elevator shaft is partitioned off to make a separate main vertical flue through which the pump and tank risers and other stacks of pipes are run, and into which foul-air flues for the ventilation of toilet-rooms are connected on different floors. In this ventilation shaft a special galvanized-iron flue is run and provided with branches extending to within 2 feet of the different water-closets and urinals, from each of which a local vent pipe 3 inches or 2 inches respectively is taken to the flue branch. The top of the main flue opens freely in an iron exhaust chamber on top of the roof, where an electrically driven 5-foot Blackman fan exhausts the air from both the toilet-rooms and their fixtures.

Figure 2 shows the location of basement pumps, tank, hot and cold water drums, and the arrangement of distribution lines and position of the valves

by which they can be completely controlled from this one point near the engineer's headquarters.

Separate metered supplies are taken from the two adjacent street mains and discharged through four 1½-inch ball cocks into a 1,000-gallon open iron suction tank, from which the two 14'x7'x10' Deane house pumps are supplied. The pumps deliver through a 4-inch pipe D to the roof tanks, and also through two 3-inch fire lines E E that have valve and hose couplings on every floor. A third similar fire line is branched overhead from pipe D, but is not here shown. Pipe D has a check valve, opening up to prevent the escape of water from the tanks. The supply and pressure from the tanks is brought through a 4 inch pipe F to the distribution drum G. 8x2 feet, that has dished heads, is tested to 200 pounds hydraulic pressure, and is hung from the floor beams above by iron suspension straps. This drum has two 8-inch handholes, and distributes the house supply of cold water through 12 1½-inch risers H, etc., which are controlled by the angle valves I I. etc., and may be separately emptied through the drip pipes and waste valves just above. A 2-inch pipe K from the bottom of the drum supplies the hot-water boiler J, whose contents are heated by two interior 3-inch brass coils, each 30 feet long, and connected to the live and exhaust steam mains. This boiler delivers hot water under tank pressure to the different groups of fixtures through the three 2-inch lines L L L that vent above the roof tanks, and are connected above their highest fixtures with ¾-inch return-circulation pipes M M M, that are branched into the 2-inch emptying pipe N, between its valve and the boiler. These pipes are commanded by valves O, and each one can be separately and independently emptied by closing its valve O and opening its valve P.

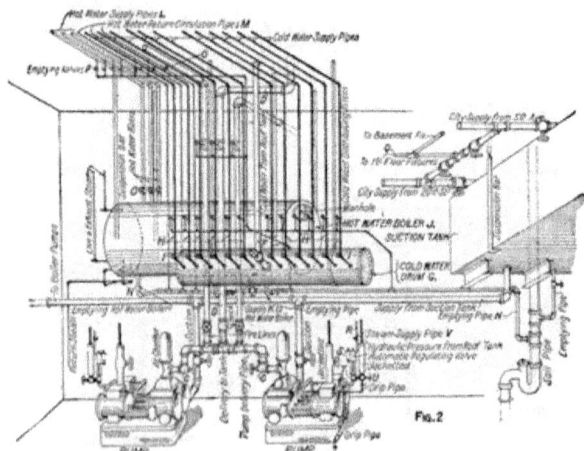

Fig.2

Each of the three systems, hot water, cold water, and tank delivery, has a blow-off above the roof tank. The pipe Q that supplies the first floor and basement fixtures with cold water is connected as shown with the mains from each of the two streets and by a valve not shown with tank-pressure system, which is usually cut off, but may be used if necessary. All hot water, to basement fixtures inclusive, is supplied under tank pressure.

Each of the duplicate Deane steam pumps receives steam through pipe V and valve U, which is usually open, and admits it to a valve T which is operated by a piston attached to its stem and continued in hydraulic cylinder S. This cylinder is connected with the roof tanks by a small pressure pipe R so arranged that when the water in the tank is more than 6 inches below the overflow it opens valve T,

upper edge inside, and two sets of nine 1-inch horizontal tie-rods hooked to inside lugs riveted to the side plates. The tanks measure 10'x12'x6' high and have a capacity of 5,000 gallons below the overflow line. The tanks are supported on 10-inch rolled-steel I beams that distribute its weight upon the columns of the building and are placed in a ¼-inch iron safe pan 3 inches deep and 12'8"x29'2", projecting 16 inches beyond the sides of the tanks and drained by two 3-inch safe waste pipes.

The connections of the pipes are clearly indicated in the illustration. All pipes have flange joints to tanks. The main tanks T T are filled through separate valves from 4-inch pump riser D, that is carried along the ridge of the roof about 8 feet above their tops and has a vent and air valve at its highest point. They are connected at the bottom by an

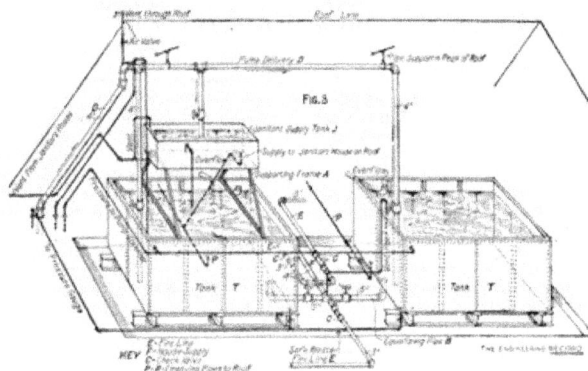

PLUMBING IN THE PRESBYTERIAN BUILDING, NEW YORK CITY.

admits steam to the pumps and starts them; when the water rises in the tank to 1 inch below the overflow line the valve automatically closes and stops the pumps, the practical result being that the pump usually works very slowly most of the time, the valve T being only a little open. Of course valve T can be fixed open and the pumps operated by hand by valve U. Another pressure pipe and dial in the engine-room, not here shown, indicates constantly the height of water in the tank by a spring gauge with its index arranged to show feet instead of pounds.

Figure 3 shows the arrangement of storage tank T T in the roof house. Each tank is made of ¼-inch steel plates, with 3x4 inch vertical T-bar stiffeners 4 feet apart inside, a 4x3 inch flange angle around the

equalizing pipe B, with valves to cut off either tank if requisite for cleaning, painting, alterations, repairs, etc., and from this pipe the 4-inch house-supply pipe F and three 3-inch fire lines E E E (see also Fig. 2) are branched, the latter with check valves to prevent fire-pump pressure from entering the tanks. All the lines are separately valved, though the valves are normally kept open and a small vent pipe suffices to promote the ready emptying of all of them. The arrangement of pump governor pressure gauge, overflow and emptying pipes is clearly shown, and a frame A made of 2-inch riveted angles rests upon the top of one tank and supports a 500 gallon tank J that is elevated to give sufficient head for the supply to janitor's apartments on the same floor as the main tanks T T.

PLUMBING IN AMUSEMENT HALLS AND PUBLIC BUILDINGS.

PLUMBING IN THE MADISON SQUARE GARDEN, NEW YORK.

(PUBLISHED IN 1891.)

THE water supply of the Madison Square Garden is divided into two separate systems. The one for the auditorium building receives water through the 2½-inch pump pipe A, Fig. 1, which delivers to the fire tank S. The latter is supported above the roof by the 6-inch iron beams R R R, etc., which are carried by the wal. W and plate-girder T.

Figure 2 is an elevation at Z Z, Fig. 1. Water enters freely from the pump pipe without any ball valve, and overflows through the 3-inch stand-pipe B to the house tank Q, Fig. 3, which is below it and conveniently placed some distance away on the upper gallery floor. When this tank is full the pipe is closed by ball cock F, and the water, rising in tank S, rings an electric alarm X, which is in tank S, not shown here. This arrangement insures the constant maintenance of the upper tank S, full of water for fire purposes, as required by law.

The height of the water in house tank Q is always indicated in the pump-room by a gauge operated by the heavy float G. The latter is a copper vessel filled nearly to submersion with sand, and then tightly sealed. The fire tank S must always be full.

The overflow is through 4-inch pipe H to an adjacent slopsink.

I is the 2-inch house supply pipe; J is a ¾-inch branch to an upper washbasin, and also serves as a vent pipe to facilitate the emptying of I when its valve K is closed, and the stop cock L is opened; M is a lead-lined safe.

Figure 4 is a diagram of the system of sprinklers placed in the roof over the stage of the amphitheater for protection, while in use as a theater. C is the 4-inch pipe from the fire tank S, Fig. 2. It is run horizontally for some distance, just under the ceiling of the upper gallery, and then descends by the 2½-inch riser O to the cellar, where it is connected with the fire line A, Fig. 5. P is a check valve, opening with a current towards O; Q is a main supplying the sprinkler branches R R R, etc. Different sections of both Q and R R, etc. vary in size to correspond with the number of sprinklers supplied from given points. All these pipes are suspended directly from the roof trusses.

The sprinkler heads were supplied by John Simmons, New York, for this job. They were placed about 10 feet apart, as indicated by the black circles in Fig. 4, and are nominally closed, but will automatically open whenever the temperature in any part of the auditorium exceeds 160 degrees. They are

PLUMBING IN THE MADISON SQUARE GARDEN, NEW YORK CITY

intended to be operated by the high pressure maintained by the pumps throughout riser O and the rest
of the fire line. This pressure closes check valve P,
and prevents waste through tank S. Should this
pump pressure fail, valve P will open and supply the
contents of tank S to the sprinklers or to any hose
cock on the fire line.

Figure 8 is a general view, and Fig. 9 is a section
of the automatic sprinkler head, which is screwed on
to the pressure mains R R. etc., Fig. 4. Figure 8
shows the sprinkler head closed, and Fig. 9 shows
it open and delivering a spray, as indicated by the
arrows. A is the supply tube carrying two slide
rods F F, on which the perforated valve cap C moves
vertically. Rods F F carry the head L, which has
the adjustable yokes M M, to which are pivoted the
catches N N.

The stem J of cap C moves freely through head L,
and is held up by a weak spring K. Ordinarily cap
C is raised to embrace collar P and make a press-

ure contact between ground edge B and copper
gasket D resting on the rubber cushion E, which
closes the tube A.

Cap C is maintained in position by the catches N N.
which secure the lower end of its spindle J, and are
bound together by the fusible sleeve O, as in Fig. 8
Tightness of the joint at B is obtained by screwing
up yoke M. If now the temperature be raised to 160
degrees the seams of sleeve O will be fused, and it
will separate into two parts and release catches N N,
as shown in Fig. 9.

The pressure of water in A against cushion E will
then overcome the weak spring K, and forcing down
cap C and spindle J, open the valve and admit water
into chamber H. From this it will escape through
the perforations I I, etc.

· The house pump and the steam boiler pump and
elevator pumps are all interchangeably connected. so
as to be able to command a 4-inch fire main A, Figs.
5 and 6, which is carried completely around the

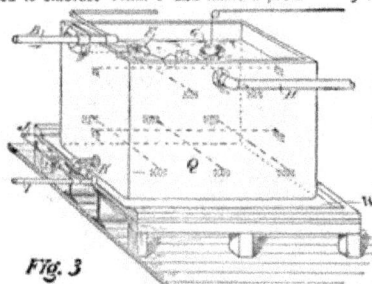

Fig. 3

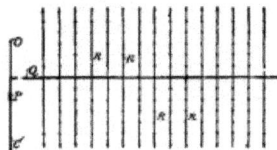

Fig. 4

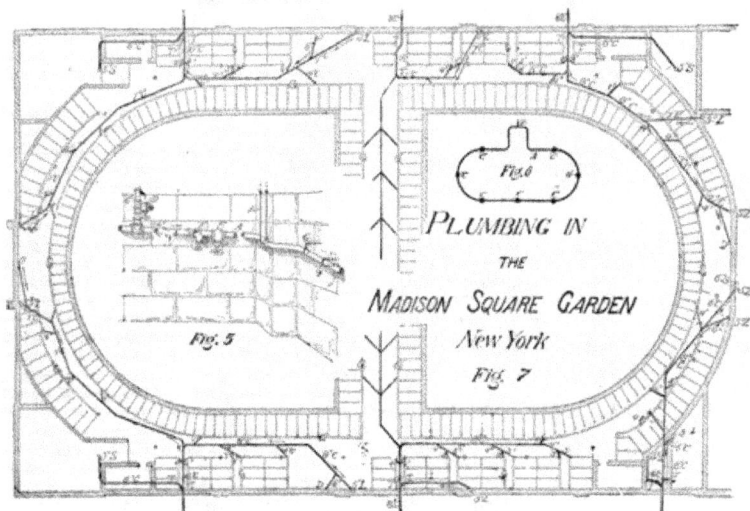

Fig. 5

Fig. 6

PLUMBING IN
THE
MADISON SQUARE GARDEN
New York
Fig. 7

amphitheater on its outside foundation walls, and is constantly commanded by the special fire pumps at Z. This pipe has several 3-inch risers C C, etc., each of which has a 3-inch hose cock D in a corridor of every floor.

The fire system commands the entire building, and serves the concert hall, restaurant, theater and tower, as well as the amphitheater.

There are also, on opposite sides of the building, two 4-inch hose cocks E, each supplied with hose enough to meet that from the other. Just beyond the cocks E E are gate valves F to shut the water off from the rest of the system if a fire occurs between that point and the pumps.

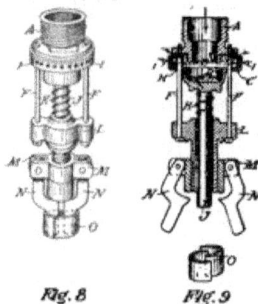

Fig. 8   Fig. 9

The main A is made without elbows, the lengths being very smoothly bent on the ground by hand to a radius of about 3 feet to fit the offsets in the wall, and so as to present a very regular and mechanical appearance. Different kinds of hangers, G, H, I, and J, are used to support the pipe according to circumstances, and are placed at intervals of not more than 10 feet. All of them, except J, are drilled and leaded into the masonry.

Figure 7 is a diagram plan showing the drainage system of the amphitheater and the horse stalls beneath it. The reference letters designate: C, cast-iron drain pipe; D, horse drinking-trough; F, drain from center stalls; G, gutter around the stalls; L, rainwater leader; S, soil pipe; T, main trap; M, branch to street sewer.

The gutter G is simply an asphalt-lined trough, covered with iron gratings, and connects through half-S traps with the drain pipe. Each trap is accessible through a handhole, and every bend of these, and all other drain and soil pipes in the building, are commanded by cleaning caps. As is shown by the current arrows of Fig. 7, almost all parts of the main drain pipes are flushed by the leaders L L, etc. whenever it rains, and by the waste from the horse troughs.

As the horse stalls and basements are unoccupied during intervals of several months each year, the gutter branch traps would lose their seals, and each one would have to be guarded by a gate valve if the stable drain pipes received also the discharge from the soil and waste pipes in the rest of the building. If this large number of valves had been provided,

employees would probably neglect to close some or all of them when the stables were empty. It was therefore proposed to make the stable drainage and the general soil and waste systems entirely separate, the former having main traps T T, etc., and the latter main traps T' T', etc., connecting them with the branches M M, etc. to the street sewers. This plan was adopted by the contractors in consultation with John C. Collins, Chief Inspector of Plumbing for the New York Health Department, and the rainwater leader and stable-drainage system is, as above described, entirely independent of the plumbing and drainage elsewhere throughou the house. All the soil and ventilation pipes were subjected to a water-pressure test.

The work was executed by Byrne & Tucker, of New York, who also fitted up the toilet-rooms and all other plumbing in the building, which, although similar to corresponding work done elsewhere by the same firm, does not present unusual features for special description here.

PLUMBING IN THE UNION DEPOT AT ST. LOUIS.

(PUBLISHED IN 1895.)

THE mechanical and sanitary equipment of the large new Union passenger station used by the numerous railways having their termini in St. Louis, Mo., has been designed to be of corresponding beauty and excellence with the elaborate provision for the accommodation of passenger movement and the rich decorations of all public rooms in the new station, and is complete and conformable to modern advanced practice. Theodore C. Link was the architect and Herbert P. Taussig Chief Engineer of the depot. Adams & Chandler were general contractors and the Abel & Gerhard Plumbing Company contractors for the plumbing, some details of which are illustrated herewith from sketches made by a member of our staff.

The main building is four stories in height, the lower or basement floor and the first floor being devoted to ticket and waiting rooms, emigrant rooms, restaurant, etc. The second and third stories contain the galleries of the main hall, and in the east and west wings over 100 offices for railroad and other purposes. The mail and express, baggage, and other departments are in adjacent separate buildings with individual plumbing equipments. In the depot itself the plumbing fixtures are designed and arranged for the greatest convenience and comfort of travelers and employees, and have been generously allotted, making some provisions not ordinarily found in similar installations. For example, in the special travelers' bathrooms, baths can be quickly furnished to men or women between arrival and departure ef terminal trains. The refrigerating system comprehends cooling all the drinking-water at one place without contact with the ice, and piping it thence through insulated pipes to the various drinking-fountains, some of which are elaborately decorated.

The plumbing substantially consists of the drainage and trap vent system, the cold-water supply,

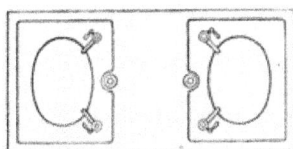

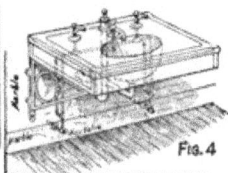

Fig. 5

Fig. 4

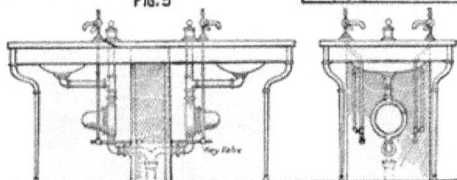

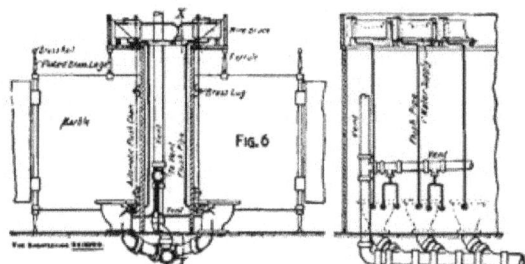

Fig. 6

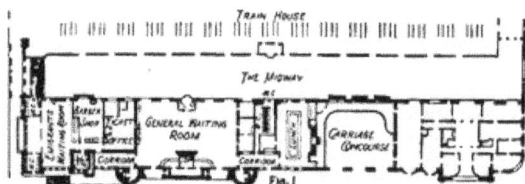

Fig. 7 SECTION AT X-X.

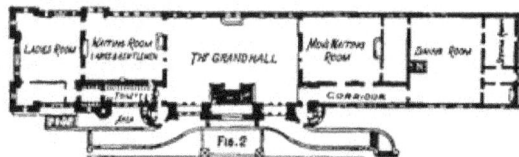

Fig. 1

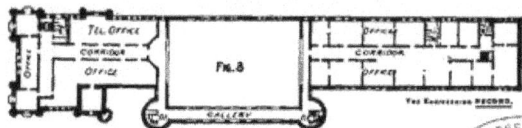

Fig. 2

PLUMBING IN THE ST. LOUIS UNION DEPOT.

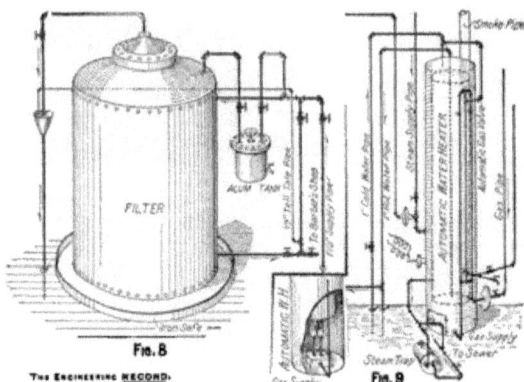

Fig. 8

Fig. 9

PLUMBING IN THE ST LOUIS UNION DEPOT.

toilet-rooms for male and female emigrants and for first-class passengers, barber shop, washbowls, men's and women's bathrooms, hot and cold water and cooking apparatus in the restaurant, dining-room, and kitchen, employees' toilet-rooms, private and public water-closets, washbowls in the offices on the second and third floors, public toilet-rooms in the upper stories, a system of drinking-water fountains, fire-protection lines, and a hot-water heating apparatus for baths and the barber shop.

Figures 1, 2, and 3 are diagrams of the different floor plans showing the location of fixtures. The second floor is arranged like the third.

Figure 4 shows the arrangement of bracket washstands, one of which is provided on each chair in the barber shop.

Figure 5 shows the center washbowl in the barber shop.

Figure 6 is a cross-section through two of the water-closets in the first-floor toilet-room, showing that passageway between the stalls behind them in which the pipe lines are both concealed and accessible. Figure 7 is an elevation at X X. Fig. 6.

Figure 8 shows the connections of the 5x4-foot Western filter through which all the drinking-water is passed.

Figure 9 shows the Seaman's automatic water heater by which a gas flame is made to raise to a uniform fixed temperature the water in the tank for the barber shop and the bathrooms. The device is arranged to cut off the heat when the temperature of the water rises too high.

PLUMBING IN THE RAILROAD MEN'S
READING-ROOM, NEW YORK.

(PUBLISHED IN 1895.)

THE Railroad Men's Reading-room is a handsome four-story structure about 80 feet square on the ground, located on Madison Avenue, New York City, near the Grand Central Station, and by reason of its architectural treatment is prominent among the sightly edifices of that neighborhood. It was erected by Mr. William K. Vanderbilt, and is maintained as a kind of club-house wherein employees of the railways having their terminus at the Grand Central Station may pass their leisure time. It is administered with a view to the healthful recreation and mental and moral improvement of those making use of its facilities, and contains beside parlor and general meeting-rooms, offices, club and committee rooms, a library, a gymnasium, baths, bowling alleys, etc.

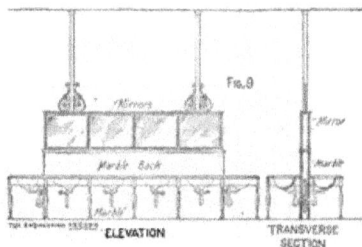

ELEVATION   TRANSVERSE SECTION

The plumbing embraces a main lavatory and toilet-room in the basement, bathrooms with tubs, shower and plunge baths and sinks, slopsinks washtrays, water-closets, urinals, and washbowls as required for convenience upon the upper floors. The soil and vent pipes are provided with cleanouts on dead ends and at principal angles, and are in lines branched from a 6-inch sewer pipe which extends under the basement floor to about the center of the building, and receives most of the discharge from rainwater leaders and floor strainers. One group of rainwater leaders discharges into an old outside masonry trapped catch-basin, and the discharge from the plunge bath is received in a catch-basin whose contents are pumped out into the sewer.

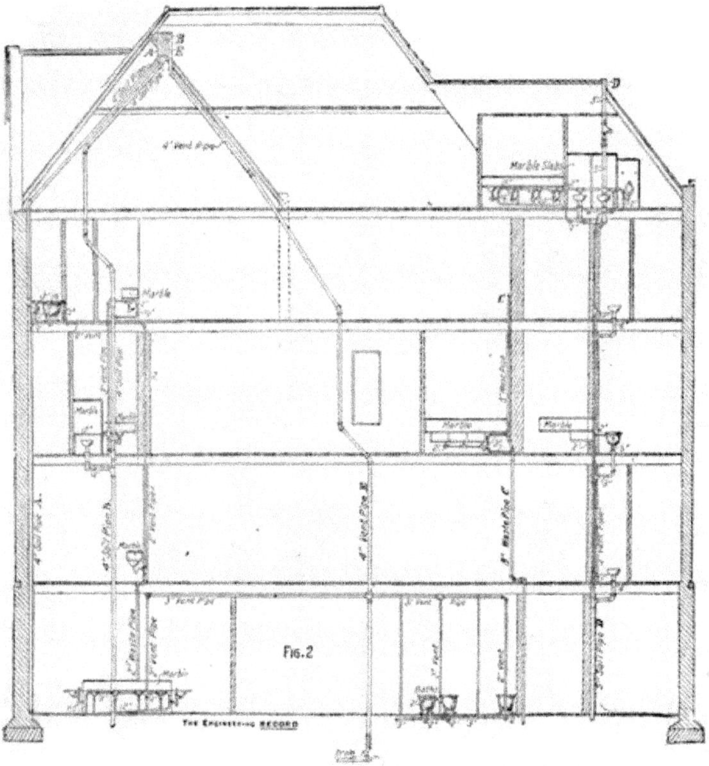

Fig. 2

PLUMBING IN THE RAILROAD MEN'S READING-ROOM, NEW YORK CITY.

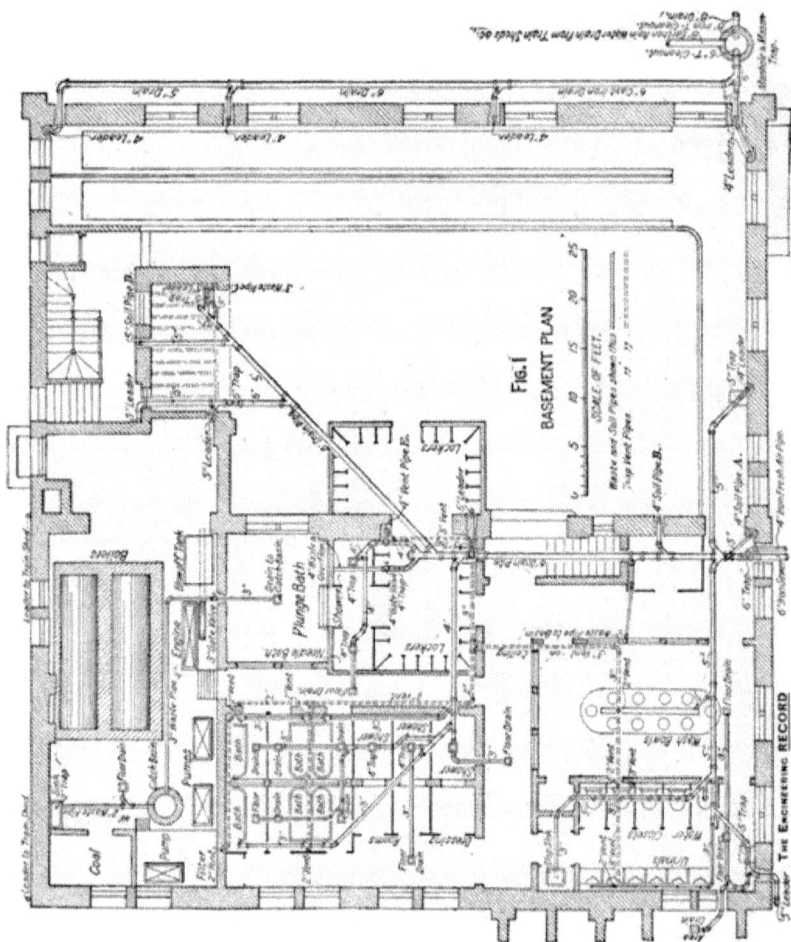

FIG. 1
BASEMENT PLAN

PLUMBING IN THE RAILROAD MEN'S READING-ROOM, NEW YORK CITY.

THE ENGINEERING RECORD

Figure 1 is a basement plan showing the arrangement of drain and trap vent pipes and the location of baths, lavatory, etc.

Figure 2 is a general vertical section and elevation showing five of the principal stacks of pipes.

Figures 3, 4, 5, 6, 7, and 8 are cross-sections showing views of portions of the lines and the fixtures at right angles to the plane of Fig. 2. Figure 3 is of line A at the third story. Fig. 4 of line B at the second and third stories, Fig. 5 of the basement water-closets.

Figure 6 shows how the vent and soil pipes of line D are run through the second, third, and fourth stories and are branched together in the mansard roof space. Figure 7 is a diagram of the basement urinals, and Fig. 8 shows the bends necessitated in line C to avoid obstructing windows and to go through the third-story partition.

Figure 9 shows the oval table, with the washbowls and handsome marble and plate-glass mirrors, in the basement.

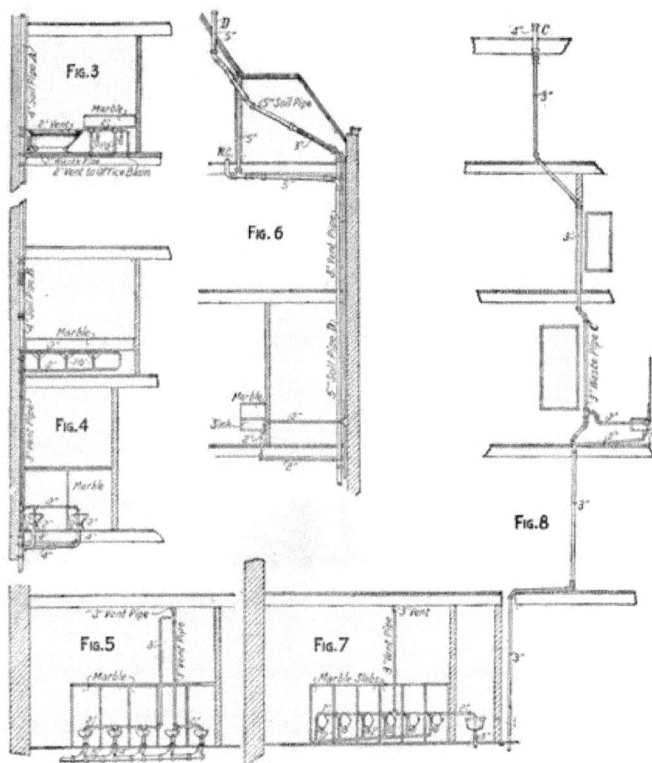

PLUMBING IN THE RAILROAD MEN'S READING-ROOM, NEW YORK CITY.

PLUMBING IN THEATERS.

PLUMBING IN THE FIFTH AVENUE THEATER. NEW YORK.

(PUBLISHED IN 1892.)

PART I.—GENERAL PLAN, DIAGRAMS OF BASEMENT, PIPE SYSTEMS, AND DESCRIPTION OF PLUMBING.

CURRENT plumbing practice, in its application to a modern metropolitan theater building, may be held to be illustrated in the work in the new Fifth Avenue Theater, in New York City. This structure is situated in West Twenty-eighth Street, near Broadway, and replaces one burned a year before. It was built according to the plans of Francis H. Kimball, architect, of New York, and all the work described in this article was designed or approved in accordance with his plans by William Paul Gerhard, C. E., consulting engineer for sanitary works, of New York. The contractors were John Toumey & Son, of New York. The fire service in this theater is notable as having been designed with the intention of conforming to the latest revised requirements of the New York Board of Fire Underwriters and of the new building law, and this is said to be the first large theater system installed since the adoption of these rules. The plan and general features of the plumbing, gas, and fire-service systems, are shown in the engravings, the details conforming in general to the illustrations familiar to our readers.

Figure 1 is a diagram showing the pipe system and general arrangement of the apparatus in the cellar. A A, etc. are 2½-inch fire-line risers. B is the gas distributer. C and C′ are 3-inch and 1½-inch risers of the automatic sprinkler system and are supplied by the independent 3-inch pipe D, forming the outside connection to which the hose from a fire engine may be connected, and from a downpipe, not here shown, from the special elevated fire tank. E E, etc. are sinks to receive overflows, drips and for other purposes. F is the suction tank. G is the Worthington compound duplex house pump. H is a 6-inch meter. I is the 20″x11:15″x10′ Worthington fire pump. J is the hot-water boiler.

Figure 2 is a plan of the gallery floor and indicates sufficiently the arrangement of the other floors, which are similar, at least, as regards that portion of the building which adjoins the stage. The riser pipes are indicated by the following reference letters: H, hot water; D, cold water; S, soil pipe; T, trap vent pipe; L V, local vent pipe; I, house-pump pressure; J, tank pressure. F is the regular set of hot, cold, circulation, soil, and vent pipes.

The plumbing equipment and fixtures comprise: On the roof, one 6,000-gallon fire tank, for fire supply only; at the top of the stage, one house tank, on the gallery floor, two water-closets, three urinals, one porcelain washbasin, one slopsink, five iron enamled washstands; on the second intermediate floor above the balcony, one water-closet, five iron enamled washstands, one painter's sink; on the balcony floor, four iron enamled washstands; on the first intermediate floor above orchestra floor, one water-closet, four iron enamled washstands; on the orchestra floor, four water-closets, three porcelain washbasins, one iron enamled washstand; in the basement, two water-closets, four urinals for men's toilet-room, two urinals and one slopsink, two water-closets under Twenty-eighth Street sidewalk, one engineer's sink and connection for future engineer's closet, connections under stage for supernumeraries' toilet-room and washtrough, one water meter, one receiving or suction tank, one fire pump, one house pump, one hot-water heater, one carpenter's sink in carpenter shop.

The 1,500-gallon cold-water house tank is on top of the loft over the stage. It is made of ¼ inch boiler-iron, with riveted joints, has three coats of Prince's metallic paint inside and out, has a 3-inch galvanized wrought-iron overflow pipe to the nearest roof gutter, a 1½-inch sediment valve and blow-off pipe, and is fitted with a cut-off connected with the house pump to automatically start the pump when the water in the tank is drawn down.

The hot-water heater J, Fig. 1, is a 200-gallon galvanized-iron closed tank containing a brass steam coil to the water. The tank is guaranteed to be tested at 300 pounds hydrostatic pressure, and is covered complete with asbestos and canvas covering.

The open receiving tank F, Fig. 1, is supplied direct from the city mains and serves as a suction tank for the pumps. It has a capacity of 1,000 gallons, is made of ¼-inch boiler-iron, and painted with three coats of Prince's metallic paint. The house pump always draws from this reservoir, as does the fire pump when filling the fire tank. In case of fire the house pump will draw directly from the 6-inch main. The 3-inch overflow discharges into a sink, the waste from which is trapped and connected with the sewer. The weight of all cast-iron pipes was specified to be as follows: Six-inch pipes to weigh 20 pounds per foot; 5-inch pipes to weigh 17 pounds per foot; 4-inch pipes to weigh 13 pounds per foot; 3 inch pipes to weigh 9½ pounds per foot; 2-inch pipes to weigh 5½ pounds per foot. The branch waste and vent pipes from fixtures were specified to be drawn D lead pipes of the following

weights: One and one-half-inch pipes, 3½ pounds
per foot; 2-inch pipes, five pounds per foot; 3-inch
pipes, six pounds per foot; 4-inch pipes, eight pounds
per foot. Where nickel-plated pipes are exposed at
fixtures they are drawn brass pipe or iron pipe size.
All pipes were tested by the usual water-filling
method.

PART II.—PUMPING CONNECTIONS, HOT-WATER TANK
AND GAS DISTRIBUTER.

FIGURE 3 is a plan of the pump connections. The
reference letters indicate pipes as follows: A A,
cold supply from house tank; B B, supply mains to
fire tank; D D, etc., fire lines; H, C, and I are respectively hot, cold, and circulation pipes to the different sets of fixtures, etc., G is the steam connection
to the tank coil, S S are pump suctions, L is a supply
to the receiving tank, K K are drip pipes for emptying rising lines, and M is the main 6-inch supply under city pressure.

Figure 4 is a view of the fire pump. Letters B, D,
M, and S have the same significance as in Fig. 3.
H is the 6-inch Thomson meter, N is a hose cock, Q
is the steam supply, P the exhaust and Q the automatic regulator.

Figure 5 is a view of the connection of house
pump G, Fig. 1. City water from pipe L is
ordinarily delivered through two 2 inch ball cocks
to tank F, whence it is delivered by the 3-inch
suction S to the pump and forced to the house
tank, about 85 feet above, through the 2½-inch
pipe B. Water may however be drawn directly
from the city mains by opening the valve U, which
is usually closed, and by closing the valves *a b*
and opening *c d* it can be pumped directly to the fire
lines and fire tank through pipe E. C C C are cold-
water supplies to fixtures on the lower floor, and O
is a 1½-inch boiler feed pipe. M is a 4-inch overflow
and N is a 1¼-inch emptying pipe for the receiving
tank. K is an emptying pipe for the fire-tank force
main. Steam is supplied to the pump through pipe R
and is exhausted through T; a 1-inch pressure pipe
P connects with the house tank and has a ½ inch
branch to the automatic governor D, which shuts off
steam when the water in the tank reaches a level
near its top, and admits it and starts the pump when
the level falls a few inches.

Figure 6 shows the connections of the hot-water
tank J, Fig. 1, which is supplied from the roof tank
through 1½-inch branch C of pipe A, and delivers
hot water under tank pressure through the 1¼-inch
pipes H and I, which supply groups of fixtures in the
north and south parts of the building respectively.
The return circulation is brought to main F, and
enters the tank through branch P, which also serves
to empty it to the sewer through pipe Q, when valves
K K are closed and valve G is opened. N is a safety
valve and pipes B B, D D, and E are supplies to the
fire tank and fire lines from the pumps, as in Figs.
1, 4, and 5.

Figure 7 is an elevation and Fig. 8 is a verti-.l
center section of the steam-pump governor D, Fig.
5, which controls the operation of the house pump to
correspond with the level of water in the tank. In

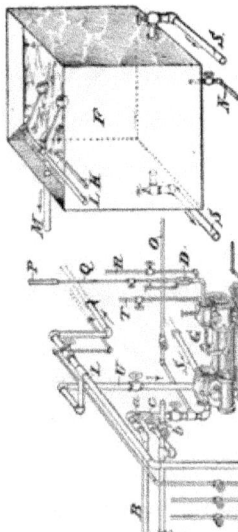

Fig. 5

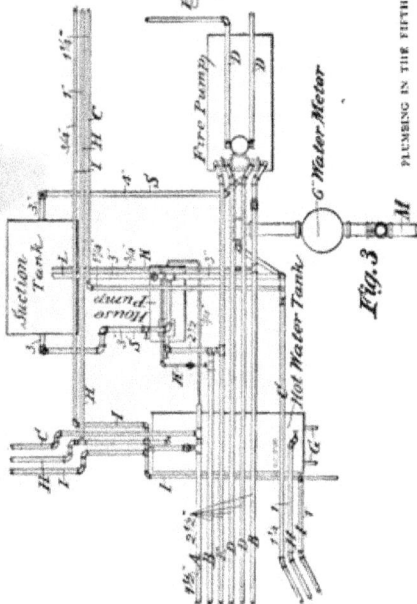

Fig. 3

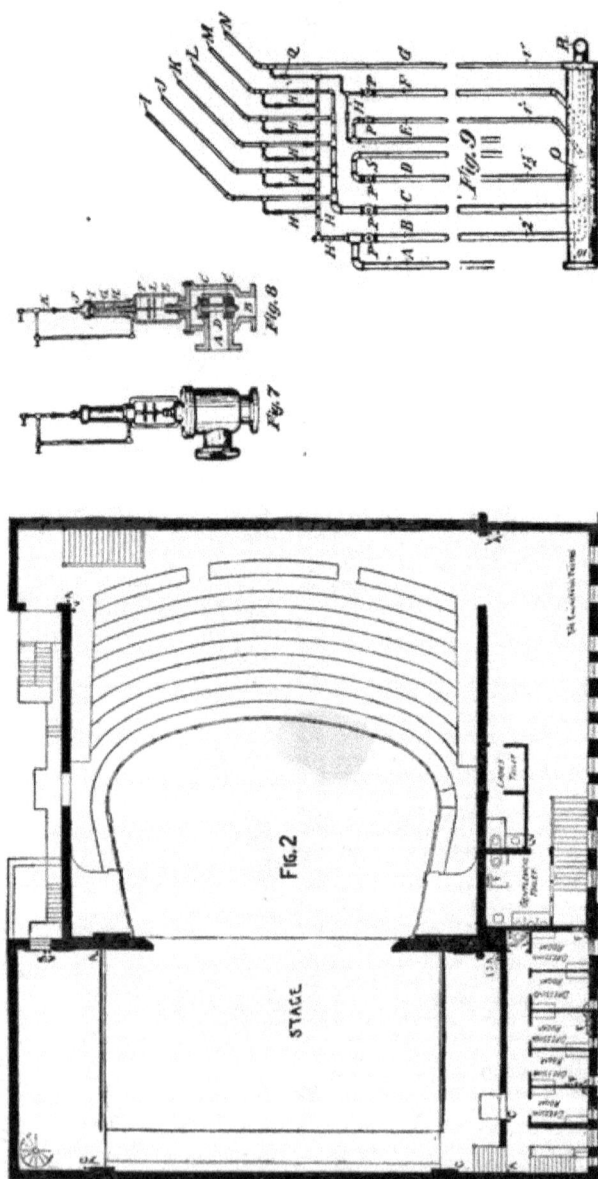

Fig. 9

Fig. 8

Fig. 7

Fig. 2

STAGE

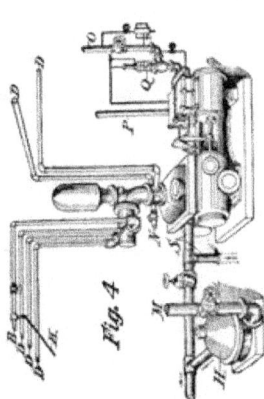

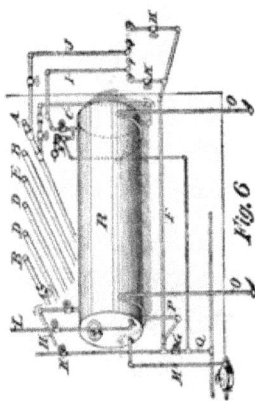

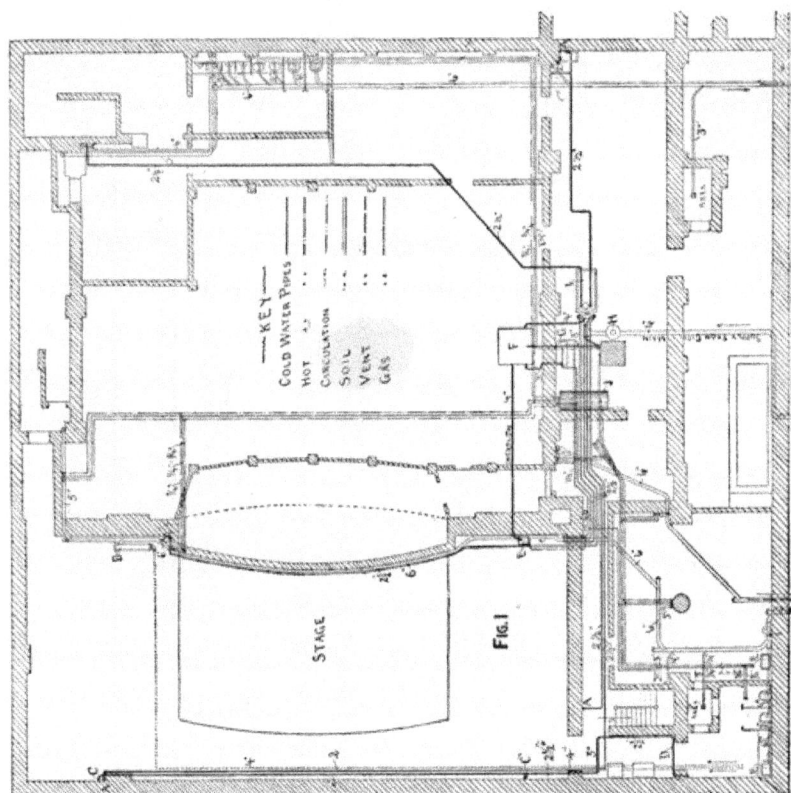

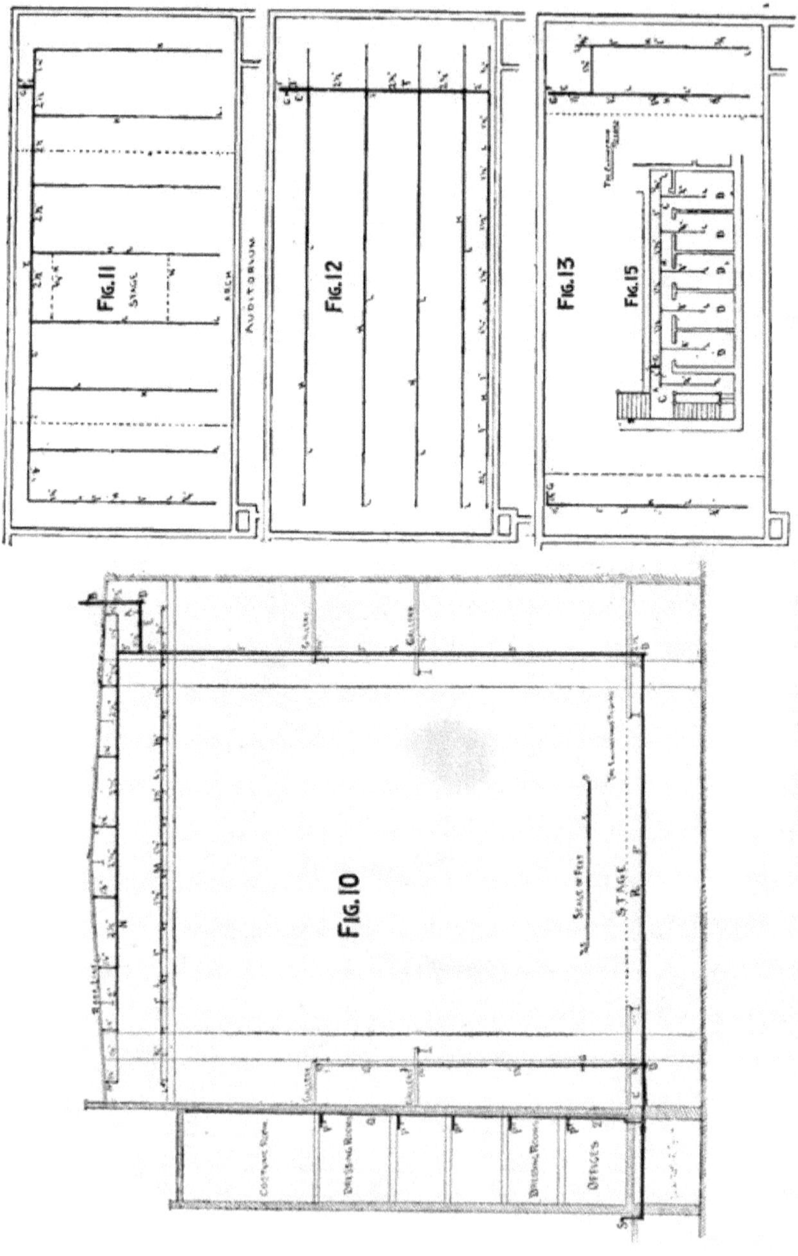

Fig. 11

Fig. 12

Fig. 13

Fig. 15

Fig. 10

PLUMBING IN THE FIFTH AVENUE THEATER, NEW YORK.

Fig. 8 the valve is shown open so that steam is admitted at A, passes through seats C C of the double balanced valve D, and enters the steam chest of the pump through the inlet B. The valve D is supported by stem E, which is connected to the rod H by the adjustable sleeve yoke F. The rods E and H work through glands, and the latter terminates in a piston head I which works in cylinder J under tank pressure from pipe K. As soon as the operation of the pump has filled the tank to a fixed level the increased pressure on piston I overcomes the resistance of spring G, and depresses and closes valve D, thus stopping the pump. Drawing off a small quantity of the water in the tank diminishes the pressure in cylinder J so that spring G raises and opens valve D, and so on. L is a lock nut and adjustment for lengthening or shortening rod E in sleeve F, so as to set the spring G at any required tension. A similar governor (Q, Fig. 4) is attached to the fire pump, so arranged as to be balanced by a constant pressure of 100 pounds, maintained in the stand-pipes, and to turn on steam the moment that pressure falls.

Figure 9 shows the gas distributer B, Fig. 1. A 4-inch gas supply P is connected to a 10 inch drum O, from which the supplies B, C, D, E, F, and G are taken for different groups of stage and auditorium lights, which can be quickly adjusted by the key valves P P, etc., which, however, cannot totally extinguish them even when entirely closed, because a ¼-inch by-pass pipe H is always open and admits enough gas to preserve the flame at all times. These by-passes are all supplied from branch Q, taken from the pipe N to the stage chandelier. Pipe A supplies the orchestra, gallery, and balcony burners; pipe S is to the ground lights on each side of the stage; pipe E is to the footlights, F is to border lights, and I, J, K, L, and M are for the rigging loft. The dressing-rooms, corridors, foyers, halls, toilet-rooms, entrance, and all stairways, etc. have a separate meter and independent 3-inch supplv. All the piping was put together without the use of red lead, and the use of gasfitters' cement was absolutely prohibited. After the piping was completed it was tested to successfully maintain for one hour an air pressure of 15 inches of mercury.

No pipe is less than ⅜-inch bore, and this size is used only for one or two bracket lights. No pipe for chandeliers is less than ½-inch bore up to four burners, or three-fourths of an inch for chandeliers with from four burners up to 15. Gas pipes were proportioned according to the following table of sizes, etc.:

⅜-inch pipe	20 feet.	3 burners.
½ " "	30 "	4 "
¾ " "	50 "	15 "
1 " "	70 "	25 "
1¼ " "	100 "	40 "
1½ " "	150 "	70 "
2 " "	200 "	140 "
2½ " "	300 "	225 "
3 " "	400 "	300 "

PART III.—FIRE SERVICE.

THE fire pump I. Fig. 1, has a capacity of 1,348 gallons of water per minute when running at the rate of 125 feet per minute piston travel, and delivers by separate horizontal lines of 2½-inch pipe to five lines of fire stand-pipes, each 2½ inches diameter, of galvanized wrought iron, screw-jointed, upon which are 25 2½-inch fire valves, as follows: 10 valves on the two lines on each side of the stage, seven valves on line in corroder along dressing-rooms, eight valves on the two lines on each side of the auditorium. Just below each fire valve the fire stand-pipe is provided with a 2½x¾-inch tee, into which an extra strong steam metal detached lever-handle ground-key bibb may be screwed to be used for filling the fire pails.

The fire valves are 2½ inches inside diameter, and are finished and nickle-plated in all places except in basement under the stage, on the fly gallery, and in the rigging loft. Each fire valve is fitted with 50 feet of best quality four-ply rubber-lined cotton fire hose, able to stand a pressure of 300 pounds per square inch, and the hose has a 1¼-inch plated fire nozzle. The hose and nozzle are supported in improved hose racks, attached by pipe clamps to the fire stand-pipes. The fire pump, its automatic attachment, the fire valve, and the fire hose were subjected to tests directed by the fire commissioners. The fire pump was operated under a water pressure of 120 pounds per square inch, which the hose and fire nozzles successfully endured. Besides this stand-pipe system the Harkness wet-pipe system of automatic sprinklers was installed to protect that portion of the theater back of the proscenium wall designated as follows: Under roof, over stage, under gridiron, fly galleries and under the stage as shown by the accompanying plans, also the dressing-rooms, property-room and dressing-room for "supers" in the basement.

The apparatus comprises a cedar tank of 6,000 gallons capacity, 227 automatic sprinklers placed so as to meet the requirements of the New York Board of Fire Underwriters, one 2½-inch and one 3-inch riser and all pipes and fittings necessary for the equipment, a separate valve for each floor or gallery, a drip pipe with suitable valve for each floor or gallery, a watchman's automatic fire alarm having one 8-inch gong placed inside and one 10 inch located on the outside of the building, a low-water alarm having an indicator in the engine-room, a 3-inch pipe connected to the main pipes of the system and carried down and through the walls of the building to the street and provided with a check valve and coupling and cap of the Fire Department Standard, a 2½-inch supply pipe connecting the water tank to the fire pump, and a 2-inch overflow pipe for the water tank. There was a steam connection made with the roof tank, and its riser was jacketed with felt to prevent the danger of freezing. Movable fire-extinguishing appliances were also provided as follows: A number of portable copper fire extinguishers, fire axes and pick heads, sets of polished axe brackets, one fire hook 6-foot pole, one fire hook 10-foot pole, one fire hook 15-foot pole, one fire hook 20-foot pole, and contracts were made for the systematic maintenance of the electrical works of the equipment and for the monthly inspections required by the New York Board of Fire Underwriters.

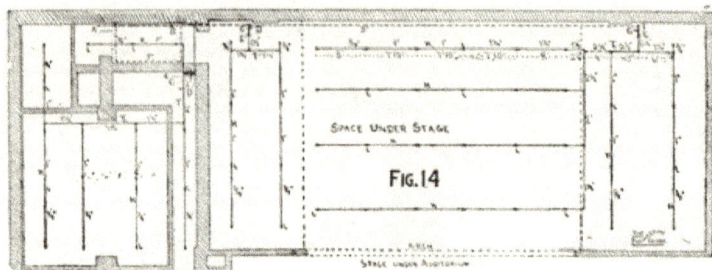

SPACE UNDER STAGE

FIG. 14

PLUMBING IN THE FIFTH AVENUE THEATER, NEW YORK CITY.

Figure 10 is a vertical section showing arrangement of the main riser lines of the sprinkler system. B is the supply from the fire tank placed on raised platform above highest roof over stage; J, K, and Q are distributing risers; N and R are the horizontal trunk mains supplying the roof gridiron and substage systems respectively; P P are branches supplying the dressing-room sprinklers; S is the outside Fire Department connection on Twenty-eighth Street; A and G are gate valves, and E is an electric attachment under the roof near the riser, which is connected to an alarm bell outside the manager's office on the stage, and one large outside bell over the stage entrance. These bells ring automatically when water moves in the sprinkler system. The fire pump is not connected automatically with the sprinklers, but with the tank only, as required by the laws of the Building and Fire Departments.

Figures 11, 12, 13, and 14 are diagrams showing the arrangement of sprinkler heads L L, etc. on branches H H, etc. of main pipes T T, etc. in different horizontal planes. G G, etc., are gate valves, D D, etc. are drip cocks. F F, etc. and A are risers, B, Fig. 14, is a check valve, and C is a 2-inch riser.

Figure 11 is the roof plan at N, Fig. 10; Fig. 12 is the gridiron plan at M, Fig. 10; Fig. 13 is a plan of the second gallery at T T, Fig. 10. The plan of the first gallery is similar to it, except that branch L¹ is omitted. Figure 14 is a plan of the system for the dressing-rooms for "supers," and for toilet-rooms in basement. Figure 15 is a plan of the system in one of the sets of dressing rooms (see Fig. 10), to which the others are similar.

A THEATER FIRE-PRESSURE SYSTEM.

(PUBLISHED IN 1890.)

THE automatic steam and tank pressure arrangement for general and fire purposes in the Broadway Theater, New York City, is shown in the accompanying diagram, where T is a 6,000-gallon iron roof tank filled through a pump pipe A. B is an overflow. C is an emptying pipe. The general house supply is through a pipe D, which pierces the tank at E halfway up, so that it can draw off the water in the upper half of the tank only, always leaving below the

level E 3,000 gallons of water that can be drawn only through the fire line F, which has branches to four other lines G G, etc. H is a check valve, closing by an upward pressure. I I, etc. are hose cocks. J is a Worthington pump with a patent automatic pressure regulator connected by a pipe K to the pump pipe F. Steam at a pressure of about 60 pounds is always kept up, received through the pipe L. A spring at N is regulated so that a pressure of about 70 pounds in the pipe will balance the steam pressure in the pipe L, and close a valve at M. If now water be drawn from any hose cock I, the pressure is diminished, the steam opens the valve M and starts the pump, whose pressure closes the check valve H and allows the pump to work at any pressure on the fire stream until the hose cocks are closed and the pressure in the pipe F becomes great enough to close the

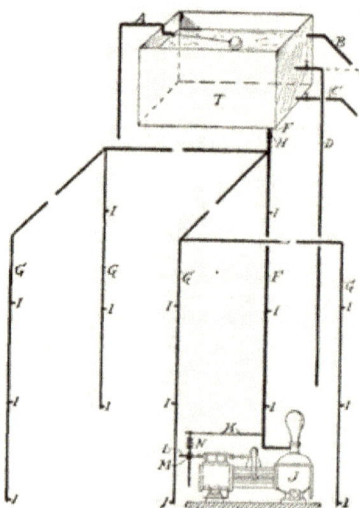

A THEATER FIRE-PRESSURE SYSTEM.

valve M. The tank water is thus shut off while the pump is working and so is preserved for use while the pump is starting if the street supply should fail.

This arrangement is illustrated from a description by W. R. Bracken, of the firm of Moody & Bracken, of New York City, who did this work, with the other plumbing in the building.

PLUMBING DETAILS IN ABBEY'S THEATER, NEW YORK CITY.

(PUBLISHED IN 1893.)

THE plumbing in the new Abbey Theater, Broadway and Thirty-eighth Street, New York City, was executed by Messrs. Rossman & Bracken, in conformity to the requirements of J. B. McElfatrick & Son, architects, and does not differ essentially from standard metropolitan work, except in some practical details of connections which have been sketched to illustrate the convenient and advantageous methods used. The system consists substantially of a pump plant, tank and street supply, hose connections, hot-

water supply, toilet-rooms for the theater, public, office and private use, and washbasins. The automatic fire extinguishers are a separate installation.

A 4-inch street connection directly supplies the suction pipes of a 20"x10"x10" fire pump and a 6"x4"x6" house pump, both of Worthington make. The fire pump is commanded by an automatic governor placed on the steam pipe between the throttle valve and the cylinder, so that, the throttle being open and steam continually on, the governor will exclude it when pressure in the discharge pipe is 100 pounds, and admit it and start the pump the moment the pressure falls in the discharge pipe. The fire riser has three hose cocks and reels with 100 feet of hose in each corridor (12 in all). Great care has been taken to assure the reliability of the fire apparatus in case of an emergency, and its magnitude is indicated by the statement that the fire pump has a capacity of 750 gallons per minute, or more than 1,000,000 gallons in 24 hours, enough to supply 50 gallons per capita per day to a city of 20,000 inhabitants. The automatic apparatus is so arranged that in case of fire anyone

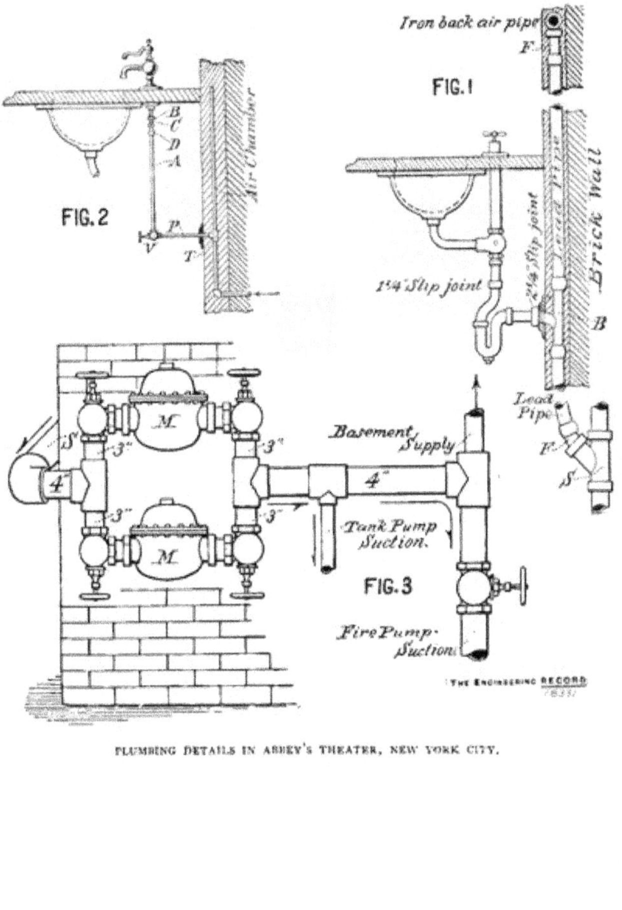

FIG. 1

FIG. 2

FIG. 3

PLUMBING DETAILS IN ABBEY'S THEATER, NEW YORK CITY.

can run out a length of hose, open the valve, and instantly turn on a fire stream, while the pump simultaneously automatically starts at full speed without signaling the engineer, thus gaining time that may be of the utmost value. The house pump supplies the house tank and steam boilers and is controlled by hand. It delivers into a 5,000-gallon iron roof tank, the supply in which is indicated by an electric high and low water alarm of Bracken's patent. The house supply of cold water is from a 3-inch tank pipe branched to distribution pipes in every story above the street which is supplied directly from the street pressure. A separate 1½-inch supply from the tank connects with the 500-gallon hot-water boiler, which is heated by a 2-inch brass live-steam coil about 70 feet long. The water pipes are all galvanized iron except where exposed in toilet-rooms and at fixtures, where they are nickel-plated brass. There are in all 90 washbowls, 35 water closets, 26 urinals, two bathtubs, six slopsinks, and eight ordinary sinks, which are installed for the public and private uses of the theater, for rented offices, clubrooms, and other tenants of the building. Washbasins with marble slabs and hot and cold water and handsome fixtures are a noticeable feature in all of the 36 dressing-rooms. When the main riser lines were run it was impossible to locate fixtures with precision, but the tees, Y's, etc., were set as nearly right as possible and closed with screw or calked caps, and a water-pressure test

was applied. Afterwards everything but the tee or Y was built in or plastered up, and when the washbowls were set they were connected up with special adjustable pieces which allowed for a variation in position of several inches.

Figure 1 shows the connection of the washbowl waste and overflow to the soil pipe S by means of two slip joints, the lower one of which is screwed on to the projecting end of a brass tee " Y " B, which is united by wiped joints to the lead pipe branches connecting it to the brass ferrules F F calked into the cast-iron hubs of the main pipes.

Figure 2 shows the connections of the basin supplies, which are all made with a ¼x½-inch angle V and short nickel-plated pipe P screwed on to the tee T, after the bowl is set. The coupling C is connected to the foot of the cock by a ground joint N, and the distance between it and valve V being measured, a ½-inch nickel-plated pipe A is cut to fit and connected up with a packing nut D, thus making an easy screwed job, claimed to be perfectly tight and durable and to be not much more expensive than lead pipe and wiped joints. One man has fitted up six such bowls in a day.

Figure 3 shows the connection of the two 3-inch Thomson meters M M to the 4-inch supply S, thus avoiding the use of a 4-inch meter and providing a by-pass to enable one meter to be cut out for repairs, etc., without interrupting the supply.

PLUMBING OF SWIMMING AND RAIN BATHS.

BATHS OF THE NEW YORK ATHLETIC CLUB.

(PUBLISHED IN 1886.)

THE accompanying illustrations show the baths of the New York Athletic Club. Figure 1 is a ground plan of the main floor of the building, which is devoted to office and bathing purposes. The second story contains a café, dining-room, reading-room, parlor, and public and private billiard-rooms. The third floor has a private dining-room, 1,024 private lockers for the members, a boxing-room, fencing-room, douche and shower rooms, lavatory, and water-closets and urinals. The fourth floor is entirely occupied by a gymnasium, the gallery of which is a running track of 21 laps to a mile, and the kitchen is in the top of the house. In the basement are four public bowling alleys and two private. A repository is also provided for bicycles or tricycles, and 84

lockers are here provided for the use of the help The building is 100x75 feet, the greatest length being on Fifty-fifth Street, and it is on this front the main entrance is. The staircase hall is a square of about 21 feet, open to the top of the building. On the left is the office, and beyond it a committee-room. Opposite the main entrance is the coat-room, 14x18 feet, and adjacent to both stairways. The remainder of the floor is devoted to bathing. There is a large swimming bath, 66 feet long by 22 feet wide, lined with glazed brick, as is also the whole room, and shown in perspective in Fig. 2. At one end the depth of water maintained is about 4 feet 6 inches, at the other about 6 feet 6 inches. To the left of this bath are the dressing-rooms, and on the right is a position for spectators. Light is admitted through skylights over the spectators' gallery. The temperature of this room is maintained at from 65 to 70 degrees in cold weather. At one end of the spectators'

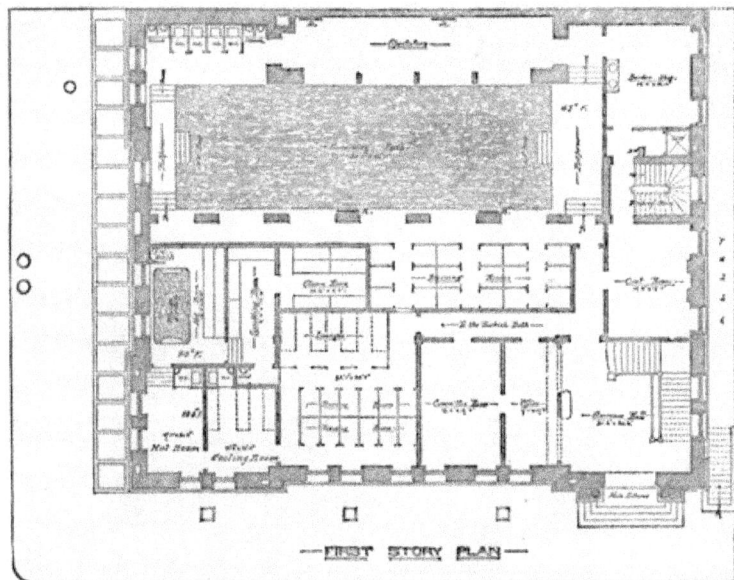

FIG. 1.

BATHS OF THE NEW YORK ATHLETIC CLUB.

gallery the water-closets, etc. are placed, and they
can be approached either from the steps of the swimming bath or the gallery. Here also are two shower
baths for the bathers. At the opposite end is the
barber shop. Gas and electric lights are both provided for evening use, and a swimming master is
always in attendance. Turkish and Russian baths
are also provided. A room 23x24 feet is devoted to
dressing-rooms and lounges; a steam-room with
marble terraces is 14'6"x9'6"; the " hot-room " is 9x6
feet, a hot-room with plunge is 12'6"x18'6", one side
of which is terraced. A spray and needle bath occupies one corner of this room, marked S N. The
temperature maintained here is about 100° Fahr.,
while a temperature of 160° Fahr., is maintained in
the dry hot room. The scrubbing-room is 18'6"x8',
and the cooling-room 12'x11'6". A study of the plan
of this floor will show the relative positions of the adjoining rooms of the baths and closets.

To warm the water in the swimming bath or
plunge the exhaust steam from the pump and electric-
light engines can be used, but the principal mode of
tempering the water for the swimming bath is to pass
it through a coil of 16 1-inch pipes, 30 feet long,
arranged in the boiler smoke flue so the waste heat
of combustion can be utilized. This is a header coil
within a long iron smoke flue, and the Croton connections are so arranged that the water in flowing to

the swimming pool, after passing a filter, can be
either passed through the coil or to the pool direct by
the manipulation of two valves in the engineer's department.

The architect was Mr. Charles W. Clinton, and the
plumbers Messrs. Locke & Monroe, all of New York.

PLUMBING OF A SWIMMING BATH.
(PUBLISHED IN 1892.)

PART I.—GENERAL VIEW AND CROSS-SECTION OF THE
BATH. DETAILS OF CIRCULATION CONNECTIONS AND
CASCADE, ARRANGEMENT OF HEATER AND PUMPS.

FIGURE 1 is a general view of the large swimming
bath in the house of the Manhattan Athletic Club,
New York City. The dimensions of the pool are
about 100'x21'x8' deep. Fresh water is periodically
supplied through pipe A, being aerated by falling
several feet into the pool in a wide sheet at B.
Ordinarily, circulation with the heating boiler is
maintained through the 6-inch pipes A and D, as
indicated by the full arrows. E is the overflow and
F is a suction pipe to a filter pump. G is an emptying pipe to the sewer. Figure 2 is a general cross-
section of the swimming bath showing its construction. Figure 3 is a partial horizontal section and
plan, showing an enlarged view of the supply and

FIG. 2.

BATHS OF THE NEW YORK ATHLETIC CLUB.

circulating connections. Supposing the tank to be full of water it· temperature is maintained or elevated by the steam-coil heater H. hot water from which circulates through pipe D and enters the pool through a grating a. O colder water flowing back to the heater to replace it through the outlet I and the pipe A. Inlet O is at the middle of one side wall of the pool about 3 feet above outlet I, which is in the bottom at one end, 50 feet away. so as to insure diffusion of the hottest water throughout the pool, while the circulation through heater H is so easy that the water returns to the tank as fast as its temperature is slightly raised, thus never becoming too hot in small quantities. For this circulation, valves K and L are open and more water may or may not be supplied at pleasure from the pump delivery pipe N. When however it is desired to aerate the water by delivery through cascade B, Fig. 1, valve L and outlet I are closed and the pump delivers through pipe M, heater H pipe A and branches J J. Figure 4 is a vertical elevation at Z Z, showing the connections at outlet O, Fig 3 The cascade was designed after the completion of the bath. so the sleeve Q was calked into the outlet elbow P of circulation pipe A. Brass pipes J J were branched from Q, whose top was fitted with a screw plug R.

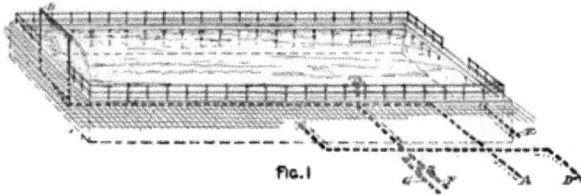

FIG. 1

FIG. 2

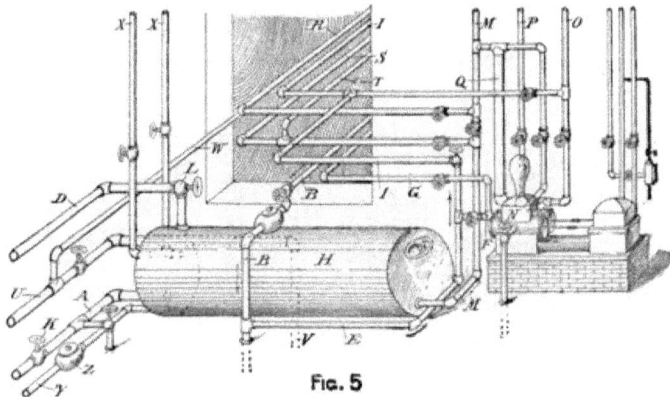

FIG. 5

PLUMBING IN A SWIMMING BATH.

Branches J J have two vertical risers T T about 12 feet long, which are connected by an 8-foot horizontal pipe B, which has for nearly its entire length a $\frac{1}{16}$-inch wide slot S (or of a total area about equal to the cross-section of the pipe), through which the water escapes, forming the cascade when plug R is screwed in. When R is removed, circulation is directly through the top I, of sleeve Q. A key wrench and guide enable the attendant to insert plug R from the surface of the water.

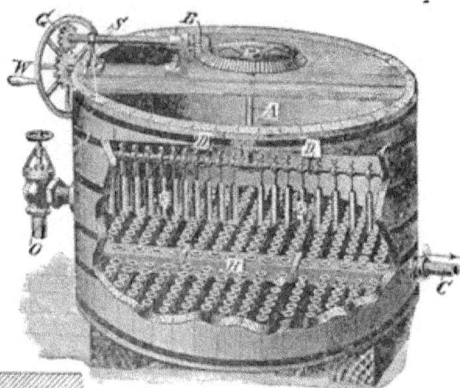

Fig. 7

Bath

Fig. 3

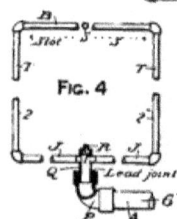

Fig. 4

Figure 5 shows the heating tank H and its connections. M is the 3-inch cold-water supply from the roof tank. A and D are circulation pipes to the swimming bath, and with their valves L and K are designated by the same reference letters as in Figs. 1 and 3. B is a 4-inch supply from the city mains. It passes through meter C and has for the pump service a branch E with the connections F to the roof-tank pump N, G to the filter pump, J and I to the bath pump K. House pump N delivers to the roof tank through pipe M, or to the boilers by O, or to the fire line by P; Q is an equalizing chamber. R, S, and T are branches

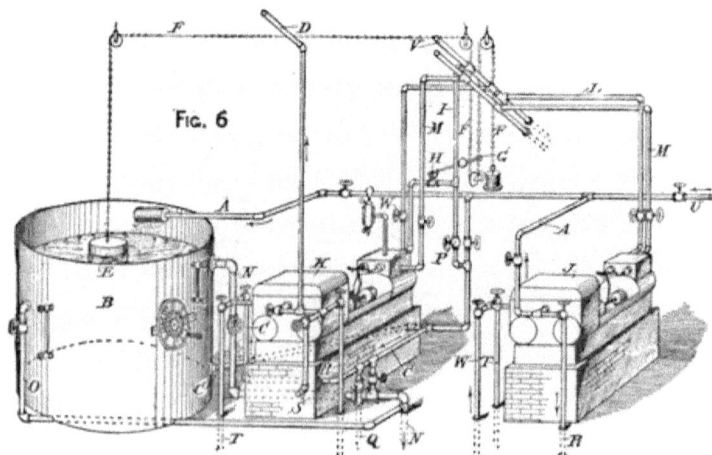

Fig. 6

PLUMBING OF A SWIMMING BATH.

to the suction and delivery pipes of pumps J and K. U is the emptying pipe of the swimming bath, and V and W are branches for sewer waste and pump suction respectively. X X are live and exhaust steam supplies to the heating coil H, and Y is the return steam pipe with trap Z.

PART II.—BATH PUMP, FILTER PUMP AND FILTER.

WHEN fresh water is admitted to the swimming bath, the dirty water, overflowing from the opposite end through pipe U, Fig. 5. may be discharged through sewer pipe V, but is generally taken through the suction pipe W of pump J. Fig. 6, which delivers it through pipe A to the Jewett gravity filter B, whence it is drawn through suction pipe C of pump K, and delivered through pipe D to the bath heater H, Figs. 3 and 5. When the water in the filter rises to a fixed height, the heavy float E, suspended by chain F, attached to the counterweighted lever G, opens the valve H, and admits steam from branch I of main V to the pump K which works until it reduces the level of the water in B and the descending float cuts off the steam and stops the pump. The pump may also be operated by the hand valve W. L is a steam pipe to pump J, and M M are exhaust pipes. N and O are overflow pipes discharging into the sewer. Ordinarily valve P is closed, but it may be opened to drain pipes V and I through pipe C into N. R R are delivery pipes to the boiler feed pipe. S is a hose connection T T are suctions direct from the city water supply, and U is a connection to the delivery pipe of the feed pump. Thus these two pumps, J and K, the roof tank pump N, Fig. 5. and the boiler feed pump (not here shown) are interchangeable throughout, and each can be connected on to the system of any of the others. When it is necessary to wash the filter, pump J forces city water from pipe T through pipe C and it passes through the filter in the reverse direction, and thus escapes into the sewer pipes.

Figure 7 shows the construction of filter B, Fig. 6. but represents the well to be made of hooped wooden staves instead of steel plates, as is really the case.

About half of the upper part of the filter and the front wall of the remainder is removed so as to show the washing and collecting apparatus in the filter chamber, from which the filtering medium is shown removed. The hand wheel W, through pinion G, shaft S, and beveled gear B, drives shaft A, revolving its horizontal arms D D, which carry cutting bars E E, which, when the filter is washed, disintegrate the filter bed. On the bottom of the filter a cast-iron collecting box H has connected to it numerous lateral pipes with horizontal inlets, through which the water is received and delivered to the supply main at C. Mr. P. Lauritzen was the architect of the club-house, and the plumbing was executed by Byrne & Tucker, all of New York City.

EUROPEAN RAIN BATHS.

(PUBLISHED IN 1891.)

THE recent agitation in New York City for public baths lends renewed interest to a number of descriptions of similar European establishments, published in THE ENGINEERING RECORD several years ago. The illustrations first published in THE ENGINEERING RECORD of October 11 1883, show a bath exhibited at the Berlin Hygienic Exhibition, by Mr. David Grove, of Berlin. The establishment was designed to afford baths at very low prices, in reach of the poorer classes, and is an extension of a plan first adopted by Mr. Grove in a bath at the barracks of Kaiser Franz Garde Grenadier Regiment No. 2, which was established in 1878. At the Berlin exhibition, baths were given at a price of about 2½ cents each, including soap and towels.

A special building of corrugated iron was erected at the Berlin exhibition for the bath. Each bathing cell in the men's department is provided with a shower bath at about 90° Fahr., and a small douche fitted to a flexible tube and supplied with cold water. In the cells of the women's department the arrangement is the same, with the addition of a similar douche supplied with warm water. The water for

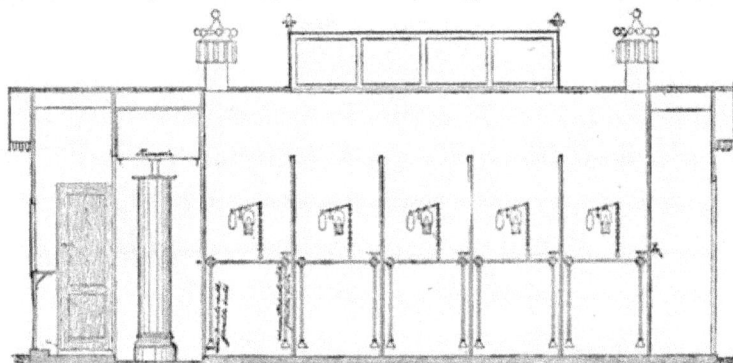

PUBLIC BATHS IN THE BERLIN EXHIBITION.

the baths is heated by an apparatus placed at one
end of the building in the center of the chamber used
for drying the towels. Either fire or steam can be
employed as heating agent for the apparatus. At the
end of the building opposite to that containing the
drying chamber and heating apparatus is the washhouse containing a machine for washing the towels.

The building has brick foundations for outside and
middle partition walls: thickness of foundation one
brick; height, 12 inches. Floors are made of one
course of bricks covered with asphalt. Each cell has,
besides, a movable grating of laths covering the
whole floor. The cells are drained by a cast-iron
pipe under the floors with a branch to each cell. The
inlets to the drains are covered with a copper sieve.
The drain pipe is trapped at its lower extremity at
the end of the building. The supply pipe for each
douche is a ¾-inch wrought-iron pipe.

The boiler of the shower bath exhibited is arranged for heating by steam only, as shown. If a
fire were placed beneath the boiler the condensed
steam would have to be carried off at the lower end,
of side of boiler, and a smoke flue could be arranged
in the center of the steam coil. The water for the
shower bath (only warm) is turned on or off by a
closet valve which allows a certain regulated quantity of water to pass through, and then closes automatically if not held open. The valves are provided with a chain and pull. The douches each have
a separate screw-down tap for cold or warm water
only. (See letters and explanation on illustrations.)

The rooms of the bath at the exhibition were not
heated, as the establishment was intended for use
only during the exhibition, which closed before
winter. The best place for fixing pipes for heating
the rooms would probably be along the foot of the
outer wall in the passages, or under the floors of the
latter. The reservoir was made of wrought iron,
and was about 59 inches long. 39 inches wide, and 30
inches high.

The cut and data of the public bathing-houses at
Frankfort-on-the-Main, Germany, were given in THE
ENGINEERING RECORD of May 25, 1889. Space,
water and fuel were limited, and it was desired to
furnish baths with soap and towel for about 2 cents
each. The roof of the house is an octagonal pyramid,
14 feet 9 inches high in the center. The outside
octagonal wall is 11 feet 9 inches high, and 13 feet 9
inches long on each face. M is the entrance for
women, and N that for men. A is the cashier's
office: B is the linen-room; C, the drying room; S,
chimney; E, the stairs to the basement where a
furnace heats water in reservoir D. This is in the
upper part of room C. The same furnace also heats
the air supplied to each bathroom through registers
K. L is the partition between the men's and
women's corridors; I I are water-closets; F F are
sliding doors to dressing-rooms G, which are supplied each with a chair, a mirror, and a wardrobe.
The floors are covered with linoleum. The partitions O are about 7 feet 3 inches high. P are water-
proof curtains for the bath closets H.

There is no full-length bath, but each closet contains a basin and a douche with hot and cold water

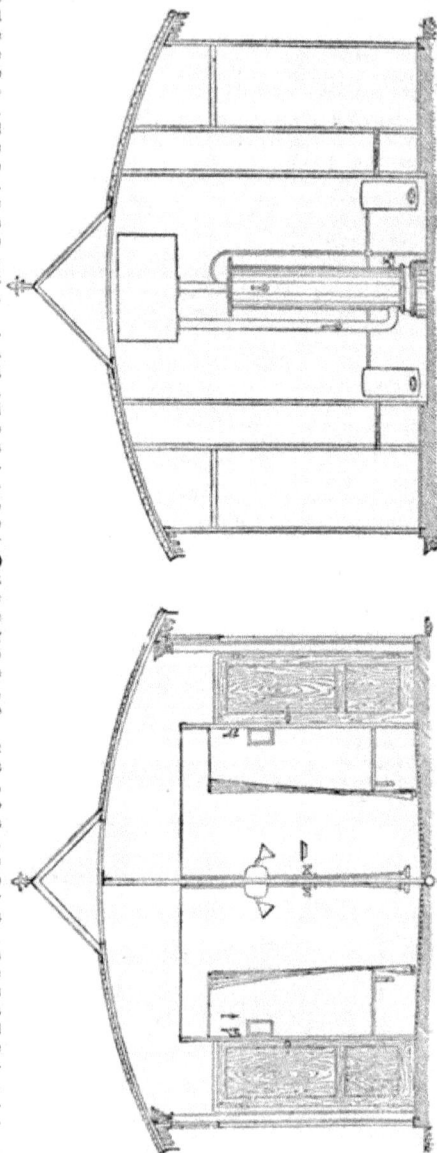

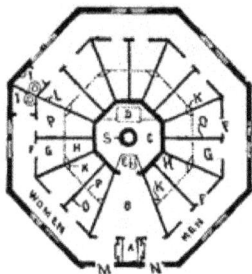

PUBLIC BATHS IN THE BERLIN EXHIBITION.

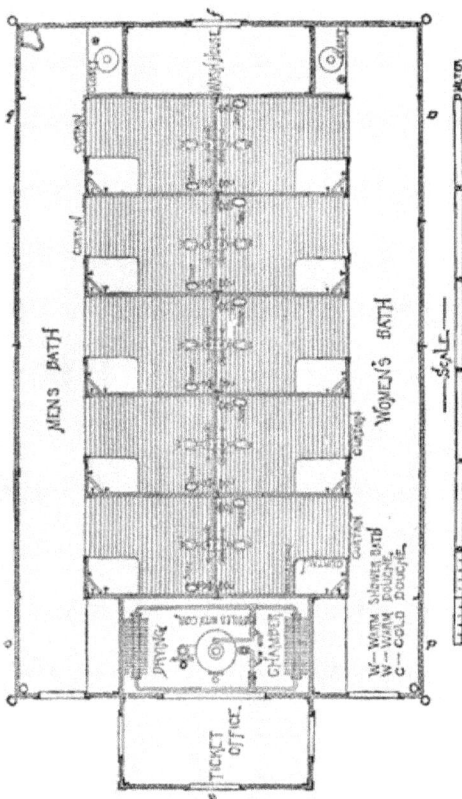

BOILER OF SHOWER BATH.

cocks, by which the bather can regulate the quantity and temperature of the water at will. The floors of the closets have wooden gratings, through which the water is drained off.

The establishment cost 25,000 francs, and it is estimated that it will cost about 4,000 francs per year to maintain it.

A public bath at Vienna, Austria, described in the RECORD of August 4, 1888, occupies the ground floor of a house, and comprises, on the right, 42 baths for men; on the left, 28 baths for women, and at the end a lavatory. Each division contains an attendant's hall, a dressing-room and the bathing-room. In the latter a closet is assigned to each bather for his clothes.

The bathrooms have sheet-iron walls, and are closed by curtains. They are 2.6 feet deep, 3.28 feet long, and each is supplied with a douche cock, which the bather can operate at pleasure. The passages between the groups of baths are 3.28 feet wide, and

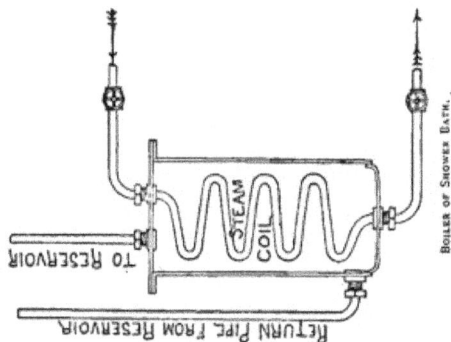

are paved with Holland tiles set in beton. Water is taken directly from the city mains to two reservoirs, each having a capacity of 3,434 gallons. The water is maintained in the reservoirs at a temperature of 68° Fahr. in summer and 104 degrees in winter by means of hot-water coils separately heated. Each reservoir is employed alternately. The rooms are heated by an independent hot-water system.

For the modest sum of 5 kreutzers each bather is entitled for 20 minutes to the use of a dressing-room, douche room, and 10½ gallons of water, besides a pair of drawers and a towel, and for each woman a bathing dress besides.

According to Dr. Lassar, Professor of Hygiene at the University of Berlin, who took the initiative in the establishment of this type of bath, from one-eighth to one-quarter of the above quantity of water would suffice. The total cost of establishing the baths was 43,855 fl., and the annual maintenance, inclusive of 5 per cent. interest on the first cost, is 17,195.5 fl.

PUBLIC BATHS IN NEW YORK CITY.

(PUBLISHED IN 1892.)

PART I.—DEVELOPMENT OF FOREIGN AND AMERICAN
PUBLIC BATHS, GENERAL DESCRIPTION OF THE BARON
DE HIRSCH RAIN BATHS, AND FLOOR PLANS.

THE subject of public bathhouses, which has been
so often represented in the columns of THE ENGI-
NEERING RECORD, has received recent increased
attention and has developed into practical operation
in several places in New York City. Some of the
features of design and construction are adapted and
modified from the practice abroad, and some are entirely of a special nature. A number of the leading
characteristics and practical details will be illustrated
in this and subsequent descriptions, which we have
prepared from the working drawings and from special sketches. A brief *résumé* of the subject will
show the conditions and circumstances now existing.
The Greeks, Romans, Egyptians, and other ancients
provided and used abundant facilities for public
baths, freely available to the poorest citizens, and in
an article of the *Dietetic Gazette* of May, 1891, Si-
mon Baruch, M. D., says that in Russia every village has its vapor bath, where the bather, after being
steamed, is well scrubbed with soap and water and
receives a massage with switches, and a shower bath.
Public baths are quite common in Constantinople
and in the interior of Turkey, the fees being suited
to the very poorest people. The same is true now in
Egypt, there being 60 or 70 baths in Cairo alone.
In Japan public and private baths are much fre-
quented by both sexes, a bathhouse being visible
every 100 paces in Yeddo.

Since the first one was established at Frederick
Street, Liverpool, in 1842, public bath and wash
houses have become both popular and cheap in Eng-
land, and they are numerous and excellent in France,
Belgium, and Germany, where the fees, although
moderate, are often beyond the means of the poorest. The adaptation of these baths to the populous tenement districts requires that they be
located near by and afford a good bath quickly, com-
fortably, and cheaply. The development of the system has tended to the abolition of the old-fashioned
tub, and the adoption of shower baths, or, as they are
termed, rain baths, for which the following advantages are claimed. First, a large economy over the
provision and maintenance of tubs; second, economy
of labor, time, and expense of filling and cleaning
the tub for every bath, the rain bath being automatic, simple, and requiring only supervision, not
attendance; third, quickness, greater efficiency, the
mechanical effect of the descending stream, and the
prevention of contact of soiled water with the body;
fourth, economy of space; fifth, economy of water;
sixth, freedom of danger of communicating diseases; seventh, stimulating and refreshing effects.

Among the requisites for public baths it is important that they should be located in the most populous
districts of laborers' residences, that they should be
neat, clean, inviting, and well warmed, ventilated,
and lighted; that they should be substantially and
economically constructed and managed, and open
every day and night. The first bath fulfilling these
conditions approximately was exhibited at Berlin
Hygienic Exposition in 1883. It was a corrugated-
tin house of about 430 square feet, and was so successful that it has since been adopted by many barracks and factories. The first public bath on this
principle was constructed in the Mondscheingasse,
Vienna, when an old building was divided into 72

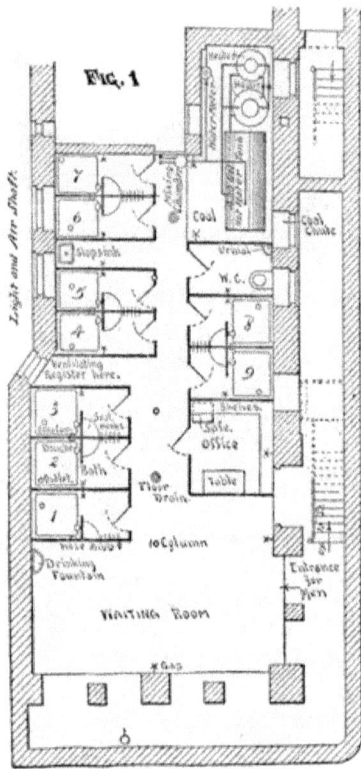

FIG. 1

BASEMENT FLOOR
THE BARON DE HIRSCH RAIN BATHS.

bathing cells, and nine gallons of warm water, soap,
and towel are furnished for about 2 cents. A public
bath association in Berlin furnishes different kinds
of baths for from 2½ to 12½ cents each. Rain baths
have also been provided for schools in Gœttingen,
Munich, and Weimar, and are used daily by 75 per
cent. of the pupils. Rain baths are also provided in
many European factories.

In this country rain baths are of very recent public
adoption. One of the first was provided at the suggestion of Dr. Baruch for the New York Juvenile
Asylum. It consists of 68 sprinklers, 20 inches

apart, placed near the ceiling above a bathing space, where companies of children can soap and rub themselves every 10 minutes at the rate of 280 an hour. The water is heated by the admixture of steam and delivered warm.

Recently the trustees of the Baron de Hirsch fund in America adopted the rain-bath system for the first of a series of free baths intended to be distributed

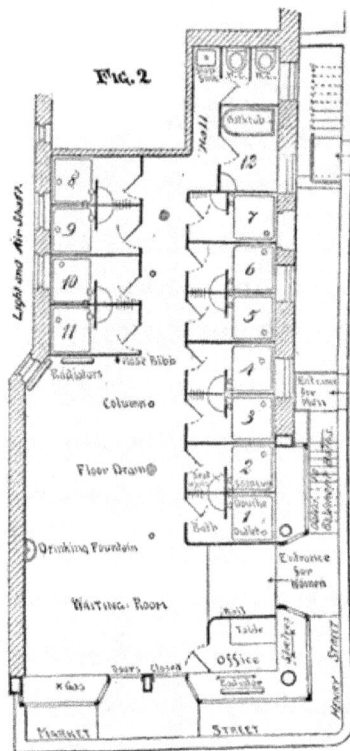

FIG. 2

FIRST FLOOR
THE BARON DE HIRSCH RAIN BATHS

throughout the tenement district of New York. These baths are located in a corner building at the intersection of Henry and Market Streets, in the eastern end of lower New York City. A basement and street floor of an apartment-house have been rented for five years, with privilege of renewal, and the place has been transformed into a bathhouse according to plans and specifications prepared by William Paul Gerhard, C. E., assisted by Messrs. Brunner & Tryon as consulting architects, Kennedy & McDermott being the plumbers, Hitchings & Co. furnishing the heating apparatus, and James Elgar general contractor, all of New York City.

Figures 1 and 2 are plans of the basement and first floors, for men and women respectively. Provision has been made for 30 douches in all, but only 20 are yet put up. The baths are open five days of the week from 9 A. M. to 10 P. M. and Sundays 6 A. M. until noon. A cake of soap and clean Turkish towel accompany every bath, for which adults are charged 5 cents and children 2 cents. The rooms have cement floors and are heated by direct radiators supplied with hot water from the bath heaters. The rooms are lighted by gas and ventilated by registers into a 16x24-inch galvanized-iron flue, inside of which the 11-inch smoke flue from the heaters is carried up above the roof. The bath apartments are partitioned by corrugated-iron walls on angle-iron frames with wooden caps, and have wooden screen half-doors. All walls, iron-work, pipes, etc. are painted five coats of special white bath enamel. The water-closets, urinals, slopsinks, and drinking-fountains are of porcelain, and the bathtub, Fig. 2, is of enameled iron with glazed rolled edge, and stands on high legs. It is also provided with movable seat inside for small children.

PART II.—DETAILS OF BATH COMPARTMENT, HEATING, STORING, AND MIXING ARRANGEMENTS IN THE BARON DE HIRSCH RAIN BATHS.

FIGURE 3 shows the arrangements for heating, storing, and mixing water in the basement. Water under direct street pressure is taken through a 1-inch trap and 2-inch service pipe to a 2-inch meter M, Fig. 1, from which it is delivered through pipe D to the 600 gallon galvanized-iron boiler B, and from it flows through the 4-inch circulation pipe E, and the 2-inch branches F F to the two Hitchings heaters A A, returning hot through the 4-inch circulation pipe G, and the 2-inch branches H H. The heaters are also connected to the hot-water radiators by the 1½-inch flow pipe I and the return pipe J. Hot water is delivered from the boiler through the 2-inch pipe K, which, entering with the cold-water pipe N direct from the meter M, delivers to the 20-gallon galvanized-iron mixing chamber C, from which the water is delivered at any required constant temperature to the bath douches through the 2-inch pipe O and its 1½-inch branches P P P. Ordinarily valves *a*, *b*, *d*, and *e*, are closed and cold water enters at the bottom of the boiler B through pipe D, and hot water from the circulation pipe G enters through branch S, but if hot water is needed quickly, as early in the morning, before all that in the boiler B is heated, valve E is opened so that the boiler is supplied most directly from the heaters. When the heaters A A are not required to heat water for the radiators, the thermal energy supplied to the boiler B will be so much increased that a means for cooling the water in the boiler has been provided—viz., by means of the branch V that is taken from the cold supply pipe D and terminates in a perforated horizontal pipe 6 feet long that is near the top of the boiler B, inside of it. Ordinarily the valve *a* being open and *b* closed, this device is not in operation, but if *a* is closed and *b* opened the cold-water supply is distributed all over the top of the boiler and instantly mixes with and lowers the temperature of the hot water supplied

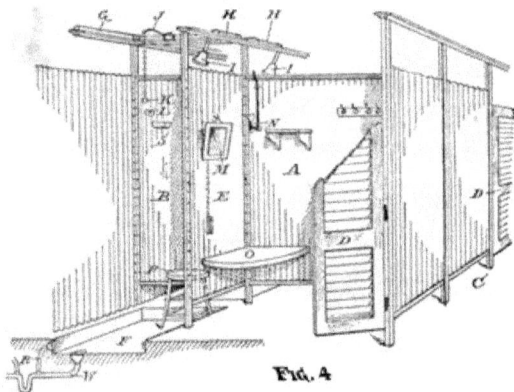

FIG. 4

through pipe K. The boiler can be emptied through the waste pipe X. W is a drip pipe to empty the rising lines. R is a cold water supply to the slopsinks and bowls and Q is a water-pressure gauge. T_1, T_2, T_3, are special hot-water thermometers made by the Hohman & Maurer Thermometer Company, of Peabody, Mass. Thermometer T indicates the temperature of the hot water in the center of the hot-water tank, T_1 indicates the temperature of the mixing of hot and cold water. T_2 indicates the temperature of the water as supplied to the douches, and T_3 indicates temperature of hot water as coming from top of hot-water tank.

The mixing is adjusted by valves ff, which are so sensitive that a very slight movement is shown on the scale of the thermometer T_1, by which the baths may be regulated, though thermometer T_2 is usually consulted after the valves are approximately set; h is a safety valve. Pipes N and K were at first connected to the heating chamber at jj, but have been found to operate more satisfactorily arranged as shown with a common delivery L. Y is a smoke flue and Z is the high angle-iron frame supporting the boiler.

Figure 4 shows the construction and arrangement of a bath compartment. C is the corridor from the wooden screen doors D D, etc., open into a dressing-room A, about 4½x4½ feet, which is separated by half-partition E from the bath place B, which is about 4½x4 feet wide, with a depressed basin F about 8 inches deep. Water at about a temperature of 100 degrees is supplied through the 1½-inch distribution pipe G, with

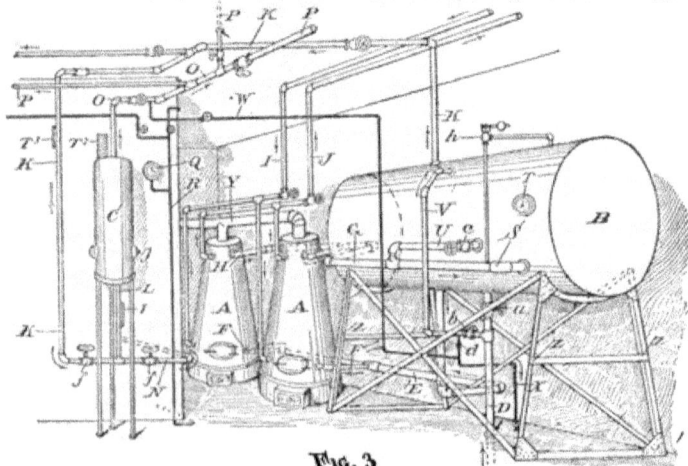

FIG. 3

THE BARON DE HIRSCH RAIN BATHS, NEW YORK CITY.

¾-inch branches H H, to the copper douches I I, etc., which are above the bather's reach and set at an inclination to deliver water upon his neck and body, but not upon his head unless he especially presents it. The douche is governed by a self-closing bibb, which is opened by a lever I and chain K, whose ring may be pulled down to hook I and secured there, so as to hold the valve open and leave both the bather's hands free. Each bath compartment is provided with a stool P, seat O, mirror M, gas burner N, comb, soap tray S, and with shelf, clothes hooks, and a bolt on the door D. The brass strainer or waste pipe W is designed to be not quite large enough to

ELEVATION WITH FRONT of BOWL REMOVED.

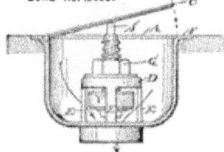

DETAIL OF VALVE.

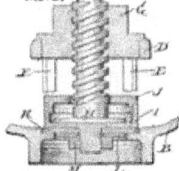

FIG. 5

Valve Removed

carry off the water as fast as it is received from the douche I, so that it rises in the basin F until it overflows through R, thus keeping a few inches in the basin F to cover the bather's feet.

Figure 5 shows the floor drainer (Figs. 1 and 2). It is essentially a brass bowl B, nearly hemispherical, and about 9 inches in diameter, which receives the house flushings, etc., and delivers through pipe P, which discharges freely into a trapped waste pipe. The flange F is set flush with the surface of the floor, which is graded to this point, and when the valve V is fully shut, the cover C can be entirely closed and rest in its seat, offering no obstruction above the floor surface. If however the valve is raised at all, its stem S will project above the bottom A of the cover

seat, and interfere as shown with the cover, thus compelling the attendant to screw it down and close it tightly in order to remove the floor obstruction. The valve stem S screws up and down in block G, which is removable from ring D, cast with legs E E solid to the bowl B. The solid head H of stem S lifts valve box I by its cap J, through which the screw passes loosely. K is a rubber gasket, L a brass washer, M a nut. This floor drainer was designed by Oliver Barret, New York, who has placed it upon the market.

PART III —RAIN BATHS AT THE DEMILT DISPENSARY, GENERAL PLAN, DESCRIPTION, SECTION OF BATH COMPARTMENT, DETAIL OF MIXING VALVES AND ARRANGEMENT OF HEATER AND BOILER.

THE Demilt Dispensary is at Second Avenue and Twenty-third Street, New York, and in its basement six rain baths and one tub have lately been placed, with provision for increasing the number. The rooms are arranged as shown in Fig. 6 and are open to men and women on alternate days, a charge of 10 cents each being made, unless the bathers are unable to pay, when free tickets are issued to them. The rooms are lighted by gas and heated by steam.

The six rain-bath compartments are duplicates and fitted the same as the Baron de Hirsch baths except that the bath proper is paneled with 1¼-inch blue-veined Italian marble slabs, and the dressing-rooms with wood painted with five coats of Aspinall's special bath enamel. The floor of the office and corridor is of wood, that of the bath being of cement.

Unlike the system adopted for the Baron de Hirsch baths, the douches at this bath have no bibbs, and are not under the control of the bather, but are operated from a valve board in the office. Figure 6 is a general floor plan showing a hot-water heater A, the 200-gallon boiler B, and the valve board C; 1, 2, 3, 4, 5, 6 are the rain baths and 7 is a room for invalids, children, etc., with a high, enameled iron, rolled-edge tub T. Figure 7 is a section at Z Z, Fig. 6. When a bather is undressed he touches the electric button and the attendant at the valve board mixes the hot and cold water and turns it in his douche, afterwards further regulating the temperature if requested to. When the bather has finished he again presses the button and the water is turned off. Figure 8 shows the heating, storing, mixing, and delivering arrangements. Cold water under city pressure is delivered to the boiler B through the 2-inch pipe E; circulates to and from the heater A through 2-inch pipes F F and is delivered through 2-inch pipe G to the 2½-inch heater H (on the valve board C, Fig. 6). Cold water is also delivered through the pipe D to the heater I, and a branch J from each of these headers unites in the mixing chamber K to form a pair for each of the six rain baths, the numbers of which are painted between them. The water is mixed by valves M M, and when the special thermometer N indicates the required temperature (normally 100° Fahr.), valve O is opened and the water is delivered through pipe L to whichever bath compartment is indicated by the electric annunciator P, which rings a bell and raises a pendulum bob attached to the corresponding numeral above. A is a safety valve, R an emptying pipe, and S S S pipe legs supporting the boiler B. T is a drip pipe for emptying lines I I, etc. Meter experiments by Mr. Gerhard showed that with both cocks open the bathtub was filled to within 5 inches of the overflow (45 gallons) in two minutes, and that the douche with both valves half open, under the same water pressure, about 20 pounds discharged about 7.1 gallons per minute, so that, allowing three minutes douche for each bather, not quite half a tubful of water is used. Probably, however, it is used much more lavishly in practice, as the attendant states that the 200-gallon boiler does not suffice well for more than six simultaneous baths.

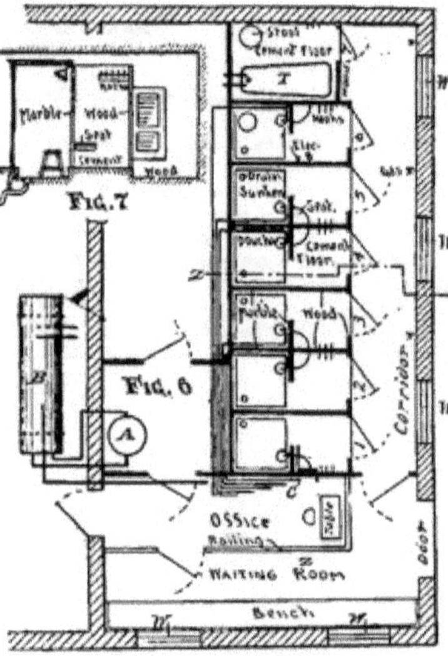

PART IV.—RAIN BATHS IN THE HEBREW INSTITUTE, GENERAL DESCRIPTION, PLAN, ELEVATION AND SECTION OF A BATH COMPARTMENT, ELEVATION OF HEATER.

THE Hebrew Institute is a newly completed building designed by Architects Brunner & Tryon, assisted by William P. Gerhard, as consulting engineer for the sanitary work, and Mr. Alfred R. Wolff, as consulting engineer for the heating and ventilation. It is located at Jefferson Street and East Broadway, in eastern lower New York. Its functions are intended to be similar to those of Cooper Institute, and it contains a large assembly-room with stage, several classrooms, reading-room and library, workshop, gymnasium, and shower baths. There are five baths (Figs. 9, 10, and 11) located in the upper story, and have a cement floor ("flintolithic pavement") raised one step above the general floor level. The bathroom also contains a large enameled washup sink, in the adjoining large dressing-room are lockers, and in another room a slopsink, a large enameled sink, water-closets, and urinals. The five compartments are about 4x5 feet in size and paneled with blue-veined Italian marble slabs 6 feet high. Each entrance is furnished with a portiere and a half screen

Fig. 7

Fig. 6

door. The rooms are ventilated by wall registers, furnished with gas and electric light and heated by direct steam radiators. Figures 9, 10, and 11 show the fittings for one of the bath compartments. Hot and cold water is supplied by pipes H and C respectively, the bather admitting first cold water by valve c and then tempering it to the required degree by admitting hot water through valve L. The water is mixed in the chamber G, where a special hot-water thermometer T indicates the temperature of the water delivered

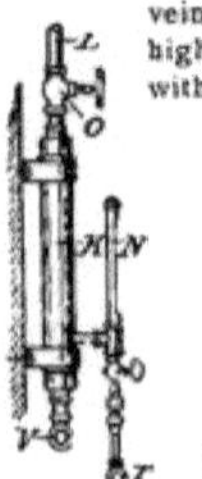

Detail of Mixing Chamber.

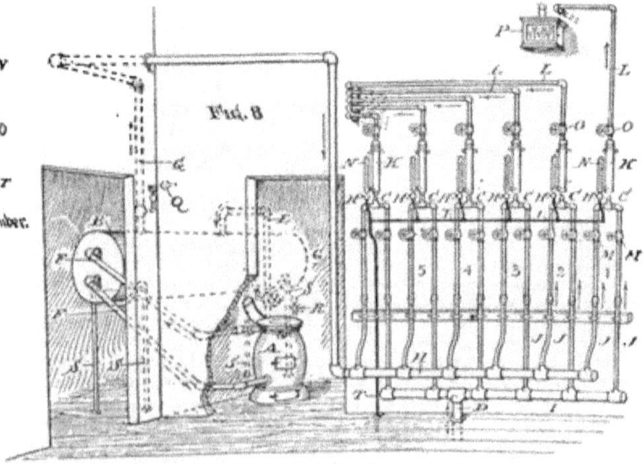

Fig. 8

RAIN BATHS IN THE DEMILT DISPENSARY, NEW YORK CITY.

through pipe P to the brass douche A. The douche is controlled by the self-closing bibb B, is operated by a chain D, which terminates in a ring E and may thus be secured to the hook F so as to hold the valve open and permit the bather to use both hands freely. I is an emptying cock, J is a towel rack, K is a soap cup.

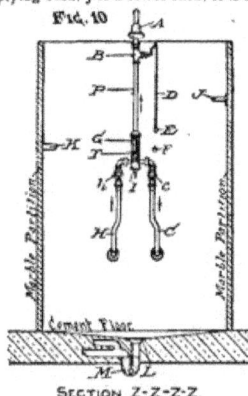

Fig. 10

SECTION Z-Z-Z-Z.

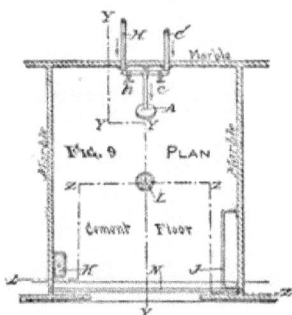

Fig. 9   PLAN

L the floor strainer, and M a waist trap, the trap screw of which is accessible from the room below. The douche and all the pipes and fittings are of brass, nickel-plated. Water is furnished from two storage tanks on the roof and the hot supply is obtained from a small Foley hot-water heater in the basement.

Figure 12 is an elevation of the heater, which is located in the basement, and is operated by steam from the steam-heating boilers. Cold water enters the main drum *k* through the supply pipe A and is heated by the steam coil B with flow and return pipes L and M. The hot water passes at C into the upper drum I, where it operates the expansion rod D and is delivered to the baths, washbowls, etc. through pipe H. D is a flexible metal rod fastened at F, passing around a loose drum E and movable at G, where it is attached to a rod bearing on the short arm of lever N. When the water cools to a certain temperature, rod D contracts and throws

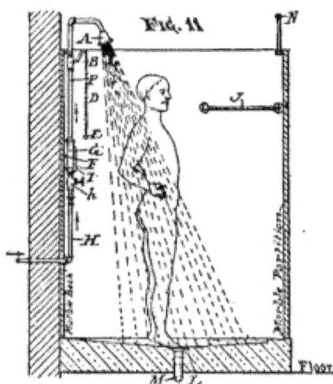

Fig. 11

SECTION Y-Y-Y-Y

the bottom of lever N to the right, carrying with it the attached stem V of the valve P, which is thus opened to a greater or less degree and admits to the coil B an amount of steam from the pipe L proportioned to the coldness of the water. As the temperature of the water rises, rod D elongates and allows the arm N to be forced back, and the valve P is closed by the action of the counterweight W attached to the long arm S of the bent lever, of which the short arm R is opposed to N. This lever is pivoted at T to a supporting bracket Q. V is a circulation pipe. O is a set screw to adjust the rod D to operate within the temperature limits desired, and it is ordinarily set to allow a maximum temperature of 120° Fahr. The valve acts positively with a difference of about 10° Fahr., and as it automatically proportions the steam used to the water drawn, it is believed to be economical, besides preventing the possibility of scalding at the baths by too hot water.

The plumbing of the work described above was executed by S. & A. Clark, of New York City.

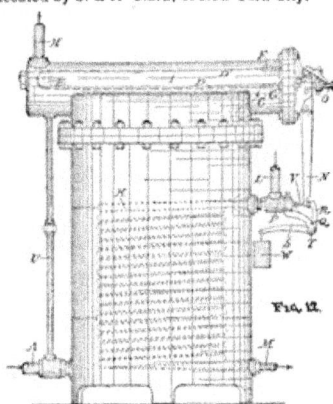

Fig. 12.

SPECIAL BATHS IN ST. VINCENT'S HOSPITAL.

(PUBLISHED IN 1892.)

In St. Vincent's Hospital, Seventh Avenue and Twelfth Street, New York City, a suite of rooms has been arranged and fitted for ordinary and special baths, as shown in plan in Fig. 1, where A A are reclining and massage rooms with chairs, sofas, tables and other furniture. B B are steam rooms, C is a needle bath, D a Turkish bathroom, E an electro-chemical bathroom, and F is a special medicinal vapor bathroom, where the patient is seated inside a marble cabinet G, within which his body is surrounded by the medicated vapor. The opening L fits about the patient's neck, so that his head is in the outer atmosphere. Figure 2 is a general perspective view

of the cabinet G as closed, and Fig. 3 is the same open. The walls, panels, bottom and adjacent wainscot are of Italian marble, and all the exposed metal-work is silver-plated. A channel H H H is cut in the bottom slab, and may be filled with water from pipe I, commanded by valve J. Any required substance may be dissolved in this water, and when the patient is properly seated, with his neck in the opening L, the panels P P P P are closed and steam admitted through the valve K to the pipe H, which is perforated on its under side, produces the vapor. The vapor is chiefly confined by the close lap joints of the marble. M is a stand-pipe overflow and waste plug, and N N are brass counterweights, lead-filled. Figure 4 shows the joints of the panels a, b, c, d, e, and g, Fig. 2. The work was designed and constructed by Byrne & Tucker, New York City.

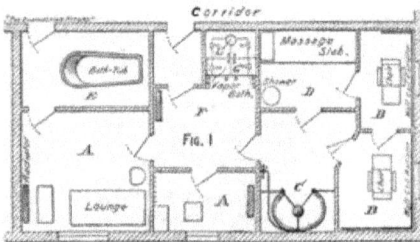

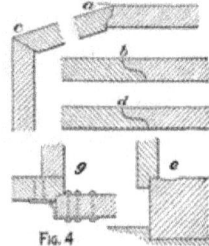

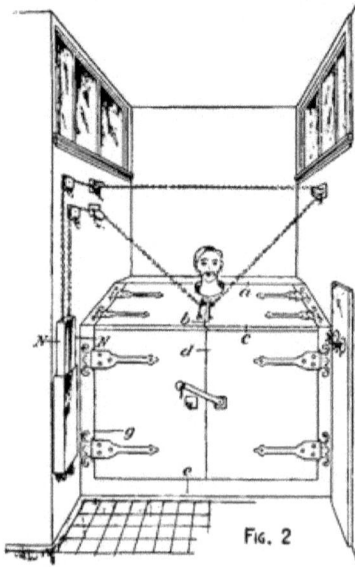

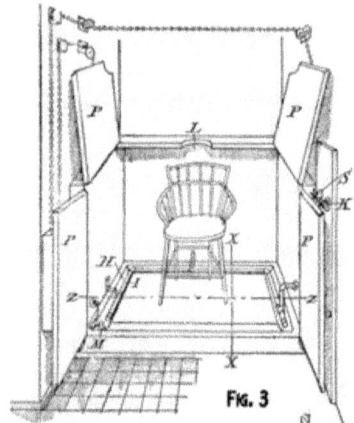

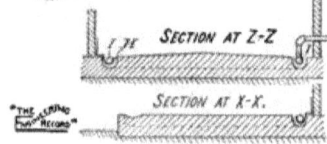

SPECIAL BATHS IN ST. VINCENT'S HOSPITAL, NEW YORK CITY.

MISCELLANEOUS.

GAS PIPING FOR BUILDINGS.

(Reprint of an editorial from The Engineering Record of September 27, 1894.)

There are few features of modern building construction which do not now receive thorough treatment when the design is fortunate enough to fall into the hands of competent architects and engineers, and yet there is one very important portion of the structural outfit, so to speak, of a building, which up to the present time receives no intelligent consideration whatever, except in very rare cases. We refer to the gas piping of buildings of all classes. The gas companies have made practically all possible advances in processes of manufacture and distribution, and while there may be unfairness in some exceptional instances, in the main the gas consumers have largely reaped the benefit of the resulting economies.

Municipal building regulations have generally prescribed fairly wise and reasonable general rules under which buildings and their various structural appointments are to be constructed, but on the question of running gas lines for proper distribution within the buildings they have been essentially silent. Architects also virtually have turned over to gasfitters, as a general statement, the whole question of fitting the buildings which come under their design and supervision. What ought to be everybody's business seems to have been nobody's business, and consequently there probably has never been any portion of the construction and fitting of a building which has exhibited more ignorance and gross blundering than the general run of the gas-pipe plans of many structures now standing, and that is saying a great deal. As is almost or quite the invariable result in such matters the purchaser and the consumer are the principal sufferers. It certainly is not creditable to the architect or engineer thus to fail to properly specify, or generally to specify at all, for so important a part of his work, and no municipal regulation can be considered as complete, either in its form or operation, unless it suitably covers a class of work which so immediately affects the comfort and health of almost every human being and the welfare of every business within its corporate limits. It is the legitimate desire, of course, of the gasfitter to reduce to his client the total cost of his work to a minimum, and he gets his work as the lowest bidder, hence the result is all but a universal decrease of size of pipes in a building far below those which ought to exist for a proper supply at the points of actual consumption. Besides, a lack of proper knowledge of design causes a very general and sometimes an utterly absurd disproportion between the main and running lines and branches, which results, in connection with the fundamental difficulties of small pipes, in excessive complaints from users in many instances, and in costly and inefficient, if usually unnoticed, illumination or heating, and very costly alterations and additions to the piping in the modern fireproof building.

Indeed it cannot be expected that gasfitting will be done carefully and efficiently or with materials and workmanship of excellent quality unless, like other branches of mechanical work, it is done under intelligent specifications, faithfully executed.

It is true that there have been a few very creditable efforts to remedy this state of things, but they may be said almost to be included in the excellent little work by William Paul Gerhard, and perhaps the unduly short printed regulations of one or two gas companies in the country, and they have not produced any apparent improvement in the general situation.

Since the advent of high buildings with fireproof floors and the demand for gas for cooking and heating has arisen, the embarrassment due to the causes cited have been more pronounced. In view, therefore, of the interests involved, The Engineering Record, in pursuance of its policy to elevate and advance all branches of building construction, submits to architects and engineers, as well as municipal authorities, a general system of specifications and rules under which buildings may be fitted with gas pipes so as to produce the greatest excellence in design and the most efficient and economical use of gas. After a very thorough examination of the whole question, and after many conferences with large firms of gasfitters, engineers of gas works, and others directly interested in the attainment of the desired end, the specifications, tables, and rules which we print herewith have been prepared. These regulations have been made essentially to agree with the few best efforts which have already been made for the same purpose; they involve no conditions inconsistent with the best interests of gas consumers, gas producers, or gasfitters, but they have been based upon such reasonable conditions as will secure in all respects the best practice to all those departments of gas interests. The tables showing the sizes required for the prescribed number of burners, logs, heaters, and ranges are based upon a very careful and thorough investigation, both analytical and experimental, in regard to the flow of gas through the pipes of the maximum lengths indicated. The resulting sizes are in some cases a little larger than hitherto prescribed, while in other cases they are not; but in all cases they will insure the free flow of the necessary volume of gas, and thus entirely avoid the annoyances and loss due to too small pipes. The slight increase of cost of piping from this source is too small to be appreciable in the total cost of any building whatever.

Should it be desired by architects or engineers, these general regulations can easily be supplemented by other clauses or paragraphs designed to cover special cases or details which it would not be proper or suitable to recognize in the concise regulations designed to meet the purposes of those which we print. We commend these specifications to the most careful and favorable consideration of architects, gas companies, and the building departments of cities. They have been carefully and rationally designed to fill a gap in building specifications and general regulations which has been the cause of most serious and widespread annoyance and loss.

ESSENTIAL REQUIREMENTS FOR THE GAS-PIPING OF BUILDINGS.

As the result of a special investigation, the accompanying table and recommendations are submitted as the basis for proper specifications for the gas-piping of buildings to meet the demands of modern requirements for lighting, heating, cooking, and manufacturing:

GENERAL REQUIREMENTS.

1. All lines of piping throughout the building, except drops, must be laid with grade so as to drip or drain back into the risers, with no depressions to hold condensation. Drips with drip pipes where needed must be provided at meters and at such other points as the plan of piping may render necessary.

2. No riser must be less than three-fourths inch in diameter in any case, and all risers must be covered up on inside partitions so as to be thoroughly protected from freezing. Wherever risers or other pipes cannot be guarded in this manner, they shall be protected from frost by special and effective coverings.

Table of Maximum Lengths of Pipe and Number of Lights with Corresponding Diameters.

The results in this table are based upon a specific gravity of 0.6 and a pressure of gas of 0.5 inch of water.

Diameter of Pipe, Inches.	Maximum Length, Feet.	Maximum Number, Lights.*
3/8	20	2
1/2	30	3
3/4	40	6
1	60	10
1 1/4	70	15
1 1/2	100	30
2	150	60
2 1/2	200	100
3	300	200
4	400	300
4 1/2	500	400

* One burner delivering 6 cubic feet per hour.

*Gas Logs and Ranges.**

Diameter of Pipe, Inches.	Maximum Length, Feet.	Maximum Number, Lights.
1/2	100	1
3/4	100	2
1	100	4
1 1/4	100	7

* The numbers for gas logs and ranges in the third column of the table refer to sizes for which the consumption in any log or range does not exceed 35 cubic feet per hour.

The sizes of piping for gas logs and ranges are for single lines run from or near the meter, or source of supply, for the specific purpose indicated.

When gas logs and ranges are supplied by branch pipes, or when any branch pipes are run from the main system of the building, the combined sectional areas of all the pipe sections must exceed the sectional area of the main supply pipe sufficiently to maintain the proper flow.

3. Wherever practicable all piping shall be exposed, but piping that must be concealed shall first be thoroughly inspected by the gas company, and the gasfitter shall give due notice when the piping is ready for inspection. Unexposed piping must be so concealed as to be readily accessible in case of examination or repairs. Wherever practicable, as in floors, the concealment shall be made by boards over the pipes, secured by brass or other non-corrodible screws.

4. In cases where extensions are made care must be taken to extend with such sizes that the rules already prescribed shall be maintained.

5. All drop pipes must be left perfectly plumb and well secured in that position.

6. Long runs of piping must be firmly supported at frequent intervals so that no sagging nor depressions can occur in which condensation can collect.

7. If pipes run across wooden beams or joists the requisite cutting, notching, or boring shall never be more than 2 inches in depth nor more than 3 feet from bearings, and as near the latter as possible.

8. Lines of piping shall not be placed under tiled or parquet floors, marble or other stone or metal platforms, or under hearthstones, unless the local conditions render such procedures imperatively necessary.

9. All pipes shall be of the best quality of wrought-iron welded gas pipe, and all fittings, including couplings, elbows, bends, tees, crosses, reducers, etc., under 2 inches diameter, shall be extra heavy malleable fittings; those of larger diameter may be of cast iron. These pipes and fittings may be plain, galvanized, or made non-corrodible by any effective method.

10. Pipes and fittings are to be put together with screw joints and red lead, or red and white lead mixed with joints made perfectly gas-tight.

11. All pipes shall be firmly and safely secured in position with hooks, wrought-iron straps, or hold-fasts, secured with screws at close intervals, so that continued use in proper line and grade may be effectively secured.

12. Meters shall be placed where they will be most conveniently accessible for reading the index and for examination and repairs, and when placed on the walls the minimum height above floors shall be 2 feet for the bottom of the smallest meters and the maximum height shall be 8 feet for the top of the largest meters. The sizes of connections shall be as follows:

3 light ¾-inch diameter.			60 light 2-inch diameter.		
5	"	¾ "	"	100	"	2 "	"
10	"	1¼ "	"	150	"	2½ "	"
20	"	1¼ "	"	200	"	2½ "	"
30	"	1½ "	"	250	"	3 "	"
45	"	1½ "	"	300	"	4 "	"

13. The completed piping shall be tested by some competent authority, who shall give a written certificate of the results before any of it is covered at any point. All outlets shall be tightly capped and the whole system shall be tested preferably with a mercury gauge, or by a low-pressure spring gauge which has been recently and authoritatively tested by a mercury column. When air is pumped into a completed system of pipes until the pressure reaches 12 inches of mercury and stands or remains stationary for five minutes, or if the column of mercury does not fall more than 1 inch per hour, the system may be considered satisfactorily tight. Otherwise leaks must be sought and stopped and the testing repeated until the preceding requirements are satisfied. When extensions to completed systems are made the same tests shall be applied to the extensions before they are put in use. In the case of large buildings the entire system may be tested in suitable sections.

DRAINAGE BLUNDERS.

[BY ALBERT M. WEBSTER, A. M. AM. SOC. C. E.
(PUBLISHED IN 1892 AND 1893.)

CHAPTER I.

THE foundation stone on which the structure of Sanitary Science and Preventive Medicine rests is the germ theory of disease. The assumption is that a specific disease is produced by the presence of a specific germ in the blood, without which the disorder cannot occur. The diseases at present attributed to germ origin are known as zymotic diseases, and it is with these that sanitary science has to deal.

The assumption of the presence of the specific germ in the blood directs attention to the channels through which it may be introduced, and to methods of intercepting and destroying it before it enters the system. Food consumed and air breathed are the general conveyors of the zymotic germs, infection through wounds excepted. Complete sterilization of these conveyors would stamp out zymotic disease; but sanitary science, recognizing the practical impossibility of this effort, directs attention also to the source of

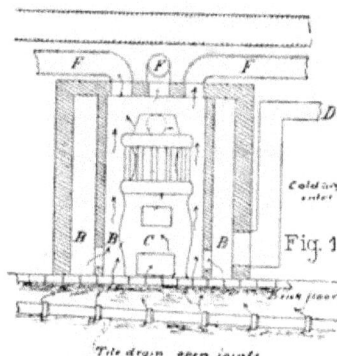

Fig. 1

Cold air inlet

Tile drain open joints

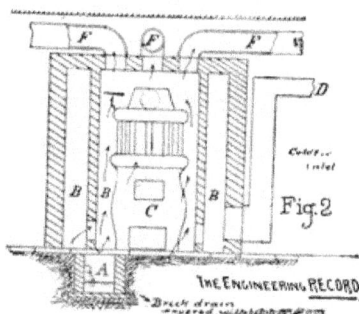

Fig. 2

Cold air inlet

Brick drain

THE ENGINEERING RECORD

the difficulty, the breeding grounds most favorable to the development of the germs; in other words, attempts the sterilization, removal, and destruction of organic filth.

Municipal sanitation directs its efforts to these fields in the aggregate, as effecting the community. House sanitation extends this work into the dwelling. Its canons are the immediate removal of all organic waste, and the insurance of pure air and food within the dwelling.

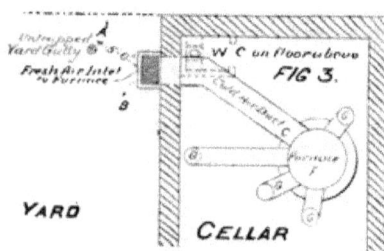

FIG 3.

YARD

CELLAR

The removal of waste is largely effected through the drains by the water-carriage system. The necessary fouling of the drains makes it imperative that the air from them should find no entrance into the household, and it is one of the purposes of modern sanitary plumbing to insure the security of these drainage channels.

The accompanying illustrations of defects in plumbing in this particular have been taken from examples found in New York houses of the better class in fash-

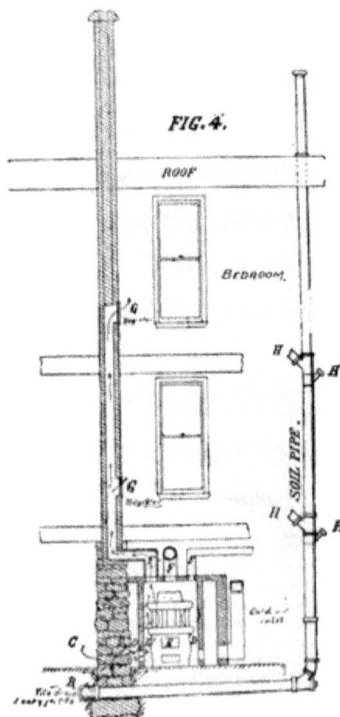

FIG. 4.

lonable parts of the city. In general they are houses in which the visible plumbing has been remodeled. In each case the defective drain was in direct air communication with the furnace, and when the furnace was in operation there was little reason to doubt that the air from the drains was drawn into the furnace and distributed through the hot-air flues to the house. The defects were discovered by the smoke test. A fan blower with furnace and appliance for producing pungent sulphurous smoke is attached to the soil pipe at the roof and set in operation. The smoke travels through the system of pipes and issues from defects or openings which may exist, either concealed or exposed.

In the case shown in Fig. 4, smoke was blown into the main soil pipe at the roof. It issued through the other pipes, extending to the roof, and also through the soil and other pipes at the roof of the adjoining house, and at the fresh air inlet of the adjoining house; arguing the connection of the drains of the two houses. The smoke also appeared at every hot-air register, in the house under test. The owner was confident that there was no connection between the furnace and drains in his house, as the plumbing had been very recently renewed at considerable expense. On opening the furnace air chamber it was found full of the smoke, which was seen to issue through defective joints in the masonry of the party wall, against which the furnace was built. The drains in the neighboring house were uncovered and found to be leaking and defective close to the party wall. Referring to Fig. 4, the course of the smoke had been through the pipe at the roof to the cellar and through to the adjoining house drain B into the furnace air chamber D by way of the defective joints in the party wall C, and from the furnace through the flues F to the rooms in the house G G.

Figure 1 illustrates a somewhat similar example, except that the drains which ran under the furnace were of earthenware and the joints defective, and there was no connection with the neighboring house. As in the first case, the smoke blown into the soil pipe at the roof issued from every register in the house.

Figure 2 illustrates a case similar to Fig. 1, except that the drain which ran under the furnace was an old brick drain with flagstone cover, the joints of which were open, and the drain itself was in foul condition.

Figure 3 is an example of the contamination of the air supply to the house, through an untrapped yard drain A opening close to the cold-air inlet B to the furnace F. When the furnace was in operation air was drawn into the air chamber, through the cold-air box B, and was distributed to the rooms of the house through the flues G. The untrapped yard drain A being so near to the furnace inlet, made it more than likely that air from the drains was at times drawn into the house.

CHAPTER II.

The examples of air contamination shown in this chapter relate in general to defects in the house plumbing proper. The first example shown in Fig.

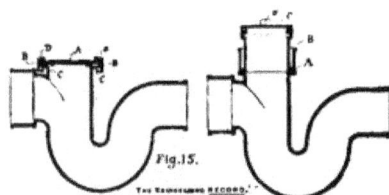

Fig.15.

THE ENGINEERING RECORD.

FIG. 8.

FIG. 9.

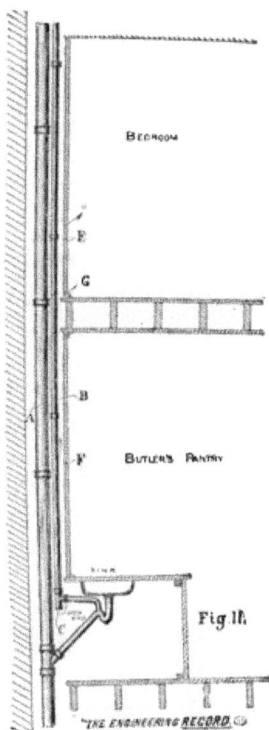

FIG. 5.

Fig 11.

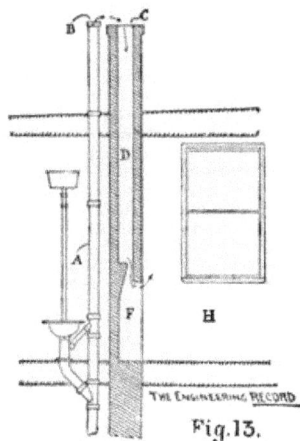

Fig.13.

FIG. 6.

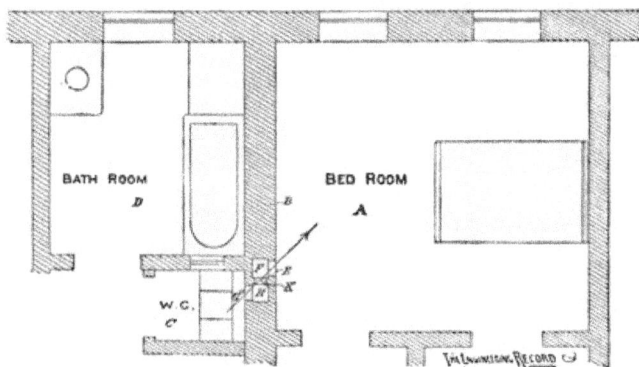

FIG. 7.

DRAINAGE BLUNDERS.

5 was found in a Madison Avenue, New York, dwelling of comparatively recent construction. The main soil pipe A extended from the cellar to the roof, and was provided with an accompanying "back-air" pipe B for trap ventilation. Both pipes passed through bedrooms on the upper floors, and were boxed in with a light wood casing F. A smoke test revealed the presence of a serious leak near the butler's sink, and on stripping the pipe it was found that the lower end C of the "back-air" pipe B was quite open, no attempt having been made to seal it. The plumber who did the work had evidently intended to connect with C the back-air pipe from fixtures on the floor below, but another arrangement had proved more convenient, and the open end C was overlooked and forgotten. An interesting feature connected with the test was the course of the smoke, which traveled from the open end of pipe C behind the wood casing F as through a flue, and appeared three stories above in the children's bedroom, issuing through openings G at the floor. As the children at the time were under treatment for diphtheritic throat affections, there appears reasonable ground to believe that the defective plumbing was the cause of the difficulty.

Figure 6 was an arrangement found in a country house where at certain times one of the rooms H seemed to be contaminated with drain air. The soil pipe A extended above the roof with the open end B close to the top C of the chimney flue D, from the fireplace F of the room H. Under certain conditions, especially where there was an open-grate fire in an adjoining room, with no fire in F, the fire in the adjoining room appeared to establish a downdraft in the flue D, and to carry with it air streaming from the open end B of the soil pipe A. The trouble was corrected by moving the open end of the soil pipe.

Figure 7 is a curious example of the possibilities of the city dwelling. It was discovered in a block of houses on Lexington Avenue, New York. The bedroom A in one of the houses was occupied by an unprotected aristocratic lady of highstrung and somewhat nervous temperament. The party wall B separated her from inoffensive but unknown neighbors in the adjoining house. All went well for a season. Suddenly the inoffensive neighbors gave way to midnight disturbances that threatened nervous prostration to the occupant of bedroom A. She was awakened from her sleep by the clanking of chains, the clang of machinery, and the roar of the rush of mighty waters. At all hours of the night these horrors would awaken her. She contemplated building a massive sound-proof masonry lining against the party wall; and in final desperation called for an examination of the wall to see what could be done. The examination revealed the conditions shown in Fig. 7. The neighbors' bathroom D and closet C were next to the party wall B in which there were two ventilating flues F and H; the former connecting through a register E near the ceiling with the bedroom H, the latter connecting with a register G near the ceiling of the closet C. The brick partition

K between the flues had been omitted or broken out by the mason near the registers; and direct sight and sound and air communication existed between the two rooms near the registers. The flush tank for the closet was close to register G, and might as well have been in the bedroom A so far as its powers of disturbance were concerned. The partition K between the flues was bricked up and the trouble ceased.

It has been found that certain features of house plumbing are so often defective under test that they should be condemned on general inspection without special trial. To this class belongs the cast-iron plate cover cleanouts so frequently used on iron traps. Figure 8 shows this cover. It is a plate A cast with two lugs B B, arranged to grip flange C C, on the under side. Two ears D D are cast on the top face to allow it to be turned. The flange C C is wedge-shape, or screwed on the under side with two vertical slots through which the clips B B pass when the plate is lifted off. A bed of putty is generally placed on the top of the flange C C and the cover pressed into it until the clips B B pass through the vertical slots in the flange. The cover is then turned by means of the ears D D, the clips B B bear against the screw under face of C C, and the plate is drawn to a firm seat in the putty. The security of the cover depends on the putty, and at its best will not stand a very mild pressure test. When the putty dries out, the cleanout is very apt to be found defective under the smoke test.

Figure 9 shows an improved form of cleanout. The trap is cast with a vertical cleanout hub A, into which a brass ferrule B is calked with oakum and lead. The brass ferrule has an inside turned thread, and is provided with a brass screw cover C with a wrench nut D cast on it. This form of cover properly set can be made to stand a severe water test.

CHAPTER III.

The foregoing chapters have been confined to defects contaminating the air supply to the building or the air in the building. The present chapter will

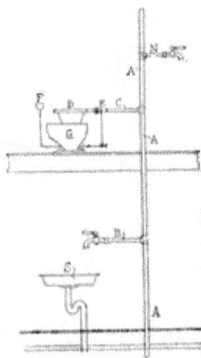

FIG. 10.

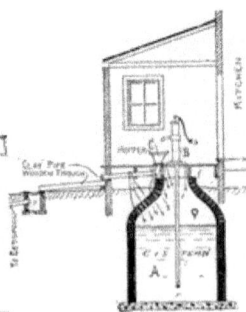

FIG. 12.

deal with the contamination of the supply of house water. Figure 10 represents an arrangement of the water pipes which has been frequently described and condemned in print, but of which many examples at present exist in New York City as the inheritance of ante health-board days. The fact that the same blunder is being repeated daily in towns and villages away from the larger cities is sufficient ground for again calling attention to it. The water supply pipe A A has branches B, C, N, etc. at various levels. B supplies a kitchen or butler's sink, where drinking-water is drawn for the tank. C supplies and connects directly with the bowl D of a pan closet or other form of direct flushing closet. A valve E on the pipe C is connected by levers with the seat handle F of the closet. The size of the water pipe A and the pressure of the water supply is such that when B is open water will not only not flow to the other branches C and N, but will draw the water out of A A and draw air into the branches C or N, if either of the cocks on those branches is open. If the closet G is in use and the handle F is raised at the time B is opened for drawing drinking-water, air is drawn directly from the bowl D of the closet and enters the water pipe A A if it does not mingle directly with the water being drawn through B. If there are germs of disease in the air from the bowl D they contaminate the water which flows through A A. To avoid this possibility, small flush tanks are required for each closet or group of closets, so arranged that this suction of air into the water pipe is prevented.

Figure 11 was an arrangement for providing for the overflow from a water tank in a Madison Avenue dwelling. It was not uncommon some years ago and may still be found in new work outside of the larger cities. As the street pressure was at first insufficient, the tank E was placed on the top floor of the building and was used for supplying water to all fixtures except the kitchen and first floors. A pump in the cellar was used for filling it, and an overflow pipe B was connected with the tank and branched into the main soil pipe A of the house. The overflow B was trapped at C to prevent air from the soil pipe finding access to the tank water. The trap was provided with a back-air pipe D to protect it against syphonage. As long as the tank overflowed at frequent intervals the trap C was supplied with water and remained secure. But improvements in the city water supply were made and increased the street pressure so that water would rise at night to the level of the tank. The pump in the cellar was done away with, and the tank was supplied directly from the street pressure, with a ball cock on the supply pipe, which shut off the flow of water when the tank was full. In connecting this ball cock the plumber had set it so that the water was shut off before the tank had filled to the level of the overflow B; in fact it could not have been otherwise arranged without continual waste of water. As a result the trap C very soon evaporated and the air from the soil pipe had free access to the surface of the water in the tank. The water was used, among other things, for bedroom basins, and a case of diphtheritic sore throat which had occurred in the house was thought to be due to

the use of the water in brushing the teeth. The tank overflow was cut off from communication with the soil pipe and an independent pipe led to the extension roof and provided with a flap valve.

Figure 12 shows the contamination of the water supply which was discovered in an isolated country dwelling. The water was drawn from a cistern A supplied by the rain falling on the roof. Under the pump B, which stood on the back piazza close to the kitchen door, there was a sink or hopper C to catch the drip from the pump. The hopper was connected with some lengths of clay pipe leading to a trough and ultimately into a cesspool. It was found that the hopper, being convenient to the kitchen door, had been used to receive much of the liquid waste from the kitchen and from floor washings. The connection of the hopper and the clay pipe and the pipe joints were found defective, as was also the brickwork of the dome of the cistern. As a result much of the waste water poured into the hopper found its way into the cistern, which in time became so foul that it was very offensive and was abandoned. Before this was arrived at two cases of diphtheria had developed in the house.

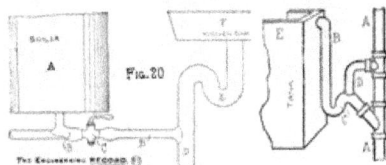

FIG. 13.   FIG. 14.

Figure 13 shows a very common blunder, found in new work as well as in old. The hot-water boiler A in the kitchen is almost always provided with a "sediment pipe" B B, for blowing out the sediment which in time collects in the bottom of the boiler. This pipe is controlled by a stop-cock C, and to provide a means of disposing of the water when the boiler is being blown off, the sediment pipe B is connected with the waste pipe D from the kitchen sink F. This connection is frequently made below the trap E of the sink. The danger lies in the fact that when the house is closed for any length of time it is very common for the plumber to turn off the water from the house and to empty all the pipes, as well as to draw off the water in the boiler through the sediment pipe B, and to leave the stop-cock C open. In this case air from the drains enters the boiler A through B B, and from the boiler connects with all of the water-pipes in the house. When the water is again turned on it receives any contamination which the drain air may have lodged in the pipes or boiler. Two cases of typhoid fever followed the occupancy of a house which had been left in the condition noted for a few months. The sediment pipe should be arranged to have a free discharge into a trapped fixture, or to be emptied through a hose temporarily laid to a yard drain.

PRINCIPLES TO BE OBSERVED IN PLANNING AND SPECIFYING PLUMBING FOR A COUNTRY HOUSE.

[BY ALBERT L. WEBSTER CIVIL AND SANITARY ENGINEER, NEW YORK.]

(PUBLISHED IN 1894.)

DRAIN, SOIL, AND VENT PIPES.

For drains, waste, soil, and vent pipes use "extra heavy" factory-tested cast-iron pipe.* Standard cast-iron pipe, which is very frequently used, is too light to permit of the joints being securely calked; in addition to which the metal in the body of the pipe is too scant, and as the core is frequently eccentric in casting, the pipe is often very thin in spots, and blowholes are frequent. As many as a dozen of such blowholes have been found in a single length of pipe under test.

The iron pipe, which extends at least 5 feet outside of the foundation wall, should be thrust well into the earthenware drain, with which it may connect outside, and a full Portland cement joint be made between the earthenware and iron. The iron drain, where it passes through the foundation wall, should be protected by an iron sleeve, or the wall should be arched over the pipe to prevent breaking the pipe in case the wall settles. The spaces about the pipe may be packed with mineral wool. The running trap, with brass screw-cover cleanout, should be placed just inside of the house wall. The outlet of the trap should be slightly lower than the inlet, and to effect this the trap may be made up of fittings as shown, or a trap of this pattern which is found in the market may be used. Just back of the trap insert a 4-inch fresh-air inlet extending outside of the house and opening above the ground not less than 20 feet from windows, and well removed from the cold-air inlet to the furnace. The fresh-air inlet should be finished with a quarter-bend and wire cage or grating, allowing a free inlet for air. The opening should be 18 inches above the ground, or enough to prevent choking with snow.

Main drain inside of the house should follow along the foundation wall when practicable or be supported on brick piers to prevent settlement. If below the floor it should be laid on a 3-inch bed of concrete in a brick trench with stone covers. If it is necessary to bury the pipe, it should not be covered until after the rough work has been tested with water pressure and the pipe to be buried has been tar-coated. A brass screw-cover cleanout should be left at the upper end, and if the pipe is buried, a pocket should be left in the cellar floor, through which a wire or cleaning rods may be worked into the pipe to remove stoppages. The vertical run of the pipes should be arranged to connect with the horizontal drain with Y branches and one-eighth bends, and should rise true to a plumb well above the roof. Avoid, if possible, offsets in vertical pipes, especially in vent and back-air pipes. If offsets cannot be avoided, they must be arranged to receive the wash from some

*The use of screw-joint wrought-iron pipe will not be discussed at present. The additional expense of wrought-iron work generally excludes its use in inexpensive small buildings.

fixture. This is to prevent the accumulation of rust scales which drop from the inside of the pipe and accumulate at bends, and in time cause stoppages. The base of all back-air columns must drop directly into a soil or waste pipe to avoid rust accumulation. See on the drawing this arrangement at the base of the main back-air and kitchen-sink back-air pipes.

The base of all vertical columns should be firmly supported on masonry piers to take the entire weight of the column. Pipe hooks should be used to keep the columns in vertical line only, and should not support the pipe, as any settlement in the lower part of the column would pull the joints apart if the upper part of the column is secured. The weight of the soil and back-air columns shown in the drawing is about 800 pounds. Where the vertical pipes pass through the roof a water-tight, flexible roof flashing should be calked into the hub above the roof, turned over and flashed under the roof covering. This admits movement of the pipes due to changes of temperature without breaking the roof joints.

Y branches should be left in the drain and soil pipe to receive drainage from all fixtures. The house should be planned and the fixtures arranged to connect them as close as possible about the vertical column of soil pipe.

The branches for all fixtures should be extended in iron as close to fixtures as possible, and kept as much as possible above floors, and open to inspection.

TRAPPING.

All fixtures should be trapped with bend traps, and all traps should be back aired from the crown of the trap. If brass traps are used, they should have ground brass couplings only. Traps with rubber or leather washers should not be used; they cannot be made permanently secure. Earthenware closets and other fixtures having the trap formed in the earthenware should have short lead bends connecting with the iron pipe. This admits of settlement in the floors without breaking the earthenware.

Connect lead with iron pipes by a wiped solder joint and brass ferrule calked into the hub of iron pipe.

All back-air branches from traps should be connected into T or reversed Y branches on the back-air column; these branches be placed above the fixtures from the trap of which the back-air branch is taken. This prevents the fixture discharging through the back-air pipe if the waste pipe becomes choked.

All lead or metal traps should have a brass screw cleanout below the seal of the trap.

After all the iron-work is completed, and traps and back-air connections and bends have been connected, all of the traps and bends and other openings must be securely sealed and the main drain and fresh-air inlet plugged outside of the house. The entire system of pipes and branches should then be filled with water to the level of the top of the pipes above the roof. The entire system of pipes should then be carefully inspected while under this water test, and any leaks discovered in the joints must be made tight by calking the lead home. Any split hubs and pipes or fittings with sandholes must be broken out and replaced by sound material, and the test repeated until the entire work stands the test without showing leaks of any description. The fixtures may then be permanently set and connected. On final completion of the work, and after all traps have been filled with water, a smoke test or peppermint test should be applied to the drains to test the connections not subjected to the water test. This applies

All pipes should be run as direct as possible and the fixtures planned to lie close to the main pipes and avoid long branch connections.

All pipes and fixtures should be left as much exposed as possible and with the least amount of joiner-work and boxing, and with the least amount of absorbent material about them. Fixtures should be located so as to avoid running water-supply pipes on outside walls where they are exposed to frost.

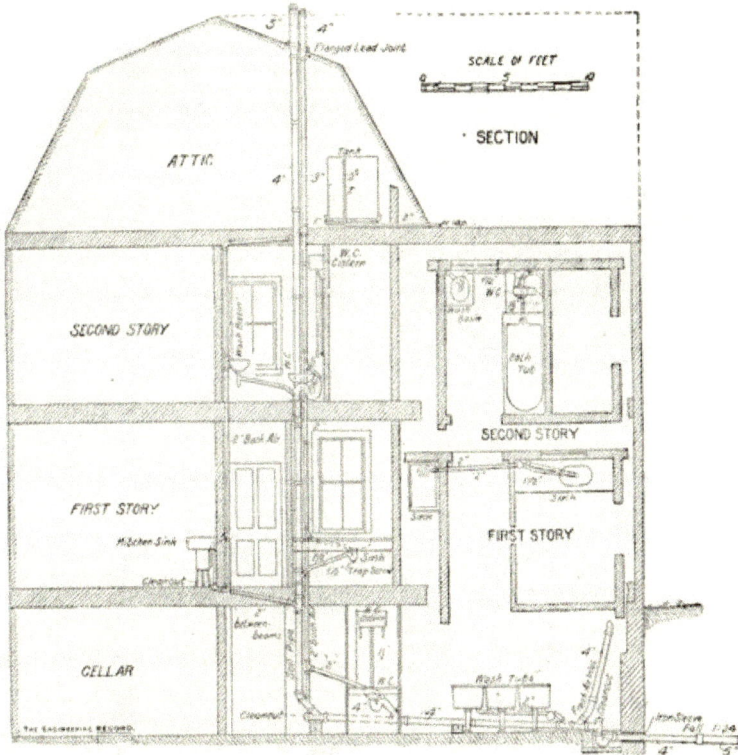

PRINCIPLES TO BE OBSERVED IN PLANNING AND SPECIFYING PLUMBING FOR A COUNTRY HOUSE.

especially to connections with earthenware fixtures and brass trap connections.

The joints in the iron pipe and fittings should be made by firmly calking a roll of oakum into the joint between the hub and spigot, care being taken to prevent the oakum entering the bore of the pipe. The joint must then be run full with pure soft lead, which must be firmly calked home when cold. The joint must be level with the face of the hub when finished.

RAIN LEADERS.

If the rain leader from the roof is carried inside of the house it should be of the same material as the soil pipe and should have an iron trap and brass cleanout on the house side close to the drain.

If outside leaders are used they should connect with an accessible iron trap inside of the house wall. See that leader lines are secure against settlement and leakage, as damp foundations and cellars frequently result from defective underground leaders.

CLEANOUTS.

Cleanouts for all large traps should be on the house side of the trap This gives added protection to the trap in case the covers are not made secure at any time.

YARD, AREA, AND COURT DRAINS.

All yard, area, and court drains may connect with the rain leader branch inside of the leader trap. When pipes pass through floors the opening should be sealed to prevent air and odors from the cellar and lower floors passing to the floors above.

SAFE WASTES.

Safe wastes, if provided, should discharge over a sink in the cellar or upon the cellar floor. In no case should they connect directly with the drains.

EARTHENWARE DRAINS.

Earthenware drains outside of the house should be laid with a fall of not less than 1 foot in 60 feet. If the line is a long one sealed T's or cleanouts should be left, to permit the removal of obstructions without breaking the pipe. Earthenware pipe should be laid true to line and grade and be well bedded to prevent settling. Pockets should be cut out in the trench for the hubs so that the pipe shall bear along its entire length and not rest on the hubs only. Pipes should be centered and the joints made full with Rosendale or Portland cement mortar, each joint to be swabbed out to remove cement from the inside of the pipe. In a wet trench during construction keep trench water down by outside draining or pumping and avoid draining through the pipe laid, as mud will deposit in the pipe and cannot be easily removed.

Seal the open end of the drain when the work is left standing during construction.

See that stones are not thrown against the pipe or come within 18 inches of it in backfilling the trench. If the trench is through rock, cut deep enough to allow 4 inches below the hubs and fill in with a bed of earth or sand for the pipe to lie on. If the outside house drain is a long one, place one or more ventilation pipes on the line, opening 18 inches or more above the ground and well removed from the house or points where the escape of foul air from the drain would be objectionable.

If the trench is through soft ground or quicksand, prepare a bed of broken stones or tarred planks to prevent settlement of the pipes and pack well under the pipes in backfilling.

GROUND-WATER DRAIN TRAP.

If the cellar is wet and foundation or subsoil drains are needed, the discharge from them must not be direct to the house drains, but must be into a tight masonry or iron box below the cellar floor, with a cover in the floor, and with a securely trapped discharge connecting with the main drain. The box must be arranged to receive an infallible supply of water from a draw cock or fixture flush pipe, or must be provided with ball cock connected to the water supply. If direct connection is made with the house drain there is danger of stoppage in the house drain, causing the back-floodage of sewage into the subsoil

and foundation drains, and consequent saturation of the earth with foul water.*

WATER-CLOSETS.

Water-closets should be of earthenware with brass floor flanges and bolted connections with white lead and putty or other secure joints. Avoid rubber or other perishable washers. Brass back-air connections should be cemented in earthenware of closet and located so as to avoid receiving the splash from the closet into the back-air opening. When the closet is flushed the back-air coupling should be of ground brass. Avoid washer connections. Each closet should have an independent flush tank supplied by ball cock from the water supply pipe. No closet should have a direct flush from the water pipes. Avoid closets with valves or concealed fouling space; and leave the closet, as well as other fixtures, as much exposed as possible, and free from unnecessary joiner-work.

BATHTUBS.

Bathtubs should be exposed, with accessible trap and with standing overflow, or overflow which can be readily cleaned. Baths may be of wood or indurated fiber lined with tinned and planished copper, steel-shell lined with tinned and planished copper, enameled iron or porcelain, the selection depending upon the available money to spend and the preference of the user. Avoid all unnecessary woodwork.

BASINS.

Basins should have an exposed overflow or standing overflow, which can be readily cleaned. Traps should be as close to outlet as possible.

SINKS.

Kitchen and butlers' sinks receive large amounts of grease, and the wastes from them are liable to stoppage. Provision should be made for the ready cleaning of these wastes. The sinks should have the least possible amount of absorbent material, and, if possible, should be set free from walls to admit of cleaning all around. Iron, copper stone, and earthenware are the materials employed. Avoid concealed overflows. Place the trap close to the sink.

GREASE TRAPS.

Where much dishwashing is done, or where the discharge from the sink has some distance to go to reach the sewer or cesspool, one of the approved forms of grease trap should be used to intercept the grease. These can only be relied upon to remove part of the grease, and they need careful attention on the part of the servants to render them of service.

SLOPSINKS.

Slopsinks should have flushing rims with a flush tank supplied by a ball-cock. They should be free all round and be set with marble or slate floor and wall slabs, or metal splashboards.

LAUNDRY TUBS.

Laundry tubs should be of porcelain, earthenware, soapstone, slate, cement, or other non-absorbent

---

* Good suggestions for ground-water drain trap and connection with sewer is shown in "House Drainage and Plumbing Problems," pp. 45 and 49, and reprinted from THE ENGINEERING RECORD, prior to 1887 The Sanitary Engineer.

material. Avoid wooden tubs and wooden or other covers. All fixtures should be placed so as to receive direct light and outside ventilation. Avoid set fixtures in bedrooms and living-rooms or unventilated and ill-lighted closets adjoining them, and do away with fixtures not needed for constant or frequent use. The water seal evaporates from the traps of fixtures not frequently used, and leaves direct outlets for drain air.

WATER SUPPLY.

The main water supply to the building should be laid below the frost line and should have a stop valve inside of the wall where the pipe enters the house. Care should be taken to see that there are no leaks in the buried pipe, as the cause of wet cellar walls has sometimes been traced to defects in the buried supply pipe. The pipes should be arranged to drain completely when water is shut off. Lead or galvanized wrought-iron pipe is preferred for general use, dependent on the character of the water used. For country houses used only a portion of the year, from which the water would be turned off, lead should be used, as iron pipes rust quickly when empty of water.

FILTERS.

Filters may be placed on the supply main in the cellar if desired. They will remove sediment and screen the water. They do not remove salts in solution or bacteria. If employed, filters must be arranged to allow of ready cleaning and washing. The waste discharge must not connect directly with the house drains. Direct supply branches should be laid to the kitchen and butlers' sinks and fixtures where drinking-water will be drawn.

If the water pressure is so strong as to cause wear of the faucets, or so light as to fail on upper floors, a tinned copper-lined or iron house tank should be placed in the attic and supplied through a ball cock or filled by a pump. The tank should have a large overflow, safe-waste, and emptying pipe, discharging upon a roof or over a fixture with a large outlet. The overflow must not connect with the drain pipes. Provide a flap valve on the end of overflow pipe. The tank should be cleaned frequently. Provide a dust-proof cover with ventilation opening protected by wire gauze and a cloth screen.

DIRECT OR MAIN TANK SUPPLY.

The direct or main tank supply should have branches to all fixtures and to hot-water boiler. The boiler may be of galvanized wrought iron or tinned copper. The boiler should have a valved sediment pipe for emptying, and this should be opened at intervals to remove the sediment deposited in the boiler. The sediment pipe must not connect directly with the house drains.

All branch lines must drain completely. The hot-water riser, if the water service is from a tank, must have an expansion pipe extend over the top of the tank and raised well above the tank.

Branches should have a stop-cock, and wastes and runs of pipes should have brass unions to admit of ready repair.

PARIS BATHCARTS.

(PUBLISHED IN 1891.)

AN American familiar with the fact that every house or apartment, renting as low as $300 per year in the United States, has its own bathtub with hot and cold water supply and waste to remove the contents of the tub, is amused if not amazed when, on a visit to Paris, he gets an idea of the custom still prevailing in that metropolis of luxury and elegant buildings.

The large hotels, some very costly private mansions and apartments, and the public bathhouses have their bathrooms as is the custom in the United States, though the French bathroom usually is much larger and is elegantly furnished with rugs, lounges, and dressing tables, etc., the idea being that if one takes a bath one must lie down and take a nap after it.

People living in apartments costing as high as a thousand dollars a year, and in the new quarter of Paris in the neighborhood of the Champs Elysées, when they wish to bathe, other than take a sponge bath in a small, portable tub, either go to the public bathing establishments or send thence to have a bath brought to their apartments. Sunday morning one sees a strange-looking, two-wheeled cart, like a very high dog-cart, on which there is a framework built over the wheels. This framework can hold three bathtubs. They are made entirely of copper and are about 5 feet long, about 20 inches deep at the end, and 18 inches on the side. The driver of this vehicle is perched up high on a small seat in front, is bareheaded and wears a blouse. On each side of him an iron ring encircles a copper-covered vessel holding about three gallons of hot water, which rests on a little shelf. He also carries a supply of dry towels and sheets. The bathing establishments have these carts, and when a patron sends word that he wants a hot bath at a certain hour, the bath is put on the cart, the kettle filled with hot water, and the cart with its strange load is rapidly driven to the building in which the apartment is. The driver carries the bathtub, as an Adirondack guide carries a canoe, on his head and shoulders from the first to the fifth floor, as the case may be, and after spreading a sheet to protect the carpet he spreads also a clean sheet inside of the tub, so that the bather does not touch the metal. Then he carries up the kettle of hot water which he has brought from the main establishment. The necessary cold water he gets on the premises, either on the same floor with the apartment or in the court yard. When the bather has had his bath the attendant removes the soiled water by dipping it out, wipes out the tub and carries it with his kettles and soiled towels downstairs to his cart. The charge for all this is about 60 cents, with the usual additional tip to the man.

BATHTUBS AND BATHCARTS IN PARIS.

(PUBLISHED IN 1891.)

H. D. Wood, C. E., of Boston, who resided at Paris a number of years, in referring to the foregoing article on Paris bathcarts, sends further information as follows:

"There were three good ordinary bathhouses within a quarter of a mile of where I lived, each having from 30 to 50 bathrooms. These averaged about 6x10 feet, and the furniture consisted of two chairs, a looking-glass, two shelves, with a comb and brush, a shoe horn, a pair of slippers, a decanter of water, and a glass, a bootjack, coathooks, and a square of cork 14x18 inches to step out on. The charge for a plain hot bath was 16 cents, with a discount if buying six tickets at once. The attendant furnished hot towels at the rate of 2 cents apiece, when rung for. Soap was extra, as in all continental Europe. A printed schedule of extras was hung in each room, including soaps, bran baths, almond-cream baths, etc., hair wash, services of barber or pedicure, also a list of cordials, lunches (hot steaks, chops, etc.), all furnished by the attendant. Besides the plain bath, there was the complete bath, in a larger room, with a sofa, lounge, a sheet spread inside of the tub, so that the body does not touch the metal, and a towel over the cork mat. Price, about 20 cents. Also one or two rooms for shower, hose, etc., baths; and six or eight rooms with iron or porcelain-lined tubs, for medicated baths. These establishments furnish Seine River water. In the cheaper districts, the bathhouses are supplied with canal water, which is not of so good a quality, nor as clean

"The public washhouses, sort of charitable establishments, where, for a few cents (nominal price) a woman may have the use of washtubs and hot and cold water for doing her laundry, and also the use of a drying closet, usually have a few bathrooms attached. These are more especially for the use of the poorer class (here a tramp may get a bath, and have his clothes washed and dried, while he waits). The object of the washhouse is to help those living in one room, as the police regulations forbid any laundry work done where a person has but one room to live in.

"The shape and form of the Paris bathtub is such that when sitting back at the head of the tub one has water up to the shoulders. That is the way the bath is prepared for the customer. If sufficient water were used here to get the same depth I fear the boiler in the $300 house you speak of would give out often, and even in the larger house, but I may be prejudiced.

"One appliance used in the bathhouses is a portable brass tube about the size of an ordinary standing overflow plug, which the attendant hooks on to the hot-water faucet when he fills the bath. The cold water being turned in at the same time, the hot water is delivered under the body of warm water mixes quicker, and no steam is thrown out in the room.

"The Paris bathtub is somewhat larger than those in common use here; they are generally of planished

copper, thick enough to stand common use without any wooden frame or boxing. The portable ones are on castors, and the stationary ones are set upon blocks an inch or two from the floor, and the outlet sets over a floor catch basin connected with the drain pipe, so that a tub can at any time be removed for tinning or repairs without affecting the plumbing; all the water pipes and faucets being made fast to the wall and located at the middle of the side of the tub. There are several shapes and styles, weighing from 70 to 88 pounds, holding from 75 to 120 gallons. In estimating for water-works, or for bathhouse tanks, a bath is called 60 gallons.

"In the more thickly built-up sections, where bathhouses are numerous, the carts for portable baths are drawn by the attendant, and carry two bathtubs. A horse is used in the less populous districts. In either case the cart consists of a barrel set on a two-wheeled frame, on top of which are placed the tubs. On a shelf in front are two copper pails with large cork floats for carrying the water upstairs, the pails being hung on the ends of a stick slung across the shoulder of the attendant. Usually on the cart each pail contains a closed doubled-lined copper cylinder in which the towels are carried and kept warm by a hot-water lining. The barrel is filled with hot water at the bathhouse tank before starting, and this water is carried up to the apartment with the bath, and cold water taken either at the kitchen sink or taken up from the faucet in the yard. When all the baths on the team have been delivered the attendant, before starting for home, lets off the remainder of hot water from the barrel into the gutter, and frequently small shopkeepers help themselves to a free supply of hot water. Thus the attendant lightens his return load, as it would be of no use taking the water back to the establishment."

ADJUSTABLE CONTROL OF CHURCH GAS-LIGHTS.

(PUBLISHED IN 1890.)

In many churches, especially cathedrals and city churches, it is desirable to quickly and definitely vary the intensity of the gaslight in different sections of the building at different periods of the service. This regulation should be done easily and certainly from one place, and should avoid danger of entirely extinguishing the lights, or any lack of uniformity. An arrangement devised to accomplish this has just been applied to the gas system in the Church of the Holy Trinity, Forty-second Street, New York City, and operates as shown in the accompanying illustrations.

The 4-inch gas main A (Fig. 1) terminates on the floor of the auditorium in a corner near the rear, and into it are tapped the 1½-inch branches B B, etc., which each supply all the group of lights in a particular part of the edifice, as the nave, the transept, the chancel, the right gallery, left gallery, etc. Each of these branches is controlled by an ordinary gate valve C, operated by a hand lever D, so that the sexton may from this point graduate any set of lights at will. Figure 2 is a side view of a valve C showing an attachment E intended to fix the opening of the gate at one or more exact positions, thus producing several unvarying intensities of illumination, or to allow it to be entirely closed when necessary. A guide bar F is screwed rigidly to the valve and has one or more rounded seats G, which engage with a spur K, attached to the adjustable bar H, which is fixed at I to any part of the valve lever and slides over bar F. When handle D is depressed from D¹ (the full open position), spur K slides easily along down on F until, in the position shown, it rides over into the seat G, which is curved so as to offer some

resistance to slipping. The lever is thus held at D as shown, unless the force is noticeably increased to push it down, when K leaves its seat G with a distinct click, and may be pushed successively into other similar seats (not shown here), or the handle D may be completely depressed and the lights shut off altogether. M is a steel spring maintaining spur K in contact with guide F. L is a set screw for adjusting the stops to any position of the valve lever, and N is a set screw for the direct adjustment of the gate stem. This arrangement has been in operation for several months, and is considered convenient and effectual. It was devised and made by Paul S. Bolger, New York, contractor for plumbing and gas work in the church.

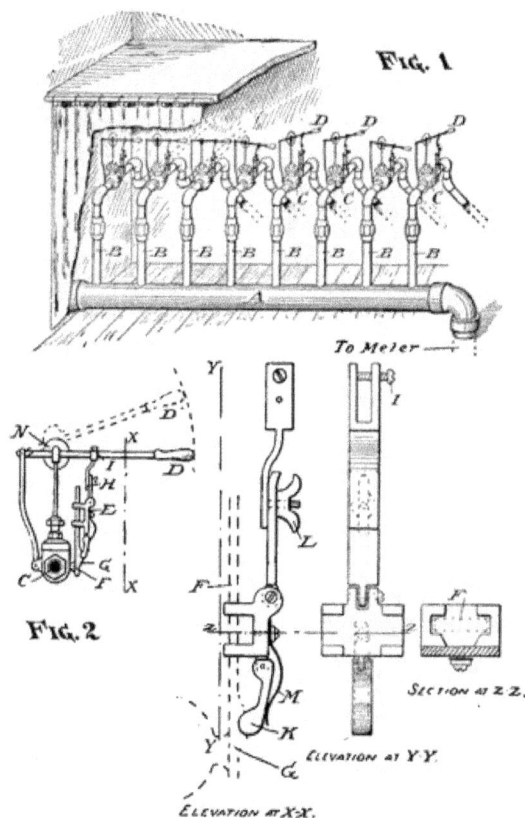

FIG. 1

To Meter

FIG. 2

SECTION at Z Z.

ELEVATION at Y Y.

ELEVATION at X-X.

ADJUSTABLE CONTROL OF CHURCH LIGHTS.

AN AUTOMATIC DUPLEX-CONNECTED
HOUSE PUMP.

(PUBLISHED IN 1894.)

In the residence of J. Pierpont Morgan, Esq., at Fifth Avenue and Thirty-ninth Street, New York City, the extensive plumbing system has been installed for years and includes pump connections for the water supply to the lower floors, roof tank, and elevator service. The general features of the original pump arrangement are conventionally indicated in

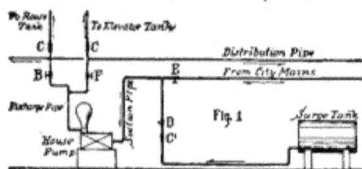

Fig. 1

the diagram, Fig. 1, which shows a hot-air pump connected up to be supplied either from the city mains direct or from the surge tank that receives the water discharged from the elevator cylinders, and to deliver correspondingly either to the house supply tank or the elevator pressure tank, both on the roof. The operation of this system, like all other double systems, requires accuracy and attention for the valves, which must always be correctly manipulated and individually changed whenever the pump service is varied. To pump into the house tank, valves B and E were opened and valves D and F were closed. To pump into the elevator tank the oily water that had already been used there, valves E and B were closed and F and D opened. Valves D and F were to be opened only when the elevator tank was to be filled.

As the control of these four valves developed upon servants who were likely to be careless or ignorant and negligent, and might often be replaced by newcomers, there was always danger of improper manipulation. It was found that the surge tank was sometimes overflowed by city water, and that the dirty water from the surge tank was at other times distributed throughout the house. Check valves C C, opening upwards, were accordingly put on the tank pipes, but of course had no effect except to relieve the pump valves. Considerable difficulty continued to be felt, and Spellman & Blair, of New York City, were employed to remedy it. They first put on a check valve C', opening up so as to prevent water from the city pipes overflowing the surge tank if valve D should be left open. But this did not prevent the possibility of pumping from the surge tank into the house system. One day when both tanks had been filled and the pump was under full headway the servant closed all its valves and it immediately burst, making a complete wreck. To prevent the possibility of a repetition of this accident and to avoid confusion with the two systems of suction and delivery it was advised to use two separate independent pumps. But Mr. Spellman objected to this and devised an arrangement for automatically operating both services with one pump without danger

of mixing the water or possibility of impeding the discharge.

An Otis electric pump No. 3 was put in and fitted up as shown in Fig. 2, with its suction and discharge pipes connected to a special valve A, which consists substantially of two three-way cocks operated simultaneously by one handle H. These cocks are separate and independent except that they are so connected that they must both be operated together by the connecting spindle S, which makes the house tank and city supply pipes register with the pump delivery and suction ports, and closes the surge tank and elevator pressure tank pipes, or *vice versa.*

Figure 3 shows the details of the valve, which consists essentially of twin chambers C C, connected by a body B and operated by a spindle S, which is commanded by a lever H. The valve is entirely of brass, with ground joints at G, capstan-headed stuffing boxes D D, asbestos packing E, and holding rings Q. The spindle S is jointed in the middle, where it is riveted to the lever head, and it has at each end a wing W, over which the ground valve V slips freely and is snugly seated by the pressure of two spiral springs A A. The pump suction and the delivery pipes are screwed on at I and J, and their ports are

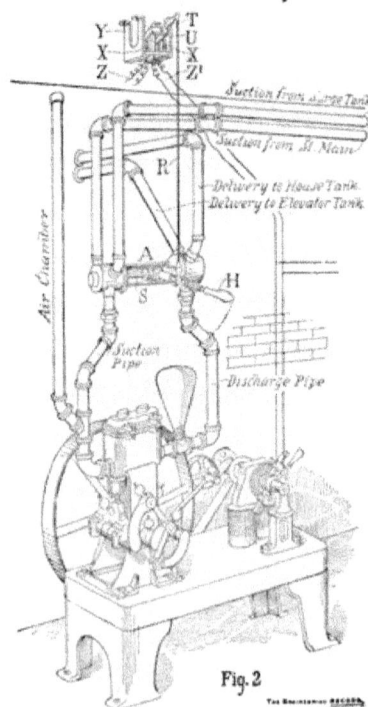

Fig. 2

always open into the chambers, which have discharge and delivery pipes screwed an at K K K K with their ports P P' commanded by valves V. The length ,M of the face of the valve V is made exactly equal to the distance N between ports P and P', so that, in changing from one system to the other, when the valve V is revolved to position V' port P begins to open as soon and as fast as port P' closes and there is always a full area of discharge ports open under all possible circumstances. When lever H is horizontal, as shown in Fig. 2, the pump is connected to city main and house tank. When it is pulled down to H' the pump is connected to the surge tank and the elevator tank. As the house tank requires filling two or three times a day and the elevator tank only about once a week, the valve is left set open to house connections, and a card affixed directing the handle to be depressed to fill the elevator tank, and then raised and left in its former position; and the pump can only be stopped by replacing the lever in its horizontal position.

The electrical connections are so arranged that when the water in the house falls to a certain depth its float operates a switch and turns the current from the street main into the motor circuit and starts the pump, breaking the circuit and stopping the pump when the float rises to the top of the tank, thus making the house service automatic. When the handle H is depressed to connect the pump to the elevator pipes, its attached rod R pulls down the lever T and makes a contact between the copper bars U U and the posts X X, which completes the circuit between the main line wire Z and the pump wires Z', and thus starts and stops the pump by opening and closing the elevator valves. Y is merely a brass frame by which the elevator switch is supported from the ceiling.

Figure 4 shows the details of the switch open, the valve making connections for the house tank. Reversing the valve so as to make elevator connections makes rod R pull handle T down to T' and closes the circuit between the main wires Z and the pump wires Z' through copper bars U U and posts X X. The operation of this valve and pump system has proved convenient and satisfactory. The designers believe it to be complete and reliable and well adapted to conditions where two different services are required from the same pump, and that it is the first instance where a house pump has been connected to two suctions and two discharges controlled by a single valve. The house tank was originally arranged to be filled as fast as the water fell 8 inches from the high-water tank, but by suspending its float weight by a spiral spring the motion of the exterior lever is reduced one-half and the pump only operates for a fall of 16 inches in level. As first arranged the elevator-tank float was made to operate at low water an electromagnet that exerted sufficient pull on the valve handle H to pull it down and open the elevator ports. When the tank was full the handle was released, allowing a counterweight to return it to its original position. As this arrangement made the entire system automatic, it was feared that no attention would ever be paid to the machinery, so it was

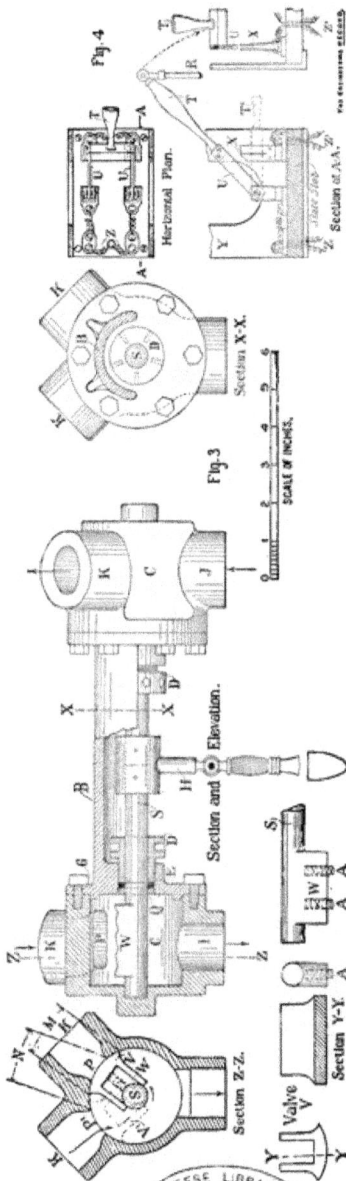

Fig. 4

Horizontal Plan.

Section at A-A.

Section X-X.

Fig. 3

SCALE OF INCHES.

Section and Elevation.

Section Y-Y.

Section Z-Z.

Valve V

AN AUTOMATIC DUPLEX-CONNECTED HOUSE PUMP.

made to require personal attention about once a week when the low-water alarm sounded for the elevator tank, and the attendant is thus reminded to clean and oil the pump.

PLUMBING OF A BARBER'S SHOP IN BOSTON.
(PUBLISHED IN 1887.)

THE accompanying illustration shows the plumbing fixtures of the barber shop in the Quincy House.

Boston. This room is very elaborate in its appointments. The floor is of black and white marble. The walls for 3 feet 6 inches high are covered with paneled white marble, finely polished; above the marble the walls are covered with mirrors. The ceiling is covered by mirrors and stained and ornamental glass set in lead frames. The chandeliers are of special design, with a one-light fixture, which hangs in front of each chair. About 3 feet from the floor in front of each chair a marble slab 16 inches wide is contin-

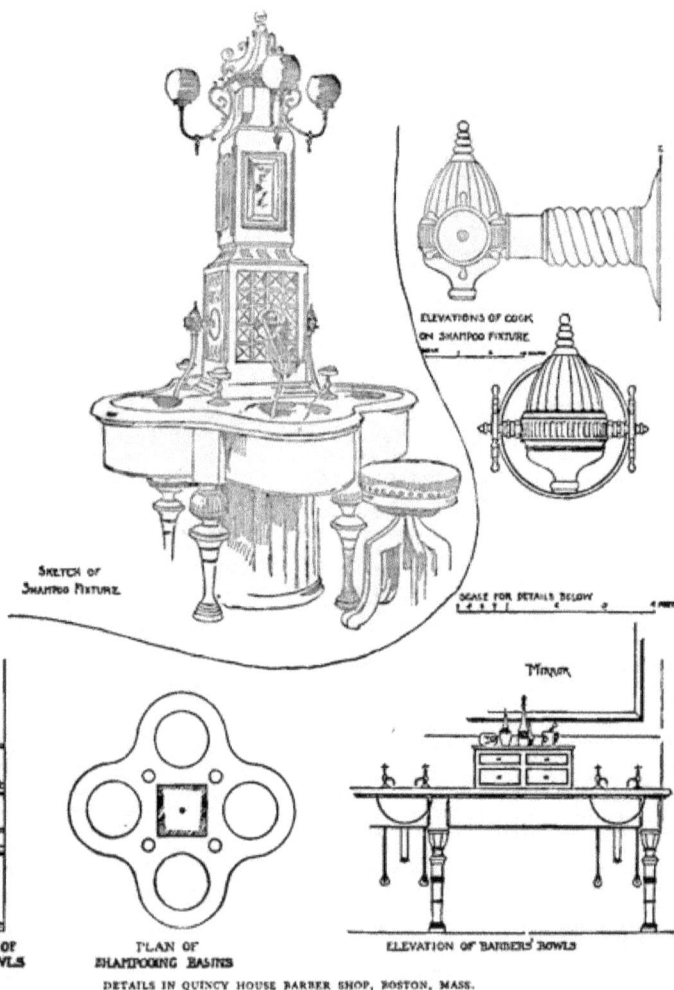

SKETCH OF SHAMPOO FIXTURE.

ELEVATIONS OF COCK ON SHAMPOO FIXTURE

SCALE FOR DETAILS BELOW

MIRROR

ELEVATION OF BARBERS' BOWLS

PLAN OF SHAMPOOING BASINS

ELEVATION OF BARBERS' BOWLS

DETAILS IN QUINCY HOUSE BARBER SHOP, BOSTON, MASS.

ued quite round the room; it rests on turned and fluted polished marble legs. In this slab a small individual basin is set a little to the right of each chair supplied by hot and cold water. The supplies and traps are of polished brass. These pipes and traps are in full view, and in a few moments can be uncoupled in case of accident. The pipes have brass plates where they pass through the marble. Back-air vents are on the other side of the wall.

In the middle of the room are four large basins set in one white marble slab which rests on four legs with brass supply and traps, the same as the individual bowls. These basins are for shampooing purposes only. The supply cocks for this fixture are from a special design. It is claimed that the hot and cold water can be readily regulated to the desired temperature without causing the customer the usual annoyance of changes from hot to cold unless great care is observed by the workman. The difficulty of ventilating the traps of the shampooing fixture was overcome by using a 2-inch wrought-iron pipe which passes from the traps through the center of the fixture and through the ceiling. This pipe has inside of it a small gas pipe which lights the fixtures by four brackets.

The architect was Mr. Samuel J. F. Thayer, and the plumbers were Messrs. Tucker & Titus, all of Boston.

AN AUTOMATIC GAS ENGINE CUT-OFF.

In a system of ventilation recently established in New York City, the fresh-air blast is produced by a Root blower, which is driven by an Otto gas engine. As there is a possibility of the flame at the engine being extinguished, the gas might escape and prove dangerous, so the arrangement herewith described was devised to insure the prompt shutting off of the gas the moment the engine stops, as it must immediately do if the jet is extinguished.

Figure 1 is a diagram of arrangement showing the operation of the cut-off without regard to scale or precise details. In it A is the gas engine, driving the blower C from the countershaft B. . D is the blast main, F is a branch hose to the collapsible bag G, H is the cut-off valve, I is the gas meter, K is the supply to the engine, and J J are equalizing bags. So long as pressure is maintained in the main D, it will be the same in F, and will keep bag G distended, thereby holding open valve H. But if the pressure ceases in D and F, bag G collapses, allowing valve H to close by a counterweight and shuts off the gas.

Figure 2 shows the operation of valve H by bag G, which rests on a shelf or table P, and supports on top a disk O pivoted to the lever L, which has a fulcrum at M and a guide at N, and raises or depresses

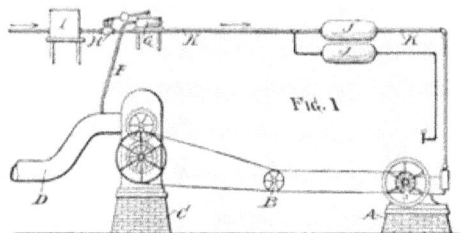

FIG. 1

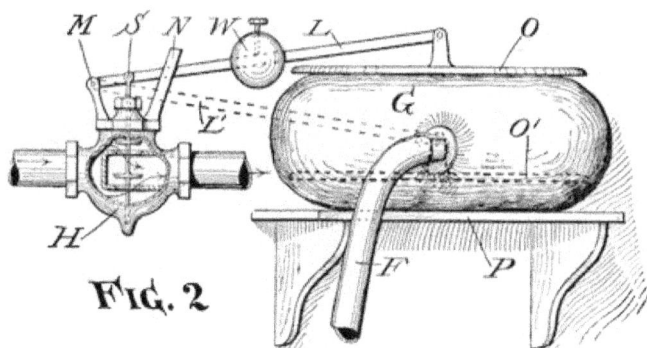

FIG. 2

AN AUTOMATIC GAS ENGINE CUT-OFF.

valve stem S. When the bag G is inflated the position of the lever, etc. is as shown, but if the bag is emptied the lever L, actuated by its adjustable counterweight W, falls to position L¹, and the double poppet valves fall to their seats, thus promptly shutting off the gas. When the engine is started it is only necessary to hold lever L up until the blower starts, when it will remain as shown.

For a description and sketch, from which our illustrations are prepared, we are indebted to W. J. Baldwin, consulting engineer, who devised the arrangement.

PLUMBING OF SWIMMING BATHS.

FUEL REQUIRED IN HEATING A SWIMMING BATH.

A. T. ROGERS, of New York, writes:

"This problem has been put to me: 'How much coal is required to heat the water in a swimming bath containing, say 85,000 gallons of water, by hot-water circulation?' The system proposed is to connect the flow pipe from the heater at one end of the tank, near the top, and the return at the bottom at the other end, thus causing all the water of the bath to pass through the heater. From data which I have obtained I have calculated that to raise this amount of water from, say 40° to 90° Fahr., would require the consumption of about 2,800 pounds of coal. Allowing a consumption of five pounds of coal per hour per square foot of grate surface, a grate containing 50 square feet of surface would be required to do the work in 10 hours. I should like to hear from the experience of others in this kind of work."

[The 85,000 gallons of water would weigh approximately 708,333 pounds. To raise the temperature of this weight of water from 40° to 90° Fahr., or through 50° Fahr., would call for the expenditure of about 708,333 × 50, or 35,416,650 heat units. Assuming that 10 pounds of water can be evaporated by one pound of coal (good ordinary practice), it will require about 3,541 pounds of coal to do the work required in 10 hours, or at the rate of 354 pounds of coal burned per hour. Allowing, say, 10 pounds of coal to be burned per square foot of grate surface per hour, which is a fair figure, 35.4 square feet of grate area would be required to do the work.]

HEATING A SWIMMING BATH PROBLEM.

X. L., from San Francisco, says:

"My swimming tank will be 80 feet long by 30 feet wide, and of an average depth of 5 feet. It will have an 8 inch inlet and the same size outlet. The temperature of the water will be about 74° Fahr., as it runs into the tank, and I would like to raise it to 84 degrees. I want also to supply hot water to 20 bathtubs. Now, how can I get the desired temperature with this large volume of water running through the tank? By answering the above information through the columns of your journal you will greatly oblige a subscriber."

[A tank of the size given will hold approximately 744,000 pounds of water when full, and if you desire to increase its temperature from 74 to 84 degrees in one hour of time, you will require a boiler that is capable of evaporating 744,000 pounds of water in an hour, or thereabouts, say, in round numbers, 250 horse-power. This, remember, is for one hour; in other words, if you want to warm that tank full of water 10 degrees in one hour, you will require boilers equal to 250 horse-power.

On the other hand, if you are satisfied to start your boiler, say the evening before, and spend 10 hours of the night in warming the water, instead of one hour, then a boiler of about one-tenth the capacity, or a little over, will do, say a 30 horse-power boiler, the extra five horse-power being sufficient to cover the loss of heat from the tank during the 10 hours you are warming it up.

You have omitted a very important item in asking this question, by not mentioning the quantity of water that runs through the tank every hour. You say, "How can I get at the desired temperature with this large volume of water running through the tank?" You omit to say how great this volume is, and also the time taken for it to run through the tank. If you mean to run the full contents of the tank through every hour, then it will require boilers of 250 horse-power, as before stated. If you intend to run it through in 10 hours it will take a 30 horse-power boiler, as before. If, however, you are satisfied with running only 1,000 gallons of fresh water in every hour, a boiler of about three horse-power will be ample, and for every additional 1,000 gallons you will want another three horse-power added to the boiler. With this you will be able to find the boiler power.

Fifteen pounds of steam per hour (one-half horse-power) condensed to water, will warm a bathtub full of water once. If your 20 tubs are used continuously, once in an hour for 10 hours a day (or for any time) you will require five horse-power additional for baths. To apply the heat of the boiler to the swimming bath you may simply circulate direct from a hot-water boiler, or use steam in a coil in the bottom of the swimming bath.]

SIZE OF PIPE FOR HEATING OF A SWIMMING POOL BY STEAM.

N. K. HOWARD, Lincoln, Neb., writes:

"I wish to heat a swimming pool 125x38 feet in size and 6 feet deep. I want to use live steam at high pressure, run through a deadener. How large a pipe will I need, what kind of a deadener should I have to insure freedom from noise, and about how long will it require to heat the water in the pool?"

[In THE ENGINEERING RECORD of November 22, 1892, an illustration is given of a "deadener" which appears to be well adapted to this case. We suggest that you make it about 4 feet long, the internal perforated brass pipe to be 2 inches in diameter and the 4-inch space about it to be filled with broken pebbles or some such substance. Lockouts should be used at top and bottom of the deadener so that it may be taken apart for cleansing. This should rest on the bottom of the tank, and have a galvanized-wire guard about it large enough to prevent bathers from touching it. With high-pressure steam a 1½-inch pipe will be amply large. The valve should be out of the reach of the bathers, so that the pipe on the tank side of the valve will be cooled when the steam is shut off. Steam should be turned off when the temperature of the water reaches 70° Fabr., as the 4,750 square feet of water surface in the tank will absorb heat from the greater heat of the air in the tank-room. The time required to heat the water will depend upon the temperature of the air in the room and of the water to be heated. The overflow pipe should be at the opposite end of the tank from the heater, as the top current when heating will naturally be towards it. If a small stream of water is allowed to flow continuously into the tank, the floating animal fats, etc. on the top of the water will pass off through the overflow, which will thus act as an effective skimmer. A coil of 1-inch brass pipe, four high, across one end of your plunge bath, the bottom pipe being capped and ¼-inch holes drilled a few inches apart, on its bottom side, will also make a very good heater, and, like the other, should be faced with a wire guard.]

NOTES AND QUERIES ON HOT-WATER SUPPLY

AIR-BOUND PIPING.

F. A. C., Chicago, writes:

" The inclosed sketch represents a defective hot-water supply in a private residence in Chicago. B B indicates the hot-water pipe and C the cold. It will be noticed that the pipe B is carried down from boiler to basement, thence suspended from ceiling and carried a distance of about 7 feet, from which point it extends up direct to second floor to washbasin in bathroom A branch not shown is taken off in the basement and extends to laundry tubs.

" When the temperature falls very low several minutes will pass before hot water flows to second floor; there is frequently no flow at all when faucet is first opened, and water is always cold. The cause of this trouble is in pipe B at A, where air accumulates, causing it to become air-bound, produced by elevation of pipe at this point. To remove the air it is

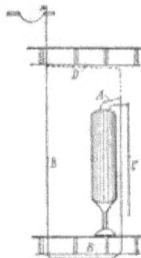

necessary to open faucet in laundry tubs, which soon releases the air, restoring flow to normal condition, permitting hot water to ascend to basin on second floor.

" The writer advised a change in the pipe B by extending it up to ceiling of first floor, connecting with ascending pipe as shown by broken lines at D. When this change is made hot water can be drawn at once in bathroom.

" I present the following to be answered by your readers, as a reply may be of interest.

" The tap and service pipe for this dwelling are each one-half inch in diameter. The head is diminished during early portion of the day, caused by an increase in the consumption at this time. To improve the supply a plumber states that the head would be improved if service pipe was increased to 1 inch in diameter. His reason for this is that a greater quantity of water would be carried into dwelling through a 1-inch pipe, thereby increasing the height of flow on second floor. Is the plumber correct?"

MORE AIR-BOUND PIPING.

A PHILADELPHIA apprentice asks:

" Will you tell me what is the trouble with the job shown in this sketch? A horizontal 1½-inch iron pipe A A runs from the pump to the lower tank, with a branch B leading off to the upper tank and a branch C acting as an overflow from the higher to the lower tank. The difference in elevation between the tanks is 15 feet and they are about 300 feet apart. When the stop-cock D is closed and the pump started up, water refuses to overflow from the higher to the lower tank. but when the cock D is open the pump forces water through all right."

[It is probable that air collects in the short horizontal overflow pipe. The whole difficulty can undoubtedly be removed by putting in a T instead of the elbow E, and running a piece of pipe with an open end about a foot above the top of the tank. The

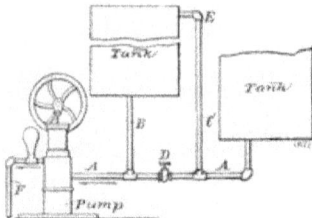

reason the water does not flow through the pipe C is that you cannot force water into a space already occupied by air, as in this pipe C. You must arrange for the air to escape, which it could not do with the bend at E as shown in your sketch.]

COIL HEATING OF A BATH SUPPLY.

N. K. HOWARD, Lincoln, Neb , writes:

" I send a sketch of a bathroom plunge bath and a tank to heat its water supply. The plunge bath is 31 feet long, 12 feet wide by 6 feet deep at one end, and 4 feet deep at the other. The tank is 10 feet long by 36 inches in diameter. I intend to try and heat the bath by using live steam in a coil in the tank made on the plan of a box coil, containing 175 feet of 1¼-inch pipe, taking steam from the heating boiler through a 2 inch pipe and returning to the boiler.

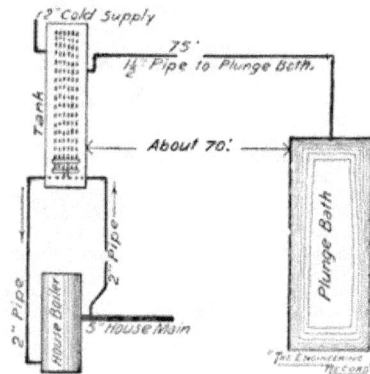

PLAN.

The hot-water supply to the bath is 1¼-inch, with a pressure of 40 pounds to the square inch (city pressure) Can I heat the water as fast as it is taken through the 1¼ inch pipe? How long will it take to fill the plunge bath?"

[A very short 1½-inch pipe, under the most favorable conditions, will deliver about two-thirds cubic foot of water per second at a pressure of 40 pounds, so that under the most favorable circumstances it would take a half-hour to put, say, 4 feet of water in the bath. The friction of your pipes, however, forms a factor which will reduce the quantity so greatly that we would not be surprised if it took four to five hours to fill the bath. At the rate of filling in five hours 17,856 pounds of water must be warmed in an hour, and to warm this, say from 40° Fahr. to 60° Fahr., is the equivalent of condensing 357 pounds weight of steam to water every hour, or, say, 10½ horse-power. Under such circumstances, we are of the opinion you have coil surface enough, but that your inlet to the coil should be about a 3-inch pipe, and your return pipe 2-inch, if you are to have a gravity apparatus.]

SIZE FOR HOUSE SERVICE PIPE WANTED.

SUBSCRIBER, Ypsilanti, Mich , writes:

" In the city of Ypsilanti, Mich., in which they are putting in water-works, the Superintendent of the Water Board claims that a ½-inch connection of common black iron pipe run from a city main—the distance of about 50 feet, and the pressure on the city main is 70 pounds to the square inch—would be sufficient to supply a house containing one bathtub, four washstands, one water-closet, one sink, and one ¾-inch street-washer. Bathtub, washstands, and sink are supplied with hot and cold water.

" The plumber wanted a 1-inch connection, but the Board of Water Commissioners laughed at him. As it was, he put in a 1-inch connection from service cock. Please state in your paper what you think about it."

[Under the conditions of pressure stated, a ½-inch round-way corporation cock would probably answer the purpose, but the service pipe should not be less than three-fourths inch, and a 1-inch service pipe is not a needless extravagance, as the additional expense is small, and the loss of head by friction is much less in a pipe of any length.]

TO KEEP COLD-WATER PIPES FROM SWEATING.

ARCHITECT writes:

" We are called upon to remedy a trouble that seems to be very general, and we would like to have your advice through your paper as to the best means of remedying it. The hot and cold water pipes in the house in question are fastened to the ceiling of the basement; the hot-water pipe is painted, the cold-water pipe is not. They are both lead pipes. Since the furnace has been stopped the cold-water pipe sweats to such an extent that the floor in the basement is wet all the time. Can you suggest any remedy that would overcome the trouble in a simple and inexpensive manner?"

[Your trouble is simply the condensation of the moisture in the air on a cold surface, the same thing that occurs on the inside of the kitchen windows in winter and on the outside of the ice pitcher in summer. The remedy is to surround the pipe with some non-conductor so that the exposed surface shall not be so cold; any non-conductor will do, but if pervious to the air like felt it will be kept damp by the moisture passing through it and condensing on the pipe inside, and will be liable to rot. A thick coat of paint will somewhat reduce the condensation and will be the neatest arrangement. If the paint is thickened with ground cork the condensation may be reduced to an inconsiderable amount, although the job will not be as smooth. Such paint is used on the interior of iron ships for a similar purpose. If ground cork is not readily obtainable probably sawdust would answer a similar purpose, though it is not quite as good a non-conductor. The increase of your trouble since the furnace fire went out has probably nothing to do with the absence of the furnace fire, but is most likely due entirely to the fact that there is much more moisture in the air in summer than in winter, and hence more of it is condensed on your cold pipe.]

MATERIALS FOR DOORS OF TURKISH BATHS.

B & C . Rochester, N. Y., write:

" Will you please let us know what they line the doors in bathrooms with when the heat is to be 180 degrees, and oblige."

[The proprietor of the Lafayette Place Turkish and Russian baths in New York says: " Florida cypress is by far the best of all woods for doors, window-casings, and sills. It should be thoroughly kiln-dried, then given several coats of the best linseed oil, and finished rough dry, *no paint, no flannel covering*. Georgia pine and other woods decay quickly. All woods will shrink a little, but Florida cypress will shrink the least and last the longest."]

THE FLOW OF WATER IN PIPES.

N. K. LUDLOW, Mobile, Ala., writes:

"Will you kindly answer this question and give me the rule to work out the same, to wit: A 1-inch iron pipe is attached to an 8-inch cast-iron water main in the street and is run in this shape a distance as is shown on the inclosed slip. [Sketch described in answer—ED.] Now, how much water would this pipe discharge per hour with a pressure of 75 to 80 pounds per square inch on the main in street? Please answer and send rule to work it out, and oblige."

[If an elastic ball is thrown against a hard substance we should expect it to rebound with the same force or velocity with which it struck, and when we find that a ball of glass or ivory dropped on a smooth flagstone rebounds nearly to the height from which it fell we are prepared to believe that, making allowance for the resistance of the air, a ball or other body thrown upward with a given velocity will rise to the height from which it would have had to fall to acquire that velocity. When, therefore, we further observe that a vertical jet of water will under favorable conditions rise nearly as high as the surface of the reservoir from which it is supplied we see that if our former supposition is correct it must issue with a velocity as great as it would have acquired in falling freely through a distance equal to the amount that the surface of the reservoir is higher than the orifice of the jet. This is in theory exactly true and would be equally true in practice but for the effect of the friction of the water through the pipe and nozzle before escaping.

Let us then first see how much the theoretical discharge of our pipe would be and then how much the friction is likely to reduce it.

To find the theoretical velocity of the flow we must know what head of water will give a pressure of 75 or 80 pounds. A cubic foot of water weighs about 62½ pounds, consequently a column of water 1 foot square and 2.3 feet high will weigh 144 pounds and give a pressure of one pound per square inch on the bottom; therefore, to get a pressure of, say 76 pounds per square inch, requires a head of $76 \times 2.3 = 175$ feet. A body in falling acquires a velocity per second equal to about eight times the square root of the distance fallen. The square root of 175 is 13¼, which, multiplied by 8, is 106 feet per second.

A "1-inch" pipe is a little more than 1 inch in diameter, but rust and roughness make it unsafe to count on more, and therefore as the area of a 1-inch circle is .7854 of a square inch, that amount multiplied by 1.272, the number of inches in 106 feet, and divided by 231, the number of cubic inches in a gallon, equals 4.32 gallons per second, or 15,600 gallons per hour, the theoretical discharge through a 1-inch pipe with

76 pounds pressure per square inch. How much the friction of the pipe will reduce this discharge is uncertain. The problem of the flow of water in pipes is most difficult and complicated, and though the ablest hydraulic engineers have long endeavored to devise a formula for it at once simple, accurate, and generally applicable, their efforts have so far been only partially successful. Your sketch shows your pipe to be 132 feet long with two elbows and a stop-cock in it. The frictional resistance of an ordinary screwed elbow is estimated to be equal to that of a length of pipe equal to 100 diameters, which in this case would be about 8 feet. A stop-cock, with its reduced opening, offers perhaps twice as much resistance as an elbow, or altogether the resistance of stop cock and elbows may be assumed equal to that of 32 feet of pipe, so your question practically becomes what will be the discharge per hour through 165 feet of 1-inch pipe with a pressure of 76 pounds per square inch. To this, some standard formulas give answers varying from 1,747 to 2,605 gallons, a difference of 50 per cent., which well illustrates the uncertainty of our present knowledge of the subject. On August 11, 1888, we published some "Notes on Simple Methods of Calculating the Flow of Water Through Pipes," by Edward Murphy, in which occurs the formula $\sqrt{(425 \times d \times p) \div l} = v$; or, in other words, multiply the diameter in inches by the pressure in pounds per square inch and multiply the product by 425, divide this new product by the length in feet and extract the square root of the quotient and the answer will be the velocity in feet per second, from which the discharge can be found as before. Applying the rule to this case $1 \times 76 \times 425 = 32,300$, which divided by $165 = 195.76$, of which the square root is 14. This velocity in feet per second multiplied by .7854, the area of the pipe, and by 3,600, the number of seconds in an hour, and then by 12 to reduce it to inches and divided by 231, the number of cubic inches in a gallon, equals 2,056, the discharge in gallons per hour. As this is reasonably between the extremes mentioned before and agrees very closely with the discharge found by the rule in Box's hydraulics, it is probably as near the truth as we can expect to get by any process of calculation. From this it appears that the friction will in this case reduce the discharge to about one eighth of the theoretical amount.

We have taken more than usual space to answer what seems a very simple question, but it is one of very general interest, and the answer here given may serve for many similar questions.]

WATER SUPPLY FOR A COUNTRY HOUSE.

A. S., West Point, N. Y., writes:

"I wish to supply my house with water from a spring 1,800 feet distant and discharging 600 gallons of soft and pure water every 24 hours. This spring is 172 feet above the ground floor of the house. What is the best method of piping this supply for culinary purposes, bathroom, water closets, and lawn sprinklers?"

[Nearly 10 years ago the late E. S. Philbrick answered a similar question in our columns substantially as follows: A ¼-inch cement-lined pipe should be used to conduct the water. The method of making such a pipe is fully described in Billings's "Details of Water-Works Construction." Care should be taken to cut the screw-thread so as to allow the ends of the pipe to abut against each other when screwed into the couplings, for if a gap of one-eighth of an inch or more is left the inside of the coupling it will rust and fill up the pipe.

If you must have a storage of water for sprinkling lawns or any exigency demanding a rapid delivery for a short time, make a small tank in your attic or as close to your house as possible, of a few barrels capacity, and draw your drinking and cooking water, not from this but by a branch tap from the main. Let the water enter the tank at the top and discharge the surplus by an overflow pipe into the open air. In this way the branch tap will never draw from the tank, but always direct from the source. Then put in a pipe from the bottom of the tank to supply the hose.]

EQUALIZING THE FLOW IN A DOMESTIC WATER SERVICE.

B. S. M., Montreal, **writes:**

"A half-inch water service is all that the water company will allow here. Introduced into a dwelling of three stories and basement, with fixtures distributed as shown on the sketch, it is desired to equalize the flow from the different fixtures so as to prevent the shutting off of the supply from a fixture in an upper flat by the opening of a cock on a flat below. The service is increased to three-fourths inch and run to the highest fixture. The cocks are graded in size for the flats below, being three-eighths inch for the basement, one-half inch for ground floor, five-eighths inch for the first floor, and three-fourths inch for the top floor. Will this arrangement shown in Fig. 2 have the desired effect? The architect pro-

poses running the main direct to the top floor, bending and dropping to the flats below, taking cocks from the drop pipe (see Fig. 1). Will that be an improvement? I propose running the ½-inch service into a closed tank, say a 30 gallon galvanized kitchen boiler, and taking a separate branch from it to each flat, Fig. 2. What would be the effect of that plan?"

[We would not advise the consideration of any of the three plans proposed. Carry your ¾-inch supply directly to a tank placed in the attic, the tank to have a ball cock and float to control the supply of water. From the tank take out a house supply large enough to supply hot and cold water to all fixtures, the pipe to start as a 1¼-inch with a ¾-inch branch to the upper floor; then reduce at the branch to 1 inch for second floor, reduce again to three-fourths inch for first floor, and from there carry a ½-inch pipe to the basement. A 5'x4'x3½' tank would give about 450 gallons, and with the piping mentioned would furnish a plentiful supply for all the fixtures. We have not sufficient data as to the length of the pipes and pressure on the main to say which of the plans would be the most satisfactory, but know that the method illustrated by Fig. 2 has been used with success. The drawback to it in your case is the small supply pipe from the street main, but if the pressure is excessively high, we believe the arrangement shown in Fig. 2 would give the best result of those you propose.]

TO MAKE A CELLAR WALL WATER-TIGHT

Subscriber, Boston, Mass., **writes:**

"Is there any way of constructing a cellar which shall be reasonably water-tight? After repeated sad experiences, I have given up any hope for one which shall be absolutely damp-proof; but it does seem as though there could be some way devised for keeping out water. My house is on a hilltop, yet the cellar is flooded regularly every spring. If you can suggest

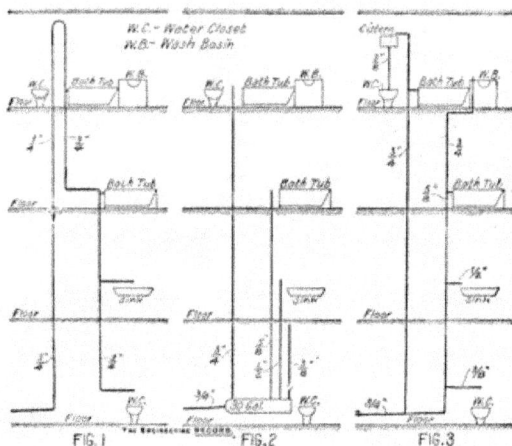

W.C.- Water Closet
W.B.- Wash Basin

FIG. 1   FIG. 2   FIG. 3

EQUALIZING THE FLOW IN A DOMESTIC WATER SERVICE.

any remedy, even a partial one, you will greatly oblige."

[Subscriber's trouble is a very old one, to which hilltop houses are nearly as liable as those in a valley. A very good plan is to dig the cellar 18 inches deeper than required for finished height, filling in with coarse, broken stone, well rammed down to prevent settling. The outside walls should be started over a similar filling, laid in a trench a little deeper than the bottom of the cellar excavation. Then over the broken stone in the cellar is laid a course of concrete 3 or 4 inches thick, the walls being laid up in cement mortar. In most localities the water will never rise higher than the stone filling, but in a clay soil, if full of seams, the cellar bottom had better be sloped slightly towards a sump at the center, or other convenient place, which should be connected with a tile drain carried far enough away from the building to discharge properly below the level of the cellar bottom, but of course not connected with any sewer or cesspool. It is possible to make tight a cellar bottom which is below tide water, but the process is usually too expensive to be considered in connection with ordinary dwellings. There are parties in New York who make a business of constructing such water-proof cellars.]

NOTES AND QUERIES ON HOUSE-DRAINAGE PROBLEMS.

BADLY DESIGNED PLUMBING IN A NEW YORK HOUSE.

THE results which may follow carelessness or incompetency of design in plumbing systems are illustrated by the condition of affairs lately discovered in a New York house where the pipes were being overhauled to make alterations for the remodeling of some plumbing which had been done within three years. It is not stated that the workmanship or material were found defective, but the arrangement, while conforming to most of the ordinary specific requirements, was so bad as to permit and produce an unsanitary condition of operation, entirely nullify the trap ventilation and effect a discharge to the sewer in an improper manner, which was never intended, and was not suspected until revealed by the

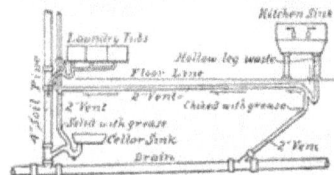

alterations. The waste from the kitchen sink was trapped below the floor, and the back-air pipe also run below the floor, nearly horizontal to its riser, so that when the kitchen-sink waste became stopped by the collection of grease below its trap the discharge was forced through the back-air pipe, and at first escaped through the foot of its riser, which also vented the cellar-sink waste, and through that waste into the sewer. After a time the vertical riser also became obstructed with grease just above the cellar-sink trap, and cutting off the discharge through it backed the water up until it escaped through another higher branch from the riser that vented the trap of the laundry tubs, through which the sink water was thus compelled to flow to the soil pipe and thence to the drain. A considerable body of dirty untrapped water was thus always standing in the vent pipe, which it filled completely up to the inlet from the laundry tubs, and the ventilation of the kitchen and cellar sinks was entirely destroyed. The use of a grease trap would have prevented the grease from obstructing the pipes, and the location of the sink vent-pipe branch above the overflow, as is carefully provided for in every case in good practice, would have made it impossible for discharge to have taken place except through the waste pipe.

HOUSE CONNECTIONS ON PIPE SEWERS.

G., of Boston, writes:

" Will you tell me what is generally used for house connections on what is known as a separate system of pipe sewers; I mean whether Y or T connections are generally used, and why one has the advantage over the other? I have generally understood that a Y branch is not so liable to clog as a T, but I heard lately that a T makes a better connection. Do you know of any sewerage system where house connections have been made with T branches?"

[For pipe sewers the general practice is to use Y branches, as offering the least resistance to flow and possible clogging. The best practice contemplates the use of a one-eighth bend, with the Y branch, the same as in house drainage. On large brick sewers in the combined system T connections are often used.]

AN EXPERIENCE WITH SEWER GAS DUE TO THE CLOSING OF THE TOP OF THE SOIL PIPE BY FROST.

H. B. in the Toronto *Globe* writes:

" I think my experience lately in regard to sewer gas, if it is generally known, will put all persons on their guard, and be the means of diminishing, if not totally preventing, typhoid and other similar fevers in the winter season.

" The facts are these: My house has a bath and water-closet connected with a cesspool in my garden, and I have taken every precaution to prevent any foul smell coming into the house, and have been successful, except on a few occasions in the winter, and I could not discover the reason until recently,

During the recent cold snaps with the wind from the east and northeast the sewer smell in my bathroom was intolerable. And I consequently made up my mind that it was caused by some recent obstruction in some of the escape pipes. In my quandary I went on the roof of the house to examine the ventilating pipes, and found the ventilating pipe from the water-closet completely filled up with hoar-frost and ice. I immediately emptied a kettle of hot water down the pipe, and at once the smell disappeared and the bathroom was as sweet as could be; but on came another smell, and on again came the smell of sewer gas, and I found again the ventilating pipe filled up with hoar-frost and ice as before, and immediately it was thawed out the smell went away, and as there has been no very cold weather since, I have had no more trouble. Now, sir, it strikes me that if the moist gas from a private house connected with a cesspool is sufficient to cause the complete freezing up of the ventilating pipe, how much more likely must be the pipes in a city like Toronto, where the amount of moist gas escaping must be enormous, and I have no doubt the ventilating pipes are often frozen up the same as I have described, and this should put every person on the alert to examine their escape pipes and keep them clear of ice."

[The above letter to the Toronto *Globe* relates an experience to which we have often called attention. It was this contingency that induced us long since to advise that the end of soil pipes should be open, without bends, hoods, caps, or cowls of any kind, and in cold latitudes that the smaller pipes should be enlarged from the roof upward to at least 4 inches in diameter, and more if experience indicated it to be necessary in a particular locality.]

DISPOSAL OF HOUSE WASTE IN TIGHT CLAY SOIL.

E. W. L., Troy, N. Y., writes:

"Please refer me to articles in THE ENGINEERING RECORD, which I have from the beginning of its publication, covering the subject of sewage disposal like the instance I have in hand.

"The lot is 300 feet front by about 800 feet deep, and the greater part inclines upward from the front about 1 foot in 10, and at the ridge the house stands. There is no creek nor any sewer to which house drains can possibly be connected, the location being quite outside of the village limits. The soil is 8 feet and 4 feet deep on the rock, and is all very compact clay. In half a dozen places I caused test holes to be dug, and all filled with water in 24 hours, which later somewhat decreased in depth; but they always contain more or less water.

"It is desired to put in complete bathroom accommodations and water supply to the house. I think the only course is to build tight cesspools. But I prefer to ascertain if similar difficulties have received consideration in your paper."

[From the facts given it would seem that tight cesspools will have to be relied on and the expense of pumping them out incurred, though the depth of the lot, 800 feet, will probably allow for good sized ones, or several. The water may also be utilized, in some degree, in the garden. If first expense be no important consideration, a sufficient amount of clay might be removed, specially prepared soil substituted, and the subsurface irrigation system adopted. This, with suitable soil, is the best thing to do. It is described fully in "The Disposal of Sewage in Suburban Residences," by E. S. Philbrick, reprinted from this journal, and in William Paul Gerhard's little book (Van Nostrand's Science Series) on "The Disposal of Household Wastes."]

THE USE OF GREASE TRAPS.

A MEM. AM. SOC. C. E., New York, writes:

"In your opinion, is a grease trap an absolute necessity in a well-planned drainage system for a country house, and if so, why not equally necessary in a city house, where the horizontal run from the kitchen sink to the sewer is 80 or 90 feet? Most country houses have a grease trap close to the house. Granted the necessity of the grease trap, why should this be nearer the house than the sewer in the city? Granted the necessity of the grease trap, what is the extreme distance from the house at which it may be placed, and sewer properly? Is it not as objectionable as a cesspool, and, with subsoil irrigation to avoid the latter, do we remove the danger if the grease trap remains? Do you know of any case where flush tanks have been used without grease traps or settling basins, the house sewage draining to the flush tanks direct, and being disposed of immediately?"

[The necessity for using a grease trap depends on several things: First, the probable amount of greasy water disposed of in a kitchen; second, the distance the greasy water must flow to reach the sewer or cesspool; and third, the length of soil pipe or earthern drain that may be buried in the ground.

A grease trap, like other traps used in drainage, must be considered as a compromise; in other words, it is used to prevent a greater evil. A small family, keeping one or two servants, may safely dispense with one, but when a number of servants are employed, and much cooking is done, and in a country house, where the waste water must pass through pipes buried in the ground where the greasy water will be rapidly chilled, we should advise the use of a grease trap.

In a city house, as in New York, for instance, the average distance of the sewer from the kitchen sink is about 90 feet, but for 60 feet of this distance the hot greasy water flows through an exposed and accessible iron pipe in a cellar which, during the winter months, when the house is occupied, is warm, so that the water is not chilled, and grease precipitated until it reaches the sewer. Besides, in properly planned work, cleanouts are provided for the removal of obstructions. If a grease trap is placed within a house, the grease should be removed from it every few days and before it becomes putrid, otherwise cleaning it is a very offensive operation. A grease trap should be located nearer the house than the sewer, and as near the source of the grease as circumstances will permit, taking into account the offensive smells when grease is removed. In country houses they are usually located in a vault just outside the house.

From the point of view of the householder, if all the grease could reach the sewer no grease trap would be needed. On the other hand, a town provided with small pipe sewers is liable to find trouble from stoppages from this cause. We believe that Mr. E. W. Bowditch, of Boston, has recommended towns where he has designed a small pipe sewer system, that they require grease traps when houses are within 50 feet of sewers, but when the private drain is more than 50 feet, to leave it optional, relying on the grease to precipitate in the drain and not reach the sewer. A grease trap is not as objectionable as a cesspool, since it is not supposed to receive anything but greasy water, and is intended to be frequently cleaned.

We do not recall any place where house sewage is delivered to a flush tank and periodically discharged from it into a sewer, and should consider such use of a flush tank objectionable. This of course does not apply to disposal by subsurface irrigation, where a settling basin and flush tank are required.]

THE PROPER SIZE FOR A HOUSE SEWER.

A. J. C., Binghamton, N. Y., writes:

"In planning a house which I am building, the architect has designed to use a house sewer only 6 inches in diameter. I claim that nothing smaller than an 8-inch sewer should be used. In the house in which I now live there is an 8-inch sewer which is frequently obstructed, causing expense and annoyance. Both architect and plumber are against me in this matter. Who is right?"

[If the house sewer is for sewage which will come to it through properly set and flushed fixtures and storm water, the architect and plumber are right; if it is for old cans, ashes, shoes, rags, and broken dishes, you are right, though you will only be right a short time, unless you make the sewer large enough to allow a cleaner to enter and remove the accumulation of insoluble matter with which it will soon be choked. A 6-inch sewer should be large enough for any private residence in the country, provided it is properly laid, with the right material, has the right kind of bends and connections, a proper pitch, and is sufficiently flushed. It must also be understood, however, that only soluble matter should be allowed to enter the sewer. We know of a large 16-room city house which has but a 4-inch drain inside and a 5-inch sewer from the house to the main sewer. This sewer is over 100 feet long. The house has been occupied 10 years, and so far the sewer has not failed to perform its functions. The house has seven water-closets, four bathtubs, five washstands, laundry tubs, slopsinks, china sink, kitchen-sink, etc. In this case the owner was willing to pay a fair price and left the entire matter in the hands of the architect, who in turn gave his instructions to a competent plumber, who carried out the details. A small sewer, sufficiently large, will scour, the flow of water and matter being more confined and rapid. A larger sewer induces a sluggish movement, tending to convey only the fluids and dissolving matters. The others remain to adhere and clog and eventually to choke the sewer.]

VENTING TRAPS INTO FRESH-AIR INLET.

Charles O'Grady, Marlboro, Mass., writes:

"I send sketch of a plumbing job of two water-closets in a basement, and one sink on the first floor. I have back-vented the water-closets into the foot vents. Is there any objection to doing so?"

[Figure 1 shows our correspondent's sketch. The practice of venting fixture traps into the fresh-air inlet (foot vent) is not to be recommended. The fresh-air inlet if open in the sidewalk is liable to become choked, in which case the traps are not protected. Circulation of air through the vent pipes is more liable to take place if the vent pipes are connected with a long upcast pipe to the roof. Discharge from fixtures, as shown in Fig. 1, would tend to cause an out-draft in the foot vent, while the traps would need an in-draft to prevent syphonage and the vent currents would be working against each other. Other faults in your design are: 1 The

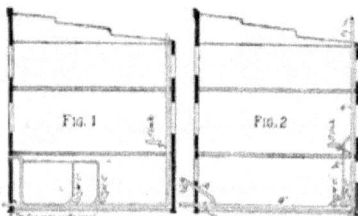

Fig. 1   Fig. 2

fresh-air inlet (foot vent) opens too close to the house. 2. The basement closets appear to be in the center of the house. They should have direct light and ventilation through outside windows. 3. The main trap cleanout should be on the house side of the trap If the cleanout cover is not tight and leaks, air from the house drain is preferable to air from the sewer. Figure 2 shows an improvement on your design, with some slight economy in piping. The bottom of the 2-inch pipe A opens directly into the crown of the trap, so that rust, scale, etc. will be washed out by the water-closet discharge. The pipe B should be given a sharp angle so that dirt and rust will run down the pipe. The upcast pipe C should at its upper end be well removed from bedroom windows. Brass cleanouts should be used at the points D D.]

SIZE OF PIPE REQUIRED TO DRAIN A FIELD.

Rodman Sands, New York, writes:

"Will a 12-inch earthen pipe be large enough to carry off the water from a stone drain, 2 feet square and 700 feet long? The drain is simply a ditch 2 feet deep and 2 feet wide, filled with stones and covered over.

"I can give the earthen pipe a fall of about 5 feet in 700.

"The field, which is 700 feet square, floods in rainy weather, and now takes from three days to a week to run off, and I want to keep it dry all the time if possible."

[A 12-inch pipe would carry off the equivalent of about one third of an inch of rainfall per hour on the area to be drained. Ordinarily, this would keep the field dry, but in very heavy rainstorms there would be flooding of the field. A 20-inch pipe would insure perfect drainage.]

TRAPS AT FOOT OF LINES OF SOIL PIPE.

PLUMBER, of Pottsville, Pa , writes:

"Inclosed please find rough sketch of county grounds and buildings to show drainage system. The dotted lines show new connections, ink circles show perpendicular run of soil pipe to be put in new. All places marked 'inlets' are not trapped, but those marked 'trap' are inlet traps. The size of drain is marked on same. The creek is open and very seldom gets as high as drain pipe. The drain is salt-glazed terra-cotta pipe.

"What I want to know is this: I have the contract to do new work, and the architect has specified running traps at the bottom of all lines of soil pipe with air vent and to continue the same to roof of buildings, each fixture to be trapped and ventilated.

"Now I claim that the best way would be to leave out the running traps and connect direct to drain; and that all foul air would at once escape above buildings and that the running traps only form a cushion whenever any of the fixtures are used, and forms foul air out of fresh-air supply. Am I not right? By answering the above you will confer a favor."

[As we understand the case from your description and sketch we concur in your opinion, that traps at foot of lines of vertical pipes are objectionable. The trap should preferably be on each drain where it leaves the building, with a suitable fresh-air inlet]

FROZEN ROOF CONDUCTOR PIPES.

J. M. G., Cleveland, O., writes:

"Each winter we have great trouble with our pipes for conducting the storm water from the roof to sewer. They seemed to begin to freeze at the foot of the pipe, after some stoppage which we cannot account for, as we keep our roof clean. They will then fill to the top of the pipes, overflowing and forming threatening and dangerous icicles, or making a disagreeable dripping, generally most annoying in the afternoon. Each spring we have to replace those pipes or pay large repair bills, many of the seams and joints being found burst. Our roof is of tin with a pitch of one-half inch to : foot. Our engineer suggests that we write you on the subject in the hope that you may suggest a remedy."

[Your complaint is a very common one in cold climates, and is especially common in New York in the winter. The fact that the dripping begins generally in the afternoons is because by that time the steam heat in your building (we assume you have steam heat, as you refer to your engineer) has begun to affect the snow and ice on your roof by passing through the wood sheathing and tin, which have been frozen hard by the colder air of the night. This action is accelerated by the sun's rays, if the temperature is above the freezing point.

If unobstructed, the water will fall through the conductor pipe to the sewer or outlet, as it cannot freeze while in rapid motion, or when the air is at a temperature that will allow of thawing. Here this water comes in contact with the cast-iron or earthen sewer pipe imbedded in the earth or stone sidewalk, which is surrounded by all the influences of frost, and which has the nucleus of an ice formation in the hoar frost on the inside, the result of warmed damp air ascending and condensing its moisture, which is readily frozen. This operation continues until the pipes are entirely closed. Then the pipe will gradu-

ally fill to the first relief point, and if the pipe has been made perfectly tight it will freeze to the top. The practice of making these pipes tight is wrong, as loose fitting joints united only, will allow the water to escape when it fills to them.

The most certain and direct cure for your trouble would be to have large conductor pipes placed on the inside of your building, where they will be removed from the influences of frosty blasts or frozen outlets. But in this case it should be of extra-heavy cast-iron pipe with leaded joints to prevent sewer gases or odors from entering the building. This is the practice adopted for public building and business structures. The pipes should be in recesses and accessible for repairs.

If you cannot change your conductors, see that the snow is removed from the roof before it has had time to freeze to the tin. This will help you, save when there is a cold rain or sleet storm. A sure method to keep the pipes open without regard to the severity of the frost is to enter a steam pipe with jet end at the lowest accessible point of the leader, arranged with a swivel joint, so that the jet may be turned up when desired to clean the pipe, or downwards when not in use, when it will not become filled with water and freeze. A valve at a convenient point under the charge of your engineer will enable him to use it as his judgment dictates in thawing the ice or frost.]

TO PREVENT TAR-COATING FROM SHOWING THROUGH PAINT ON SOIL PIPES.

HARTFORD, CONN., April 10, 1886.

SIR: One way to prevent discoloration of paint over tarred pipe is to paste firm, non-absorptive paper over the pipe and then apply the color.

M. P. HAPGOOD, Architect.

ARRANGING FRESH-AIR INLETS TO PREVENT FREEZING OF TRAPS.

R. HADDOW, Winnipeg, Man., writes:

"With reference to the correspondence on fresh-air inlets in THE ENGINEERING RECORD, the experience of another plumber in Winnipeg might not be out of place. Bearing in mind that the thermometer goes down to 40 degrees and 50 degrees below zero, I never have found a single instance wherein a fresh-air inlet froze over with hoar frost with properly adjusted drain and soil pipes. I think Mr. Hughes must be making a mistake, because it is not the fact of cold air rushing in a pipe that freezes up the mouth of the pipe; it is the warm vapor coming out of the soil pipe at top that freezes up. But even that I believe can be obviated, by enlarging the pipe at its terminus, and have as little pape as possible after going through roof, leaving mouth of pipe clear and open.

"But the greater trouble of all with us is to get a fresh-air inlet that will not freeze up the main trap, and that I think you cannot have where you enter your pipe immediately over the trap. Mr. McArthur's idea of the cold air striking the bend before entering the trap I fail to see, because the cold air never has time to strike on that point, from the fact that the suction of the soil pipe is so great. If it was forced in, then it would be another thing. I think it it is more from the fact that he put the inlet further from the trap than it was before that kept it from freezing, so that I think by having as much fresh-air inlet pipe inside of house (as is workable according to circumstances) as possible, and entering one or even two lengths of soil pipe from trap, will keep the trap from freezing. I use generally 2-inch pipe and carry it 7 or 8 feet up the wall; this is to prevent it getting covered over with snow. These few remarks are drawn from practical experience."

[Letters based on experience are always welcome. Our correspondent seems to indorse the views frequently expressed as to location of fresh-air inlets in cold climates.]

BACK PRESSURE IN A SEASIDE COTTAGE DRAIN.

COTTAGER, New York City, writes:

"I have hired a cottage for the summer close to the shore, where the tide rises or falls 7 or 8 feet. The waste from the kitchen sink and water-closet is carried in a drain of earthen or iron pipe some ways down the beach to about half-way between high and low water, where it terminates in a hole filled with loose stones. This is not exactly in front of the house, and as it is well washed out with salt water twice every day it does not seem to be objectionable. The other night, however, happening to be in the kitchen when the house was quiet, my attention was attracted by a bubbling in the trap of the sink, which I at first could not understand, until I remembered that the tide was about three-quarters high and rising, and concluded that perhaps the rising water was forcing the air out of my drain though the trap.

"I would like very much to know if you think my conjecture correct, and if you consider the inflow of air from such a drain as likely to be dangerous, and what is the best way to prevent it."

[From our correspondent's account we have no doubt that his conjecture is correct; certainly the cause he suggests is abundantly able to produce the effect which he describes, and no other seems probable.

The remedy is very simple: a hole in the drain pipe anywhere above high water mark, even if only big enough to admit a lead pencil, will relieve the confined air and avoid the trouble; the only points to be observed in making such a vent are to place it where it will not be liable to become obstructed, and where the escaping air will not be an annoyance or danger to anyone.

As to the danger, it is hardly probable that under the conditions described the sewage would have sufficient opportunity to become putrid, and give off unwholesome exhalations, and seaside cottages, during the time of their occupancy, usually have doors and windows so widely open that such exhalations would be very greatly diluted; still it is much better to be on the safe side, and even if it should be safe, it certainly is not nice to have such vapors pumped into one's house twice a day.

Perhaps the most objectionable feature of the arrangement is that in winter, when the traps dry out for lack of use, or are emptied to prevent freezing, then unless the drain is closed in some way, the air from it has free access to the house, the tide maintaining a forced circulation, so that the whole building has a chance to become pretty well saturated with the emanations from whatever filth may have been left in the drain from the previous summer's occupancy.

VENTILATION OF A SOIL PIPE.

D. G. ADELSPERGER, Baltimore, Md., writes:

"I have a contract for plumbing, etc., etc., of the university now building at Washington City. Mr. E. F. Baldwin (who is the architect and one of your subscribers) and myself, have had a consultation about the construction of the soil and vent pipes for water-closets, etc., etc. He has an opinion one way, and I another, so we agree to write to you for your opinion. Inclosed is a sketch showing the soil pipe running the highest at ridge of roof, and traps of closets all emptying into it and vent pipe the lowest which would cause a current of air to come down the vent pipe and follow the water, etc. from the closets and vent the closet pipes and syphon up the soil pipe at highest point and make current of air circulate down the lowest or shortest pipe and up the longest.

"Now suppose the soil pipe to be placed where the vent pipe now is and would thus be the shortest, and place the vent pipe where the soil pipe now is and reverse the closet pipes and run them into the shortest pipe, the water from the closets and traps would then run into the shortest pipe, thereby running against the current of air coming down the short pipe. I say coming down the short pipe because it would not do otherwise according to natural philosophy, the other pipe being the highest would cause the current of air to come down the shortest and lowest and go up the highest pipe which would be the water from closets running against the current of air.

"Will you please give me your opinion on the subject, which plan will ventilate the pipes best?"

[We have reproduced the sketch as sent, which shows no traps, although they are mentioned in the letter.

Our correspondent is mistaken in supposing that the air will necessarily flow down the shorter and up the taller pipe.

The strength of a draft of a chimney or any other flue is proportional as much to its temperature as to its height, and in this case in whichever pipe the product is the greater of its height multiplied by the number of degrees that it is warmer than the outside air, the draft will be the stronger and the air will flow up. So that in this case if the shorter pipe is enough warmer the air in it will flow up and draw down in the taller. If two pipes or flues are equally high and warm no current will start, but if started it will continue, because the descending current will cool one flue, and being somewhat warmed before it escapes from the other the difference of temperature necessary for the flow will be maintained.

If the pipes shown in the sketch are all equally warm the air will circulate as our correspondent supposes.

The discharge from the water-closets will not, we think, materially affect the flow of the air in the pipes above the highest range. When one of the upper closets discharges, it will for the moment reverse the currents of air in all the pipes below it, but they will resume their normal direction as soon as the flow has passed. A discharge from one of the lowest closets would accelerate the air currents in the lower part of the vertical pipes, but would probably reverse the currents in the upper cross pipes and produce little if any effect above them. A discharge from the middle range would produce a combination of these two results.

If it were possible to get a fresh-air inlet at any point below, a much more efficient ventilation could be obtained.

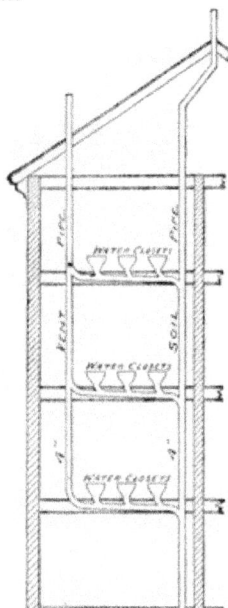

If the lower inlet cannot be had, the arrangement shown is perhaps the next best thing, and we do not think it makes much difference which way the closets discharge as far as the ventilation is concerned, as the effect of their discharge will be only local and temporary.

It should be remembered that in taking in fresh air from above, the two pipes act as flues pulling against each other, and only the difference of their respective drafts is available for ventilation.]

OBSTRUCTION OF A VENT PIPE.

V. N. S., West Newton, Mass., writes:

"A peculiar case of obstruction of a 4-inch vent pipe was brought to my notice to day. It was novel to me and may be of interest to others of your readers. According to the rules of the Board of Health, a soil pipe in the City Hall was extended full size (4 inches) through the roof several years ago. In order to get around roof timbers the stack was carried horizontally for a few feet under the roof, then by a quarter-turn was passed through the roof by a hub into which the flashing was turned, and a 3-foot length of pipe calked in the arrangement being shown by the accompanying sketch. Recently this

piece A rusted off and fell over into the snow guard. When the plumbers went up to replace it, they found the section of pipe B passing through the roof cracked, and in removing it found the quarter-bend C at the bottom entirely closed with rust. The horizontal section D laid over some ceiling joists, and from appearances may have had a slight fall towards the rusted elbow, but certainly not sufficient to hold much water. It would seem as if the small amount

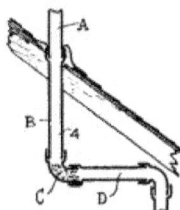

of rain that from time to time fell in the open end caused the rust to start, and as there was never any flow to carry it off or scour the pipe, it gradually accumulated till it closed the pipe. This winter snow accumulated, melted, and finally the water thus formed froze and cracked the pipe.

" This shows the great objection to any horizontal runs, especially above fixtures on a soil pipe where there will be no flushing, and that especially near the roof outlet where direct run is not practicable, one-eighth bend should be used."

[Such stoppage from accumulations of scale rust are frequent in badly planned work in both vent pipes and extensions of soil and waste lines above the highest fixture connections. But it can be entirely prevented by following the New York regulations on this point that all such pipes where offsets are required must be run at an angle not greater than 45 degrees with the vertical, or in other words, one-eighth bends must be used instead of one-quarter bends. At this angle the scale rust will slide down and be washed away. It seems difficult to make some workmen comprehend that a vent pipe through which no water runs ought to have a much greater pitch than is necessary for waste pipes. There is no excuse for this defect in modern work, and such arrangements of piping have not been permitted in New York for some years. The work described was certainly not "according to Board of Health rules," as we know them.]

TRAPS ON HOUSE DRAINS.

WALTER H. RICHARDS, Engineer Sewer Department, New London, Conn., writes:

"After reading the discussion recently published in THE ENGINEERING RECORD regarding the propriety of dispensing with a trap between the house and the main sewer, it appears that a plain statement of the whole question is desirable. It is well known that a current of air through all sewers is desirable, and is conceded by all to be necessary in the house pipes. A more complete ventilation of the street sewers can be secured by omitting the trap and ventilating through the house. A more complete ventilation of the house pipes can be secured with the trap and fresh-air inlet. A partial ventilation of the street sewers may be obtained by perforated manhole covers, and, if the sewers are reasonably well built and flushed, this without offense or danger to the community.

"In some cases, and possibly in all cases, if plumbing and traps were perfect and could be kept so, the trap on the main drain could be omitted without danger arising from the fact that by so doing all houses on a line of sewer are connected together. The question is then a practical one. Can the house plumbing be always perfect? With the most careful planning traps sometimes syphon or the water in them evaporates because of non-use of fixture, and with the most thorough inspection plumbing is sometimes defective or becomes so after inspection. A thoroughly constructed system of sewers may suffer from neglect after a few years I think that the advantage of cutting off each building from the main sewer and adjoining buildings and confining any danger to the sewer air and disease germs in the building itself, more than compensates for a more thorough ventilation of the main sewers or for the slight saving in expense effected by omitting the trap."

AS TO MAIN HOUSE TRAPS AND SEPARATE SEWER CONNECTIONS.

MR. O. P. DENNIS, office of Proctor & Dennis, architects and superintendents, Tacoma, Wash., writes:

"I wish to ask you a question or two, trusting you will give me the desired information.

"First, I will state our city plumbing ordinance provides that there shall be no running trap (or any other kind) in sewer or soil pipe between street sewer and fixtures in house (that is not mentioning traps connected with fixtures). But the ordinance provides that conductors from the roof must connect with the sewer. This I think all right. What I wish to know is if it is not better to have a running trap at a point just outside of the building, and then at a point between sewer and trap connect the conductor pipe. Lastly, what are the rules in New York City in regard to making sewer connections for a double building, or rows of connected houses? Is it the custom to make separate connections, or is one connection allowed for two or more where they are owned by the same owner?"

[As often explained in these columns, it is safer to place the main trap on the house drain between the sewer and the house system either just outside or inside of the house wall, as may be most convenient. A running trap is not, however, a good form of trap. It is better to use a half-S trap with a tee branch, so that the bottom of the inlet is a few inches above the water seal in the trap. (See illustration on page 218 of "House Drainage and Plumbing Problems.") If it is desired to ventilate the town sewers by pipes running up on the outside of the house walls, the leaders from the roof may be connected outside of this main trap. These pipes, however, should be of cast iron and made perfectly tight, especial care being taken that no leakage of gas can occur near windows.

The regulations of New York require that each house shall have a separate sewer connection. This is a wise provision, and should be insisted on everywhere. This regulation is without regard to the size of the house, and is intended to protect each householder, so that no single householder who may keep his drains in order shall be subject to annoyance or risk from the negligence of any other householder. It is also a wise provision, from the fact, that though one owner may build several adjoining houses, they usually in time become the property of different owners.]

TRAPS ON HOUSE DRAIN AT NEWTON, MASS.

T. M CLARK, Boston. Mass., writes:

"I should like to know how long it is since the health authorities of Newton, Mass., began to discourage the use of house traps, as your Pittsburg correspondent states in your issue of December 2, 1893. Five years ago I built two houses in that town. I agree with you in every point on the subject of house traps, believing that, with good plumbing inside, a detached country house, draining only into its own cesspool, is better off without a main trap, which is sure sooner or latter to be stopped with grease or rags, and flood the basement with sewage. Accordingly I laid the drains of these houses without main traps. The Board of Health made a complaint against me, and although they were polite enough to consider the arguments that I submitted to them, compelled me to dig up the drains, and put in main traps, both of which have since been choked with rags so as to discharge the sewage from the upper stories over the basement floor by the way of the basement water-closet, causing a great deal of trouble in both houses. Within the past two years the city sewers have been begun and are partly in use. Is it since then that the Board of Health have begun to discourage the use of main traps? If so I shall come into collision with them again if I build any more houses there. Objectionable as it is to have main traps get stopped, I consider that the risk of this, where houses drain into a sewer, is less to be feared than the risk of infection from disease germs entering the sewers from other houses. I read to-day in the papers that there was an epidemic of scarlet fever and diphtheria in Newton. If the Board of Health has really encouraged or allowed the omission of traps between the sewers and the houses, and schoolhouses, I am not surprised."

[We presume the regulations were modified to suit the new conditions arising from the introduction of a sewerage system in 1890-91.]

TRAPS ON HOUSE DRAINS AT NEWTON, MASS.

ALBERT F. NOYES, West Newton, Mass., writes:

"In answer to the questions asked by Mr. T. M. Clark in a communication headed "Traps on House Drains at Newton, Mass.," as published in THE ENGINEERING RECORD under date of December 16, 1893, I would state that your editorial comment was correct.

"The plumbing ordinances were changed so that the requirements now are that there shall be a fresh-air connection with the soil pipe and a running trap between it and the house drain, except when connected with a public sewer. In that case it is optional with the householder whether he has or has not a running trap and fresh-air inlet.

"This recommendation was made in my report to the City Council on a plan for a system of sewers for the city of Newton, only after a very careful study of the best practicable plan for the ventilation of the sewers, and after receiving the approval of several of the best sanitary engineers who were consulted upon this as well as other details in the design wherein it seemed desirable to depart from the common practice.

" I might here state that the plumbing rules of the Board of Health are very complete in their requirements, and the plumbing is carefully tested by the inspector with either the water, smoke, or peppermint test, and a careful superficial examination of every fixture or trap, before any portion is covered up.

" All traps are to be ventilated and all soil pipes and vent pipes are to be carried full size through, and at least 3 feet above the roof. The sewers are carefully laid, so that every section between manholes can be and is frequently inspected by the use of mirrors reflecting light through the sewer so every joint of the pipe can be seen.

" With the conditions as above described, I doubt if Mr. Clark has ever known of a case where there was a flow of air from the soil pipe into the house. On the contrary, my experience has been that should an opening occur in a ventilated soil pipe the flow of air is invariably into the pipe and not out of it.

" The introduction of a large volume of fresh air into the sewer and through the soil pipes creates conditions not only unfavorable to the life and generation of disease germs, but favorable to their destruction. It also produces conditions unfavorable to the formation of what may be known as sewer gas in any condensed or concentrated form. If the gas is created at all, it is so diluted as not to be offensive or dangerous to health.

" I have had a canvass of all the cases of scarlet fever or diphtheria which have occurred in Newton as reported to the Board of Health since August 10, 1893, and find that out of 41 cases of scarlet fever but four cases have occurred in houses connected with the sewer, and out of 25 cases of diphtheria but two cases have occurred in houses connected with the sewer.

" The cases have also occurred in sections where the sewers have either not been laid at all or laid this season only, and I do not find a single case of sickness of a person attending a school where the fresh-air inlet or running trap has been omitted."

DOES DISCHARGE OF STEAM INTO EARTHEN DRAINS INJURIOUSLY AFFECT THEM?

W. M. Dexter, East Providence, R. I., writes:

" Can you give me any information upon the effects of steam from the waste of steam heating (coming direct from a steam boiler, pressure, say 60 to 80 pounds) upon vitrified drain pipe cemented at joints thereof, the pressure in the drain being from two to six pounds, perhaps?

" There is a manufactory which is turning steam, such as above, directly into drain pipes with ordinary traps, without trying to condense same in water. Will the use of steam in the manner above be injurious to drain pipes in the long run or not ? "

[William Webb, foreman of the New York Bureau of Sewers, has observed that iron pipes having steam discharged into them become weakened, displaced, and more liable to break and leak, and that tile pipe becomes saturated with moisture, crumbles away, and breaks easily, and loses the elasticity and clear metallic ring that new pipes have when struck; that cement in brick sewers is entirely disintegrated by the action of steam, and the loose brick may be easily removed, only dry sand remaining.

Mr. Webb does not think the cement used in the joints of tile pipes is sufficiently exposed to be injured, and has never observed any indications of its destruction by contact with steam.

The sewer regulations of this and other cities require that steam shall not be discharged into sewers. Besides the question of its effect on the materials of which the sewer is constructed, the sudden presence of it makes it unsafe for men to enter sewers for purposes of inspection, and the smell in them is rendered much worse than it otherwise would be.]

ARRANGEMENT OF TRAP VENTS.

G. F. J, Denver, Colo., writes:

" Will you kindly settle a question in dispute. I send two sketches of a kitchen sink with connections, Fig. 1, as placed by the plumber, Fig. 2, as contended by me to be better, as there is no contraction in the size of pipes when continued direct as a ventilating pipe."

[Figure 2 is preferable. The bend at the foot of the 2-inch vent pipe, as shown in Fig. 1, is liable to stoppage from falling rust if the pipe is of wrought iron. Likewise the bend at the foot of the vent in

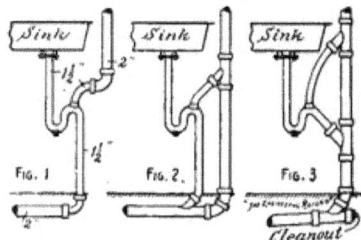

Fig. 2 although in this latter case it might not affect the trap. We add Fig. 3, as better practice than either of the others. It will be noticed that the Y branch for the trap vent is put in just above the bottom of the sink, so that in case of stoppage in the waste pipe below, it would be indicated by the waste water not running off. In Fig. 2, if a stoppage occurred in the waste, the trap could discharge through the vent pipe without its being known. The discharge, as shown in Fig. 3, helps keep the bend flushed out and the cleanout is desirable in such positions for the removal of obstructions.]

RUNNING A VENT PIPE INTO A SMOKE FLUE.

J. Reynolds & Son, Philadelphia, Pa., write:

" The Engineering RECORD contains an article headed ' Objections to Running a Vent Pipe into a Smoke Flue.'

" We ventilate apartments, etc. into the smoke flue, but not into a plain chimney. Our method is to run a cast-iron pipe from the cellar to a few inches above the chimney coping in the center of said flue. For example, if a flue measures 12x12 inches in the clear, we run an 8 inch cast-iron pipe for smoke immediately up the center of said flue, properly securing each length of 5-inch by iron stays; we then use the brick flue surrounding said pipe for the introduction of all ventilating ducts. It works like a charm and there is no conflict."

[In this case the smoke flue is an iron pipe, and the brick flue, possibly built for a smoke flue, is made to serve the purpose of a ventilating flue. There are doubtless many cases where the conditions were favorable, and satisfactory results were secured, yet it might be well to consider a variety of conditions where satisfactory results would not be secured. We therefore should not advise laying down hard and fast rules in an ordinance to cover all cases.

Vent pipes from the bowls or seats of water-closets can pretty safely be connected with any warm flue in the walls of a building, provided said flue is used for no other purpose and connected with no other flue, unless under conditions mentioned below.

When the waste heat from a furnace or boiler chimney is carried through an iron pipe within a flue, the flue becomes an aspirating shaft, and into it the water-closet rooms may be ventilated; but should there be another vent shaft in the building, then the object of the separate vent shaft for the closets might be vitiated, and the necessity for bringing them together at the top made apparent.

If there is a systematic exhaust ventilation (by fan or aspirating shaft) in the building, this flue should enter the main outlet duct just below the fan, if one is used, and pretty near the top of the heated aspirating shaft—certainly above all other ventilating flues to the shaft which lead from rooms. In such a case, presumably, it is best to vent the water closet room into the same flue as the vent from the bowl or seat, as thereby the possibility of drawing air down one flue and up another is prevented, which might be the case with exhaust ventilation should there be two flues from the same water-closet room, and which in all probability would be the case if they went separately to the coping or outer air.

If the ventilation of the building is plenum ventilation, then the flue, presumably, is best when run to the atmosphere direct, with a separate flue for each purpose, unless indeed there is a combined plenum and exhaust system, when the flues are best arranged as though they were for exhaust ventilation alone.

Houses or buildings warmed by furnace, taking air from outside or by indirect steam radiation with or without a fan may be considered " plenum ventilation," while all warmed by fireplaces or stoves, having the air drawn out of the rooms by waste heat or otherwise, are vacuum or " exhaust ventilation." With a condition of plenum within a house, chimneys and flues " draw," and all flues will show an outward current of air unless they are proportioned too large. With vacuum ventilation, if there is more than one outlet or aspirator (fan or otherwise), the stronger is apt to draw the weaker, and will do so unless there is means of admitting air below or through windows sufficient to supply the outlets which have power to draw.

Small aspirating shafts are made by providing a flue on each side of the kitchen range flues, and boiler and furnace flues, with partitions of one brick between, or by using earthen or iron pipe, or special earthen flues with partitions, within a large brick flue.

The **cast-iron pipe** within a flue makes the best **aspirating shaft, but** we know of a case where a

building was so tight that the air to supply the boiler furnace was drawn down the aspirating shaft, up which the very same boiler smoke pipe ran, until the boiler was supplied with air directly from the outer air.

In our modern kitchens it now and then happens that the flue running parallel with the range flue for the purpose of drawing the hot air and fumes of cooking from the range under the hood, works the wrong way; in fact, cold air comes down to supply the fire unless a window or door is open, some of the buildings being so air-tight. In such a case it would not be a good thing to have water closets vented into the same flue]

THE BACK-VENTING OF CLOSET TRAPS.

CANADIAN, Berlin, Ont., writes:

"In a small house the plumbing fixtures usually consist of a bath and closet upstairs and a kitchen sink down stairs. The traps of the bath and sink are usually required to be vented, but is there any valid reason why the closet trap should be vented? Even if the bath waste were connected with the lead bend under the closet, which is not now considered good practice, the discharge from the bath would hardly cause the closet trap to syphon, and of course the discharge from the sink below cannot affect the water in the closet trap when the soil pipe is continued through the roof. A vent to the closet trap would hardly be required to prevent it syphoning itself dry while being discharged; that would scarcely be possible with a 4 inch trap and a 1½-inch supply, so that I am at a loss to know why, here in Canada, we have not in the market a single porcelain washout closet without a horn for the trap vent."

[There are undoubtedly cases where the traps are few in number, and their relative position and particular connections to the soil such as to make back-ventilation of traps an unnecessary precaution, but the cases where back venting can be omitted with safety are so rare and the uncertainty of how the traps may be affected is so great that the rule for the greatest safety to the greatest number must govern, and back-venting be considered a factor of safety well worth introducing into the work. It is like all generalizations not applicable or requisite for certain isolated cases. It has, however, value for air circulation and ventilation aside from the mere protection of the trap against syphonage, but in the case cited these considerations do not appear to apply.

Referring to your immediate inquiry, the form of closet and its relation to soil pipe and the method of venting the soil pipe at the roof, would have some influence on the question of syphonage. We have seen the seal of hopper water-closet traps affected by a strong wind blowing across the open end of the soil pipe. The actual seal of water in the trap is about 1½ or 2 inches, representing a pressure of 1.15 ounces per square inch. The atmospheric pressure is 2.40 ounces per square inch, and any influence increasing the pressure 1.15 ounces on one side of the trap or reducing it 1.15 ounces on the other would unseal the trap. The conditions are so delicately balanced that special circumstances of construction would outweigh theoretical considerations. A practical test repeated under various conditions of atmosphere, coincident discharge of fixtures, etc , would best answer your question for the particular case named It has been found experimentally and in a vast field of practice that back-venting properly done practically insures the seal of the trap, and as a matter of insurance the outlay incidental to it seems to be well invested against the hazard.]

WHICH IS THE BEST METHOD OF CON-NECTING HOUSE DRAIN TO SEWER.

Jay, of Riverside, Cal., writes:

" This city is provided with a system of pipe sewers ranging in size from 6-inch to 12-inch. Flush tanks on laterals Main sewer flushed from canal every week in addition Rainwater excluded. Manholes with perforated covers at intervals As I understand it, there are the following well-defined methods of making house connections, it being understood that the house drain is extended through the roof in all cases: (1) Having no trap on the house drain, and ventilating the public sewer through it; (2) having a trap outside the building and a fresh-air inlet on the house side of the trap; (3) same as (2), but having, in addition, a ventilating pipe extending from a point between the trap and main sewer to a point above the roof outside the building.

" Will you kindly tell me which of these three methods would be considered preferable in the case of a system of sewers such as I have described? The practice here heretofore has been No. 3, but there are not many buildings plumbed that way. The plumbers favor No. 1, while I might favor No. 2 under other circumstances; here, however, we have many detached closets and sinks in which any other method than No. 1 is hardly practicable. The climate is semi-tropic, and all houses are well ventilated by open doors and windows, and there are few large buildings "

[In our opinion, the choice lies between systems 1 and 2. We have usually preferred No. 2, simply because an ideal condition of the sewers of the average town cannot be permanently depended on.]

ARTIFICIAL HEAT IN VENT PIPES.

H. G., Melbourne, Victoria, Australia, writes:

" Referring to your answer to a correspondent you say that you prefer artificial heat for local ventilation. Where should a gas jet be inserted to give the best results? Would it answer just as well if it is put in the upcast pipe on the bottom floor or any of the intermediate floors, and would not an atmospheric burner be the best to use, as they give off a much greater heat?"

[As the efficiency of ventilation by artificial heat depends mainly upon the height of the column of heated air, it is obviously important to put the burner or other source of heat as low as possible, for the same reason that a fire in the basement usually draws much better than one on the top floor.

As to the use of an atmospheric or Bunsen burner, we should say that provided the gas is completely consumed, which is the case when the flame is clear and smokeless, it makes little or no difference what kind of a burner is used, the total heat resulting from the combustion will be the same.

The ordinary burner is more readily obtained and has the incidental advantage that its light shows more plainly than the other whether it is burning or

not. It is true that it radiates more heat than the Bunsen burner and that the heat thus radiated is most of it lost as far as heating the air currents is concerned, but we should not think the difference sufficient to warrant any extra trouble or expense in procuring atmospheric burners

The ordinary burner is very easily made into an atmospheric one by slipping over it a slightly tapered sleeve, say 3 inches long and enough larger than the burner to permit a current of air to flow up all around it, the upper and smaller end of the sleeve extending slightly above the top of the burner. Such an attachment, called " Dare's burner," is, or was, patented in this country and could probably be obtained of any dealer in gas fixtures.

Our correspondent could readily determine the relative efficiency of different burners in producing air currents by a very simple experiment whose results we should be pleased to learn and publish. Let him place in or over the outlet of his vent pipe, or in any place where it can be seen and at the same time moved by the current of heated air, a wheel of thin metal, pasteboard, or even of paper, then if he will count the number of turns it makes in say one minute with each kind of burner, he will learn which of them gives the strongest draft.]

METHOD OF TESTING PLUMBING IN MINNEAPOLIS.

We are indebted to Mr. J. M. Hazen, Assistant Plumbing Inspector at Minneapolis, Minn., for a blueprint of an apparatus he has designed for testing soil and waste pipes, from which the following illustration is made. Mr. Hazen describes his apparatus and his method of making a test as follows:

To prepare new work for this test, all the iron soil and ventilation pipes must be roughed in, running trap and fresh-air inlet included. If the water-closets have a trap, then calk in a 4-inch lead bend; otherwise, calk in the lead trap and solder a piece of

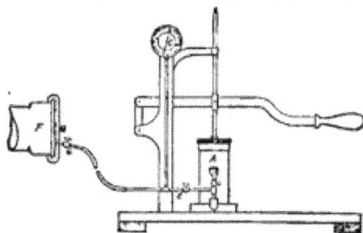

APPARATUS FOR TESTING SOIL PIPES.

heavy sheet lead over the top. Wipe on the trap ventilation and connect with vent pipes. Calk in ferrules in all openings for waste or ventilation, with a short piece of lead pipe wiped on. Pinch these ends and solder them. If the ventilation connects with stack before reaching the roof, then you only have to close the bottom and top of stack, put the proving apparatus on the fresh-air inlet, and apply the test,

Thus we have the whole system of plumbing under test at the same time. Whereas, if only the soil pipe were tested, there are yet three joints to make for every fixture, which will never be under proper test. Instead of a short piece of lead waste or vent pipe being calked in, I would, if circumstances would admit, connect up the entire waste with trap attached, take out the crown vent and connect it with main ventilation pipe. Then, if the work stands a pressure of 10 pounds to the square inch, and holds up to that, it is absolutely tight beyond question, for all fixtures that go on are outside of the traps, now under test.

I consider 10 pounds air pressure ample test for ordinary plumbing. If the work stands at that pressure, it will, as a rule, stand 15 or more.

I was told at first that cast-iron soil pipe could not be calked with lead and oakum, to stand an air pressure of 10 pounds, but that theory has vanished, and good workmen have no trouble in making their work as tight as a glass bottle. The greatest danger is in calking around brass ferrules, and great care should be taken lest they " buckle in."

To test a job of plumbing with the proving apparatus, in the absence of a three-eighths nipple to connect the rubber hose to, it is necessary to have a 2 or 4-inch iron plug with iron gasket to fit on the shoulder of the pipe in the hub " F," held in place by a clamp over the end of the hub, with a set-screw in the center to screw down on the plug. Into one side of this plug, screw in a short nipple and cock " G." To attach a hose from the pump close cock " D " and open cock E. Work the pump until the gauge C shows five pounds pressure, then close cock E. If the work is absolutely tight the indicator will remain at five pounds; if defective the indicator will go down. Now unscrew tap of ether cup B; open cock D and let the pressure off from the pipe; close cock D; put one ounce of ether in the cup; screw on cap; open cock D, to let the ether down, and at the same time begin to work the pump; close cock D; pump up to five pounds pressure, and close cock E. The ether will indicate where the leaks are, which the plumber will at once calk tight. Test the work again at 10 pounds pressure, and if the indicator stands at that the work is absolutely tight.

To test the pump, put on 10 pounds pressure, close cocks G and E, and if the indicator stands the pump is tight. A little soap and water put on the leaky joints with a brush will show the exact location of a leak by the formation of bubbles. Plumbers, architects, and builders here are perfectly satisfied with its ability to make a thorough test. That, together with its simplicity, commends it at once.

JOINING AN IRON AND EARTHENWARE DRAIN.

DEPARTMENT OF PUBLIC WORKS,
CITY ENGINEER'S OFFICE.
DULUTH, MINN., November 25. 1892.

To the Editor of THE ENGINEERING RECORD.

SIR: Will you please discuss in your department of " Notes and Queries," the different methods of joining a 6-inch tile house drain to a 4-inch iron soil pipe. such as the introduction of the end of the soil pipe a foot or so into the tile and cementing well outside, use of a tile reducer 6 inches to 4 inches, use of an iron reducer 6 inches to 4 inches, use of a 4-inch iron "double hub," fitting into the body of the 6-inch tile pipe, and any other methods which you may know of, giving the advantages and disadvantages of each, with your judgment as to the best method

X.

[In answering our correspondent's inquiry we shall assume, to prevent any misunderstanding, that by " soil pipe " be means the main drain of cast-iron pipe inside the house walls, receiving, through various branches the drainage from all the plumbing fixtures and, as a rule, running horizontally for a greater or less distance through the cellar before passing out at the front wall. This main pipe is in New York known technically as the " house drain," to distinguish it from the vertical lines called " soil " and " waste pipes," and also from the continuation of the main

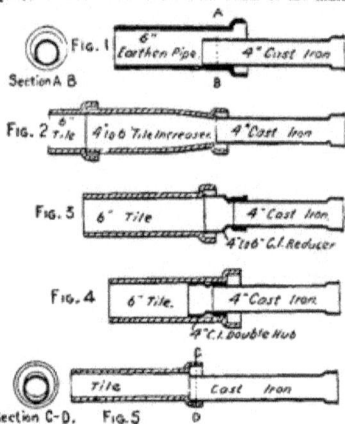

drain outside the house and extending to the public sewer, which is known as the " house sewer." The "house drain," whether above or below the cellar floor should always be constructed of iron pipe carried through and several feet beyond the house, vault, or area wall. From this point hard, salt glazed earthenware pipe may be safely used under proper conditions as to foundation, etc. If, however, the practice in Duluth corresponds more nearly to that of Chicago than New York, and the " house drain," as defined above, is commonly laid with earthenware pipe, so that the connections in question are made inside the house, we should answer that such arrangement is no longer considered good practice in this country, the safe rule being to use only iron pipe within the walls of a building.

Presuming, therefore, that it is the connection of a cast-iron " house drain " with the earthenware drain outside the walls, the accompanying sketches will explain the methods suggested by our correspondent. In Fig. 1 the end of the 4-inch extra-heavy cast-iron pipe is introduced into the earthen pipe for any desired distance and the space above the iron pipe and the entire hub carefully filled with good cement mortar. This gives an easily made and strong connection, with a good and fairly continuous line of flow, and applicable to all cases where the iron is at least one size smaller than the earthen pipe. The objection that the pipes are not properly centered seems of little practical importance. It may be noted in passing that under ordinary conditions either the iron pipe would be larger or the tile smaller than those given in the inquiry.

Figure 2 shows an earthenware "increaser," with the connection made between pipes of the same diameter. However, as the earthen hub is much larger than a corresponding one of iron, there is always a danger that the unskilled workman will lay the pipes as indicated in Fig. 5, where the discharge will be partially obstructed by the projection of the earthen pipe above the line of flow, even if no burr of cement has been left inside the pipes. In addition to this, the ordinary earthenware hub is not deep enough to make a very strong cement joint with iron pipe. The "increaser" is, of course, simply another length of earthen pipe, which it might perhaps be inconvenient to set at a proper distance beyond the walls.

Figure 3 is a cast-iron "reducer" calked on the end of the 4-inch iron pipe and cemented into the hub of the 6-inch tile. Though more expensive in labor and material, there seems to be little advantage in this method, while there are the possible objections noted in Fig. 5, and also a greater difficulty, because of the reducer, in removing any cement which may be forced inside the pipe in making the joint.

In Fig. 4 a 4-inch iron "double hub" has been used, as suggested, giving a connection similar to that in Fig. 1, but at an increased expense. To do this, however, the so-called "standard" or lightweight cast-iron pipe must be used, as this fitting in the "extra-heavy" grade measures about 6¼ inches outside. The former grade of pipe has been very generally discarded in favor of the heavier and safer material which is now commonly required by the city plumbing regulations throughout the country. It is, however, quite possible to modify this method by calking a ring of 5-inch iron pipe on the end of the 4-inch so as to make the pipes more nearly concentric when introduced into the body of the tile. This would make a good connection with slightly more labor than No. 1.

All things considered, therefore, it would seem that the method shown in Fig. 1, while apparently less careful and accurate than the others, will under the ordinary conditions of careless workmanship and the like, give a thoroughly good and strong connection at a less expenditure of time and material than with any of the others suggested.]

SUBSOIL DRAINAGE OF HOUSE FOUNDATIONS.

THOMAS LLOYD BETTON, of Kansas City, Kan., writes:

"In preparing plans of foundation for a building we have arranged drain tile running around foundation. Tile is hubless, such as is used on farms for draining

farm land. It runs about 8 inches above footing, and the architect wishes us to connect same to sewer, which has to convey water from two water-closets, two washstands, two bathtubs, two sinks, as you will see it shown connected in our plan. We say it is not proper to connect it to main sewer, for in case trap should become stopped up in any way everything from house would be forced back into the hubless tile, and would be the means of saturating the ground around building before it would be discovered. Please give us your opinion as to whether the tile should be laid above or below footing."

[Two-inch cylindrical land tiles with slip collar joints should be used with Y branches and curved pieces for change in directions. The tiles should be placed on a plank bottom laid to grade of about 6 inches in 100 feet. The invert of the tile should be level with the bottom of the footing course at the front of the house (or at the point of discharge of the tiles), and should rise in both directions from this

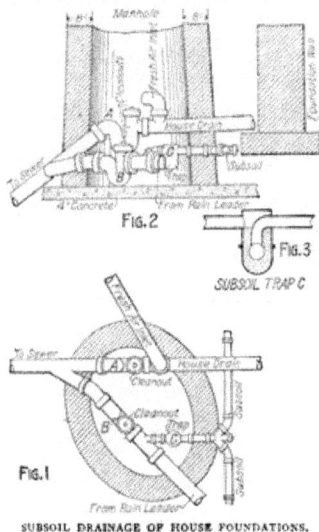

Fig. 2

Fig. 3

SUBSOIL TRAP C

Fig. 1

SUBSOIL DRAINAGE OF HOUSE FOUNDATIONS.

point to a "summit" at the back (or opposite side) of the house. Care should be taken to prevent mud from flowing into the drains while being laid. The joints in the tile are sometimes made with cloth or tar paper wrapping instead of slip collars. It is a good plan to fill in the trench with broken stones to a depth of 6 inches over the pipe and to cover this with salt hay or straw before filling in the earth. The outside wall of the house should be thoroughly plastered and troweled smooth with Portland cement mortar (1 to 2) and allowed to set before the tile is laid. This plaster coat should extend from an inch or two above the surface of the ground to the bottom of the footing course. A coating of bitumen laid on hot outside of this will make the foundation walls doubly proof against water and dampness. The discharge from the subsoil drains should not connect

with the sewer if it can be otherwise disposed of.
In case it is impossible to provide other means of
draining the subsoil line, it may connect with the
house sewer under the following conditions:

First —It must be independently trapped against
the entrance of drain air.

Second.—The trap must be supplied with water
from an unfailing source, and had best be connected
with a rain leader or yard drain. The ground-water
supply alone should not be relied on to furnish seal
for the trap, as during dry seasons the subsoil drains
may carry no water.

Third.—The subsoil drains should be effectually
trapped against the back flow of sewage, which may
be occasioned by a stoppage in the main drain beyond
the junction of the subsoil line. For this reason it is
perhaps best to connect the subsoil line on the sewer
side of the house drain; as stoppages in the main
house trap, which are not infrequent, would not then
affect the subsoil line. For the purpose in view, the
prevention of the back flow of sewage, a mechanical
trap with a floating ball should be used. A failure
in the operation of the mechanical trap, in case of
stoppage in the house sewer beyond the subsoil
junction, will result in flooding the subsoil drains
with house sewage which is not liable to be discovered until some of the house fixtures fail to
operate properly or until the flooded subsoils become
offensive. This is a risk which cannot be overcome,
and it is essential, therefore.

Fourth.—That the traps be located where they
can be frequently and readily inspected and
cleaned.

A convenient arrangement of traps to cover the
points enumerated is shown in Figs. 1 and 2, in connection with which it will be noted that, when working properly, the trap B is supplied with water from
a yard drain or rain leader. The trap C, Fig. 3, is a
ball float mechanical trap allowing the subsoil water
to flow into B, but preventing back flow of water or
sewage when the ball is seated. If C runs dry, B
prevents the back flow of drain air, and, in case of
the back flow of sewage, the first entrance of sewage
into C would float the ball and seat it. The rain
leader should be carried up with iron pipe and tight
joints to the height of the fresh-air inlet, or above
the level of a yard drain, so that sewage backing
into the leader may not escape until it has found
its way out at some opening, where it will be a
nuisance and be brought to the attention of the
householder.

The traps, as shown in the drawing, can be grouped
in an oval manhole 4x3 feet.

From what has been said it is evident that you
were right in saying it would not be proper to connect the subsoil drain as shown in your sketch. The
correct method of laying the tile has been indicated.
If laid below the footing course, it may undermine
it and crack the wall; if laid above the footing
course it is not draining the subsoil to the greatest
available depth. Plans for removing ground water
from a house to sewer are also shown on pages 45
and 46 of "House Drainage and Plumbing Problems."]

CAUSE OF STOPPAGE ON A HOUSE DRAIN.

ALLAN M. BARROWS, architect, Chicago, Ill., writes:
" Will you kindly furnish me through your valuable
journal with a comment upon the scheme of drainage
as shown by the sketch inclosed and described here-
in. A contract for plumbing has been recently executed in a dwelling at Oak Park, Ill., in accordance
with accompanying sketch. A few days after its
completion and during a heavy rainstorm water was
found to be overflowing the basement water-closet.
An examination quickly made resulted in the discovery that the running trap in the chamber had
been chocked full of paper; as also the vent pipe
from the trap on the house side. The toilet paper
had accumulated at the first bend and the flush from
one closet at so short rise had failed to discharge it
past the seal dip into the sewer. This stoppage had
caused the pipe to fill with rainwater from the
leader and back up, overflowing the closet as described.

"The owner in his efforts to locate the blame had
called upon the plumber, who, in response to the

BASEMENT PLAN
Showing horizontal lines of pipes.

owner's inquiries, said that he had followed the architect's drawings and specifications, which required
the installation of running trap chamber, vent, etc.,
though contrary to his ideas of good plumbing.
He further said that he deemed a trap of any sort
placed in the soil drain an unnecessary obstruction,
and if this were his job he should remove the
trap chamber and all, and insert in its place a straight
section. In my next interview with the owner I was
subjected to considerable reproach, and it was only
through the suggestion that for the present we remove the chief source of danger that the whole
system was spared destruction. The rainwater
leader was disconnected, and then I suggested that
in fairness to me he should consult an expert in this
matter. Time has now passed and he has failed to
do this, so I have written you asking your opinion
which I am aware he will consider authoritative, for
my own justification."

[Theoretically, every trap used is an obstruction,
fixture traps included; their use is deemed the lesser
of two evils. The position of THE ENGINEERING
RECORD on the need of a trap on main house drains
is again explained in the issue of December 2, 1893.
From the sketch submitted we should look for the
cause of stoppage either in a drain laid without
sufficient incline or in some foreign substance caught
or lodged in the main trap. Or the basement closet
may be set too low and be inadequately flushed, or
the drain beyond the rainwater leader connection
may be too small to serve the roof area drained in
case of a heavy shower. In that case there would
be a back flow at the basement closet. Assuming
that this is a case where a trap is only a proper safe-
guard, we should look for the trouble then either in

defective details of construction, insufficient fall in pipe or insufficient size, or improper use of the drain, this latter a very common cause of stoppages in new work. We see no cause for stoppage in the general design submitted, but prefer a form of trap in which the drain enters a little above the dip.]

DEPOSIT OF RUST IN THE BEND OF A SOIL PIPE.

NEW YORK, April 30, 1887.

SIR: The following unusual experience which came under my notice recently is worthy of record. The drainage and plumbing of a five-story dwelling in the city was submitted for examination and revision. There were two vertical columns of "standard" iron pipe; the one marked A in Fig. 1 connecting with bathrooms and sink on the top floors was carried by a 4-inch pipe close to the roof and there pieced out and extended above it by a 3-inch lead pipe drawn together at the top and perforated for ventilation. The column B, which was 4 inches at the base, reduced to 2 inches for the top floors, connected only with hand-basins, and was also pieced out and reduced by lead pipe drawn together and perforated above the roof. There was no trap on the main drain and no ventilation of fixture traps.

To "back-air" the traps and ventilate the drains two columns of "extra heavy" 3-inch iron pipes were put in parallel to the soil columns and all traps vented into them. The soil and back-air columns A, Fig. 2, joined near the roof and were continued by a 5-inch pipe above it. Column B, Fig. 2, was to have been similarly arranged, but it was found more convenient to carry the 2-inch soil and 3-inch vent above the roof independently, each being increased to 4 inches before passing out. The main drain was trapped at the front wall and vented to the street. Upon completing the work the smoke test was applied at the newly inserted fresh-air inlet at the curb. Columns A and C, Fig. 2, filled at once and poured out streams of smoke, but none came from B, even after A and C were plugged up. Two pails of water poured down B showed there was a stoppage, the water rising in the pipe above the roof. The top of the column was broken out at once and cut down to a point (*e*, Figs. 1 and 2) 5 feet below where the old 2-inch pipe had been increased to 4 inches. It was then seen that in passing through the top floor an 18-inch offset had been made with quarter-bends. The horizontal internal of 2-inch pipe at this point was found completely choked for several inches with a closely compacted mass of iron-rust flakes, which had fallen from the original 10 feet of 2-inch pipe above the elbow, or from a surface of about 5 square feet. The total amount of rust scales taken out weighed something over a half a pound, some of the scales being as large as a dime. The pipe did not appear seriously corroded.

How long the pipe had been choked and without ventilation is not known. The stoppage being above the highest fixture, did not speak for itself, nor would the peppermint test have necessarily shown it. The

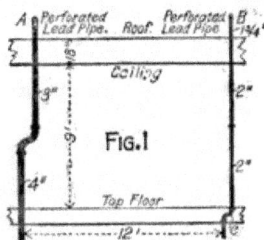

FIG.1

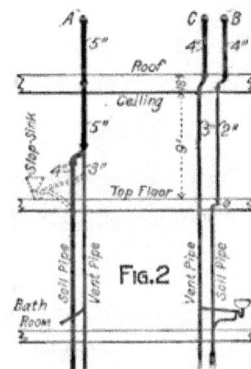

FIG.2

moral of direct lines without elbows is emphatic, and it is not improbable that many of the roof vents in old work are now inoperative from the cause here discovered. ALBERT L. WEBSTER.

METHOD OF POURING LEAD IN MAKING IRON WATER-PIPE JOINTS.

BIG RAPIDS, MICH., April 29, 1886.

SIR: Will you please be kind enough to inform me what will be the best material to use for running lead joints in 12 and 4-inch iron pipes, such as water mains? I generally use putty or clay for iron pipes up to 8-inch. I didn't know but that they use something different for large water mains to run lead in.

Yours truly, C. B.

[The best practice seems to be as follows: For pipes up to 12 or 14 inches diameter a luting of moist clay is used for covering several yards of oakum so as to form a roll for closing the exterior of the annular space excepting at the point where the lead is to be poured.

For larger pipes a wrought-iron clip of annular form, hinged at the bottom, and with tightening screw at the top, is used for covering the space. A luting of clay is used with these, also for tightness.

In cases where in large pipes a solid joint is poured (no oakum gasket being inserted in the hub) an iron ring is placed inside the pipe to cover the joint, and the joint is calked after pouring both inside and outside.]

EXPANSION OF SOIL PIPE LINE.

W. ROGERS, Philadelphia, writes:

"I would like to ask a few question regarding the plumbing of tall buildings in a cold climate like that of Chicago. Is the expansion and contraction in the soil pipe sufficient to break out a straight connection as shown at B in the accompanying sketch? This represents one floor of a 12-story building, the soil pipe being about 175 feet high, hot and cold water to be used, and the plan of other floors being identical with that shown. Would a swing joint like that at C be enough if the closets were not fastened to the floor? Would two swing joints be required if they were fastened to the floor? Is there anything better for an expansion joint than this?"

[We assume that the inquiry refers to wrought-iron pipe. A pipe of this material 175 feet high, exposed to a change of temperature from 30° Fahr. to

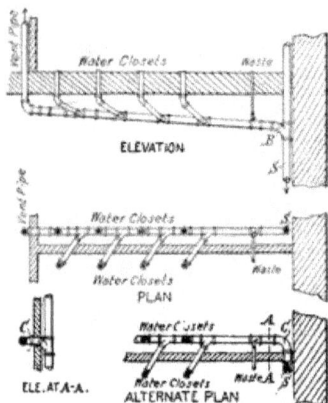

ELEVATION

PLAN

ELE. AT A-A.

ALTERNATE PLAN

110 degrees, or a difference of 80 degrees, would elongate about 1 inch. If supported at the bottom only, the worst arrangement, the top would move through a space of 1 inch between the extreme temperatures. Assuming the normal temperature to be 70 degrees, there would be a movement at the top of the pipe of one-half an inch upward and one-half downward from the normal position of the pipe. If the lateral connecting with B were absolutely rigid there might be danger of splitting B or springing a joint. As a fact, however, there are modifying conditions. The range of temperatures assumed is probably never experienced; presumably not much more than half the change in temperatures actually occurs. The vertical parts of the building are subjected to very much the same change in temperature as the pipe, and the vertical parts of the entire building move together. The long lateral for the water-closet branches would spring enough to take up the movement in B without injury, and the water-closet branches, if made with lead bends, as is customary

and desirable, would take up part of the movement and protect the earthenware of the closets. A further protection can be introduced by supporting the vertical pipe by a clamp half-way up the building, and allowing the expansion to move in both directions from this middle point instead of supporting at the base. This arrangement reduces the extreme movement at the ends with the range of temperature noted to one-half inch, or one-fourth inch up and down from the normal. The swing joint C referred to seems unnecessary if the laterals are connected with a chance to spring and the verticals are hung in the center of the columns. With the hot-water pipes it is advisable to use about two expansion loops or expansion fittings in the vertical run. We note a bad rust bend in the drawing at the base of the vent pipe, which should be removed.]

ARRANGEMENT OF A GROUP OF CLOSETS.

J. W. BARNETT, City Engineer, Athens, Ga., writes:

"I send herewith a sketch (Fig. 1) of a plumbing job proposed to be done for the University of Georgia in this city and would like to know your idea as to the size of the main vent. There will be 11 water-closets set in the basement of a three-story building, flushed by one seven-gallon tank each. It is quite probable that five of them will be in use at the same moment. Is there danger of syphonage with a 2 inch main vent?"

[Assuming the simultaneous discharge of five closets to be the maximum tax put upon the pipes and that the individual seven-gallon tanks will empty in 10 seconds, the rate of discharge of the five tanks combined would be 23 cubic feet a minute. A 4-inch pipe laid with a fall of 1 in 25, or about one-half of an inch to 1 foot, will discharge about 30 cubic feet a minute with a velocity of about 6 feet a second when running full. We assume that the lateral soil branch will practically run full and that air must be supplied through the main vent at the rate of 23 cubic feet per second. The area of a 2 inch pipe is one-fourth of the area of a 4-inch pipe, and the velocity of air in

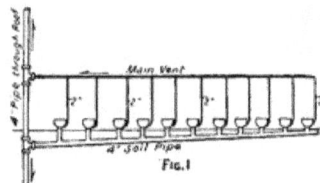

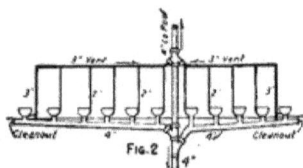

Fig. 1

Fig. 2

245

the main vent to supply 23 cubic feet a minute would be four times the velocity of discharge in the 4-inch branch, or about 24 feet a second, or 1,440 feet a minute. The resistance in a 2-inch pipe to this air discharge is about one ounce loss of pressure per square inch. The pressure, or weight per square inch of a 2-inch water seal, is about one ounce per square inch. The seal would therefore theoretically just bear the strain. The discharge of additional closets would theoretically syphon adjoining traps. In practice there are other considerations. The sudden discharge of the closets will send a volume of water into the soil branch, which will act like a piston and compress the air in front of it. This will find relief in the lines of least resistance, probably in part through adjoining 2-inch closet vents and in part through the 4-inch soil extension, and when the discharge is past, the air current will be reversed to supply the vacuum at the closets discharged, and an oscillatory movement will follow. In addition to this, the rush of water will act on the ejector principle and carry air with it, and bends and angles in the run of vents will retard the prompt supply of air. These unknown quantities modify the theoretical considerations so materially that the sizes figured can only be taken as a general guide of minimum sizes. From experience we would advise a 5-inch soil branch, a 5-inch soil extension to the roof, and a 3-inch main vent, especially with syphon closets. The ends of the soil branch should connect with the vent main.

Another arrangement is shown in Fig. 2, using 4-inch soil branches and 4-inch soil extension to the roof if a straight run of pipe can be obtained. This is a preferable arrangement if practicable.]

TROUBLE WITH BACK-WATER.

H. R. Richardson, of Hackensack, N. J., writes:

" Will you inform me of some way to prevent back water from sewer beside using a back-water trap or valve? The accompanying sketch is the plan of some work which I have done, and it gives me considerable trouble. When the work was first put in I did not use the valve A and the water backed up in the cellar drain, causing a considerable nuisance. I then put in the valve A, which worked very well, as

nothing but water passes through it. But I still found another nuisance; as I had stopped the water from entering cellar drains, it took another course, following the main soil pipe and coming out of servants' closet C. Now I thought of putting another valve at B, the same as I did at A, but as water-

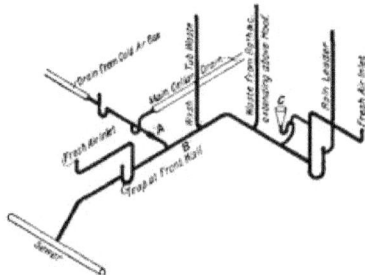

closet refuse is to pass through it I am opposed to using a valve like A, as there is a danger of its stopping or preventing it from working. If you can inform me of some other plan you will confer a great favor to a constant reader of your valuable paper."

[It is impossible to keep back water out of a cellar or house drain without the use of some sort of a valve trap, and we should advise the substitution of such a valve for the running trap shown on your sketch near the front wall. You will find some backwater traps described in our advertising columns. It is a mistake to connect the cellar drains which are to remove ground water from the house with your sewage drain as appears to have been done. At seasons when there is no ground water direct communication is made with sewer pipes and sewage matter, which should be avoided. In " House Drainage and Plumbing Problems," which is a collection of articles from this journal, you will see two plans illustrated on pages 45 and 47, under title of " Ground-Water Drainage of Country House," and " Ground-Water Drainage, City House," that will suggest a safer method of removing ground water.

Finally, cannot your town authorities prevent the water from backing up in their sewer? Possibly this action may be due to some local defect that could be corrected.]

DOMESTIC HOT-WATER SUPPLY PROBLEMS.

HOT-WATER BOILER WITH HOT-WATER HEATING COIL.

A STAUNTON, VA., correspondent sends a sketch of a bath boiler, here reproduced, explaining that in that section of the country they have a very strong limestone water. What he wants to know is whether with this form of arrangement the stopping up of the water-back with lime will be prevented; also, whether the water in the boiler can be heated in as short a time as in the ordinary form of boiler; and, finally, whether the water in passing through A B will not find its way into the pipe C.

The principle of the boiler is the same as that utilized in many forms of water heaters now on the market, the pipes within the heaters, however, serving for the circulation of waste steam instead of hot water, as in the present case. These heaters, as our correspondent undoubtedly knows, all work more or less successfully. In the particular case under consideration the water-heating capacity, or the rapidity with which the water in the boiler D will be heated by the circulation of hot water in the pipes A B, is simply a question of heating surface in the pipes A B and in that part forming the water-back. With sufficient surface, relatively, to the amount of water in the boiler D, we do not see why good results should not be reached. The tendency of hot water from the pipes A B to enter the pipe C is slight; not worth considering, in fact, unless the connection were of so very large a diameter as to admit of circulation within the pipe itself. The chance of this is remote. Some heat will go to the cold water in the inlet pipe by conduction, but this will be slight.

The theory in this case of the prevention of lime obstruction in the water-back obviously is that the water which is once admitted to the pipe system A B and the water-back extension practically remains there, unchanged in quality, except so far as that change is concerned which occurs after it is heated for the first time by circulation through the water-back. This first heating would have the effect of causing a separation of some of the scale-forming impurities held in solution, after which there would be no tendency to form a further deposit. The pipes A B and water-back would probably not be stopped up by this comparatively small amount of initial

scaling, though in the course of time, if the water contained in them be drawn off occasionally and fresh water let in from the cold-water supply pipe, new deposits would be formed, which would ultimately, if this drawing off and refilling be repeated often enough, clog the pipes.

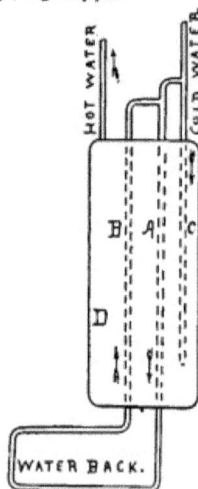

HOT-WATER COIL FOR BOILER.

The scale-forming matter in the water which is heated in the boiler D by the pipes A B would be deposited in the boiler itself, and not interfere with the working of the water-back.

INDIRECT HEATING FOR A LARGE KITCHEN BOILER.

IN the kitchen of the National Home for Volunteer Soldiers at Hampton, Va., hundreds of gallons of hot water are required about meal time for cooking, dish-

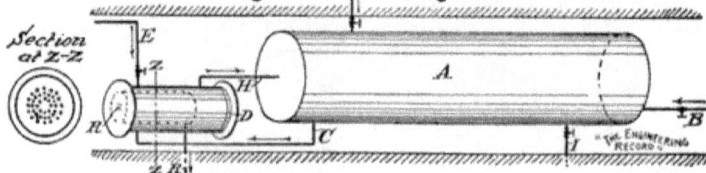

INDIRECT HEATING FOR A LARGE KITCHEN BOILER.

washing, etc , and usually only a small quantity during the intervening hours.

To meet this demand a tank A, about 4x16 feet, was provided and receives cold water through pipe B. An ordinary circular radiator R receives steam from the house-heating through pipe E and returns it through pipe R. This radiator is inclosed in a water-tight iron jacket or drum D, about 2x5 feet, which is connected with boiler A by the hot and cold water circulation pipes H and C. Steam being turned on to radiator R, the water surrounding it is heated, and rises through pipe H to boiler A, while the colder water flows out through pipe C and replaces it in drum D, and so on. G is the hot-water supply and I is the emptying pipe.

The apparatus was made by Bartlett & Hayward, of Baltimore, Md., and is said to work satisfactorily, although it was desired to place the drum D in a vertical position instead of horizontally, as shown, and as was necessitated by the limited height of the room and the position of boiler A.

AN INTERCHANGEABLE HOT AND COLD WATER SYSTEM FOR KITCHEN AND LAUNDRY.

George B. Hayes, Buffalo, N. Y., writes:

"Noticing your sketch in issue of August 8, 1891, called 'A Kitchen and Laundry Boiler System,' I am induced to ask if you have ever published a description of a perfect interchangeable system of hot and cold water for kitchen and laundry with a boiler in each."

[On page 153 of "Plumbing and House Drainage Problems" reprinted from The Engineering Record, and published at this office, is an illustra-

INTERCHANGEABLE HOT AND COLD WATER SYSTEM

tion showing how to secure circulation between boilers in different houses. A modification of this can be applied to the case cited. We cannot call to mind a recent interchangeable system that appeared in our columns, but we can see no reason why an apparatus arranged as per sketch will not do when one boiler is over the other.

A is the kitchen boiler, arranged in the ordinary manner, and B is the laundry boiler. When the cock *a* is closed the upper apparatus may be run past as though the lower one was not in existence, as no water will circulate in or out of the lower boiler and cannot flow from the lower boiler into the upper pipe, for the simple reason that *a* is closed. In like manner, when *b* is closed and the lower boiler only is in use, the boiler A is inert. Of course when the two fires are in operation the two boilers may be run together by having cocks *a* and *b* both open.]

A QUESTION OF HOT-WATER CIRCULATION.

G. C. Woods, Lawrence, Kan., writes:

"How can I secure a circulation to the sink and washbasin from the boiler shown in the accompanying sketch (Fig. 1)? Doors interfere with running

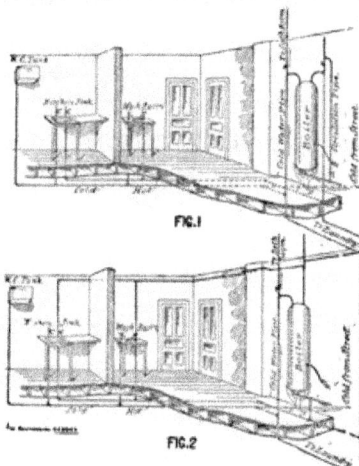

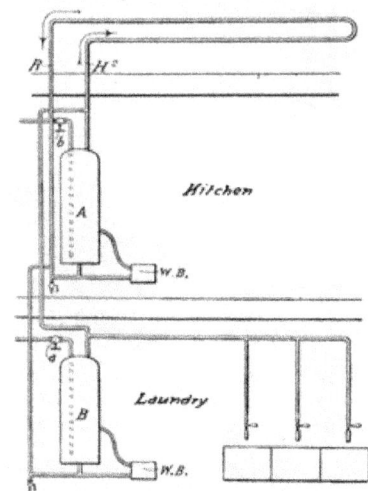

FIG.1

FIG.2

the circulation pipe along the wall level with the water-back in the range. Can I run the circulation pipe along the ceiling and down under the floor and then come up to the range, which is about 2½ feet high? Will this work right? How shall I run a circulation pipe to the laundry tubs? I have your book, 'Plumbing Problems,' but see nothing like this in it."

[The arrangement you have shown in your sketch will not work at all. Carry your hot-water pipe around the wall above the boiler as shown, dropping to the fixtures and below them for return circulation below the floor. It is very seldom that return circulation is wanted for laundry tubs. They will draw hot water pretty quickly any way, but if you think it important to have it, then connect them up on the same principle, supply above and return below, all as shown in Fig. 2, making it as short a vertical distance as possible for the return to rise into the bottom of the boiler.]

CONNECTION OF KITCHEN BOILERS TO PREVENT SYPHONAGE.

BRANION & FRIDAY, Schenectady, write:

"We inclose a rough sketch of a dwelling having hot and cold water supply for two families, each floor having independent stops in the cold-water pipe in the cellar. Each boiler has a stop cock in the cold-water pipe. The hot-water pipe on the first

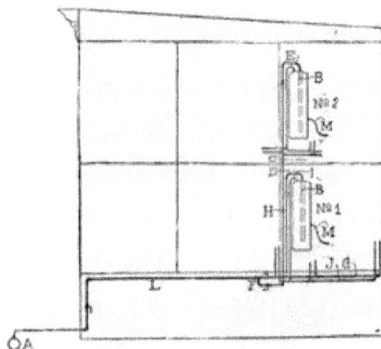

floor leads down by the side of the boiler to the cellar and branches are taken out for the sink and wash-trays. The second-floor hot-water pipe leads down beside the boiler through a partition to the bathroom, which adjoins the kitchen. The complaint made is that the water in the boiler syphons out through the cold-water pipe back into the main. Both boiler tubes are vented as shown in the sketch. The house is situated in a high part of the city and the pressure is very light at times, or not enough to supply the house, hence the syphon. We understand that boilers sometimes syphon out, but do not see how they can below the vent in the boiler tube. Will you kindly suggest a remedy for the trouble, as we find nothing in your 'Plumbing Problems' that explains the case?"

[Assuming, to begin with, that there is a good pressure on A, your city water main, that both hot-water tanks are full and that the vent holes B B are free, so long as there is head enough to send water into No. 1 it can be drawn from at any tap below it. The same is true of No. 2. When the head has been so much reduced that the water will flow only to C or D, or just sufficient to fill No. 1, the drawing of cold water on the lower floor will, if a hot-water tap is opened on the top floor, start a syphonage from boiler No. 2, down E and into G at the point of connection F, and so to the point of delivery; or if a hot-water tap was opened on the lower floor, other conditions remaining the same, the water from No. 2 would pass down E, making the circuit at F, pass up H into No. 1, out of No. 1 into I, and to points of distribu-

tion through J, though in either case the syphonage would be broken when the water in No. 2 had lowered to vent hole B in circulating tube inside boiler. When the head had been still further lowered to a point below any hot-water tap on the lower floor, upon opening it the syphonage would start from boiler No. 1 down H into G and out to the main through L. But, as before, it would be broken by the supply of air when the water had been lowered to B or earlier, should the hot-water tap be closed. In no case will the water in the boilers be exhausted below B B, but the entire contents of the boilers can be evaporated through B B into the pipes E and H. If a continuous service is required, place independent open tanks in the attic if there is room; if not, then on the roof, housing them well to prevent freezing. Carry the main supply pipe to tanks ending with a ball cock, which will shut off when the tanks are sufficiently full. Then from the bottom of the tanks connect back to the two boilers. If the water will not rise to the open tanks at all times, set independent pumps to supply them, unless it is some one person's duty to see that they are kept full. If so, one pump will suffice. This is the practice in New York, and we think will govern your case.]

TOO MUCH COIL FOR HEATING THE BOILER.

C. S , Freeport, writes:

"Enclosed you will find a sketch of a hot and cold job which gives me trouble, and as I am a subscriber to your paper, I would like you to inform me what the trouble is.

"The boiler is regulation make, and the hot-water pipe is above the cold, and has a rise all the way from cold-water pipe in the bottom of the boiler, and

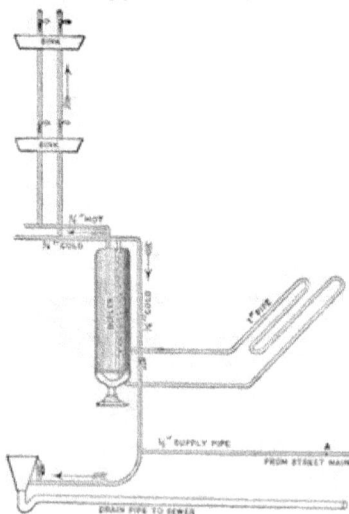

has no trap at all in it; but the coil lies on a level, and each pipe of 10 is to feet long. The water heats part of the time and gets the hottest in the cold water clear to the boiler, and when you draw water at the hopper it hammers and draws tepid water from the boiler to the hopper. It also makes noise when you draw at the sink. The boiler and hopper are in the basement. The sinks are on the next floor. I mark with arrows the direction the water takes. Please enlighten me."

[If we understand you rightly there is 100 feet of 1-inch pipe in the coil and the coil is practically level. If this length of pipe is exposed to the heat of an ordinary fire, it is too much for a boiler of the size you mention, and in any case 100 feet of pipe connected continuously with return bends or couplings cannot make a good heater. To understand this, imagine the cold water flowing into the first few feet of coil—say it enters at 80° Fahr.—and that before it has traveled 10 feet it has a temperature of 212 degrees or higher. What then can be gained by passing it through a longer coil? Why, nothing—at least for the purposes under consideration; and if we consider what the effect of a considerable length of coil beyond the point at which steam forms will be, we are forced to the conclusion there will be a repulsion and a tendency to drive the water out of the coil at both ends. If the coil is on its edge and not too long, with a good rise in the intended direction of the floor, it may find all its vent upward and be a "good water-back." If it is flat, level, and very long, it is as likely to react into the bottom of the "boiler" as to go forward, and it will be only a short time until the coil will burn out near the middle of its length.

The tepid water at the hopper or at the first sink is accounted for by the smallness of diameter of the supply pipe, or a partial stoppage within it, or both. There is sufficient pressure in the street to send the water to the highest faucet in the house when no water is being drawn elsewhere in the house. The pipes, then (both hot and cold), above the level of the boiler are small reservoirs. When the hopper is flushed the pressure in the supply pipe A is lessened, and it has then not sufficient pressure to hold the water on the upper ends of the lines of pipes, and this water must flow backward, all in the hot-water pipe going into the boiler and forcing an equal amount out through the boiler inlet (cold pipe), while the water in the cold line simply runs down the line for some distance and meets the hot water coming out of the boiler through the inlet pipe. This mixture (warm) must then pass down the cold pipe, as there is no other way for it to get out, and passes to the hopper warm or tepid. This of course cannot last long, unless it is a steam pressure from the boiler that is forcing its way down the cold pipe. The noise when water is drawn at the sink would indicate the formation of steam.]

TEMPERATURE OBSERVATIONS OF HOT-WATER PIPES.

M. C. F., St. Louis, writes:

"Do you know of any way in which I can find out the temperature of the water in the pipes of a hot-water job without breaking the pipe line to put in a thermometer cup?"

[Place the bulb of a thermometer against the pipe and put a lump of putty over the bulb so as to press the bulb against the pipe. You might further prevent radiation from the bulb by putting cotton waste outside of the putty. String can then be wrapped about the whole so as to hold it in position. If this is carefully done the thermometer will register within 1 degree of the temperature of the water in the pipe.]

CAN A BOILER BE SUCCESSFULLY CONNECTED WITH A STOVE AT A DISTANCE BY RUNNING THE PIPES ON THE CEILING AND BELOW THE FLOOR?

H. C. H., Corning Water-Works, Corning, N. Y., writes:

"I have never noticed in your valuable journal any information touching a problem in hot-water circulation which came up to-day in our practice, and I write for information thereon.

"A customer of ours desires to locate a hot-water boiler some 6 or 8 feet distant from his stove, but does not want the hot-water pipe to run directly from the stove to boiler. The question is, can the hot-water pipe be run to the ceiling, and then down to the boiler and connect at the usual place, and cold-water pipe be run under the floor to the stove, and in this way have the water heat properly in the boiler?

"If not as above, is there any way to arrange a boiler so that the water will heat properly 6 or 8 feet from the stove?"

[You may do as you describe, provided you put a "spud" on the side of the boiler high up for the hot-water flow pipe to connect with, and arrange a means of taking the air away from the top of the syphon formed above the boiler. An air-cock, or a very small pipe with a cock in it, run from the top of the syphon or loop to the hot-water pipe, rising a little all the time, so that the small pipe will not get air-bound, can be used. The cock in this pipe must be choked down so that only a very little water can circulate by this way.

If you raise the boiler to as near the ceiling as possible on a suitable high stand it will help matters.

On principle we do not advise these arrangements; but where they are necessary we should proceed as above.]

THE JOINTS OF PIPES TO A KITCHEN BOILER.

F., Denver, Colo., writes:

"I have had considerable trouble with the joint of a lead pipe connecting the water-back of a range and a kitchen boiler at the point where the lead pipe runs into the boiler. The joint is a wiped one, but I find that it leaks. What is the cause?"

[The leak is caused by the unequal expansion or contraction of the three metals, the lead, solder, and copper, of your boiler when subjected to the changes of temperature of water in the boiler. The reason for this is that these metals do not expand equally for equal changes of temperature. The solder will thus have to conform to the changes in the other metals, and a disintegration of the joint will follow, the solder becoming granulated and cracked. Use a brass or copper pipe for the hot-water connection between the boiler and water-back.]

FITTING UP A KITCHEN BOILER TO PREVENT SYPHONAGE.

YOUNG PLUMBER, Charleston, W. Va., writes:

"I am a reader of your valuable paper, and note your request for more reading matter from the craft. I have just put in an old-style range boiler in a new way to me: Putting cold water in at side A and hot water in at top B, placing stop-cocks at A and C, that water-back and pipes may readily be drained without emptying boiler when it is not cold enough to freeze a hot boiler in one night's time. D D are stop and waste cocks for controlling water on upper floors. I find that in this manner of boiler fitting the circula-

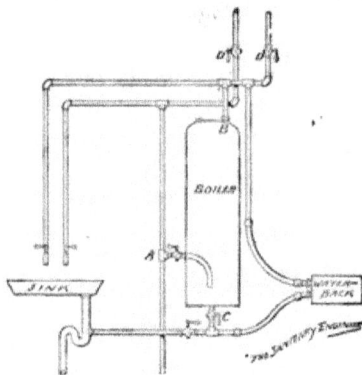

tion is perfect, no thumping or chinking, and also admits of draining pipes without emptying boiler. And we get hot water in half the time it takes the old way, as the cold water does not pass down through hot to chill it, nor does hot water have to pass through the cold to reach the proper place of storage.

"If there is any objection to this manner of boiler fitting, I would like for some member of the craft to let us hear from him."

[There is some risk of losing the water from the boiler through the cock and connection A into the street mains should the pressure in the latter become light or be drawn off for repairs while they are in their present position, as shown, or into the cellar through the stop waste should the stop be closed. A steam pressure would also drive the water from the boiler by the same connection, and unless there is a hole in the side of the inside pipe, which drops to

near the bottom of the boiler, the water may be driven so low as to allow the water-back to be burned. We would prefer to introduce the cold water to the boiler in the usual position. Otherwise we consider the scheme a very good one, and can see no objection unless it would be the possibility of some one's leaving the cock C closed when fire is in the range.

TROUBLE WITH A WATER-BACK.

M. W. NELLIGAN, of South Boston, writes:

"I have rather a peculiar question and come to you for advice. One of my customers is a baker and in his oven he has used a water-back such as is used in set ranges, and it lasted two years, then he put in another of the same make which lasted two weeks and burst in the same place that the first one did. I advised him to put in a brass coil, but he said it would work all right as far as heating water was concerned, but it would make a great deal of noise within the boiler, placed in the position it is.

"He says if the boiler was placed so as it would be all above the water back it would work all right. Now, I would like to know why it would not work all right in either position. He has had trouble with a coil before, and that is the reason he will not put one in now until I convince him that it will work. The boiler that is in is very near the ceiling."

[Our correspondent has made on his letter a rough sketch of the water-back that broke, which is V-shaped and apparently of cast iron. Another sketch shows the position of the boiler, which is about as usual, the top of the water-back being about level with the top of the lower third of the boiler and distant from it about 5 feet.

If the baker has already had trouble from a coil, it is not wise to advise him to use one again, since a coil may cause noise where a cast water-back would not.

If the oven is hot enough to make steam, the resistance of a coil to the flow of the water may detain it in the coil long enough to permit steam to form and by its sudden condensation to cause the shocks due to "water hammer" in the pipe. The steam when it is formed will momentarily force the water from the coil and the colder water as it enters again will condense the steam and make a noise. The same thing may occur in a water-back, but not to as great an extent, other things being the same. The burning of the water-back would seem to indicate that something of this kind had occurred in this case.

Put in a water-back and enlarge the connections throughout their entire length, and the trouble is likely to cease, as the resistance to the flow of the water will be lessened so that it will get through the water-back before it can be converted into steam.

We have so far assumed that the water-back of which you speak is of cast iron, of the usual pattern. If, however, your "water-back" is simply a coil of iron pipe it would not improve matters to make it of brass unless you enlarge the diameter of the pipe and connections.]

CIRCULATION FROM KITCHEN BOILERS.

A SUSPENDED KITCHEN BOILER.

EDWARD W. LOTH, Architect, Troy, N. Y., writes:
"I wish to suspend a hot-water boiler from the ceiling by means of iron clamps and rods passing through the beams above. The boiler to be used is a comparatively new one and in making alterations in the building I would not like to discard it for a new one. For various reasons the boiler should be on the side of the room opposite to the range containing water-back. The piping is to be brass and hung from the ceiling. The illustration will show, I think, clearly what is desired. In the figure A is the cold-water supply to the boiler, B the

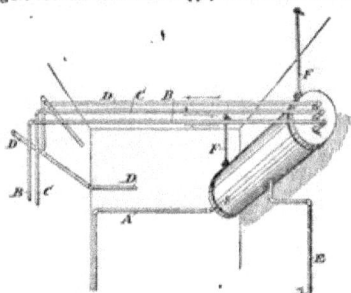

cold-water supply to the water-back, C hot-water pipe from the water-back, D hot-water supply to house fixtures, E cleanout pipe, and F F are suspension rods. Your opinion is solicited as to the efficiency of the arrangement, not alone as regards circulation, but safety as well."

[We see no objection to the arrangement, and the water should circulate if the pipes are properly run without traps. Numerous instances will be found in THE ENGINEERING RECORD of boilers being suspended from the ceiling or supported on brackets.]

PLACING AN ORDINARY KITCHEN BOILER IN A HORIZONTAL POSITION.

PLUMBER, Trenton, N. J., writes:
"A party in Trenton is going to have hot and cold water in his house, but he has not much room in his kitchen, and the only place he can put the boiler is over the top of the range, with only 18 inches of room between the top of the range and ceiling, as shown in diagram. The boiler is a plain 35-gallon one. Will you please let me know through your paper if it will work, or if there is another way to fit it up, and oblige."

[There is nothing unusual in suspending a boiler horizontally above a range, and in either "Plumbing Problems," page 203, or in our issue of December 24, 1885, may be found two very well arranged sets of horizontal boilers in connection with ranges.

With the arrangement you show in the diagram you send (the accompanying figure) the circulation

will be rather feeble. This may be improved somewhat by putting a "spud" on the boiler at *a*, taking the return circulation back to the water-back in that manner, as shown by the dotted lines. Then the cold water to the boiler can be fed into it through the usual connection, as shown at *b*, instead of into what is usually the bottom spud, as you show. The inside cold-water pipe will then be in its usual place, or it may be dispensed with, possibly to advantage, as then the cold water will not be admitted so near the point of outlet. In the pipe *b* a stop and check should be used, the latter to prevent hot water from being drawn through the cold pipe at the sink, which might follow under some conditions. In a galvanized-iron

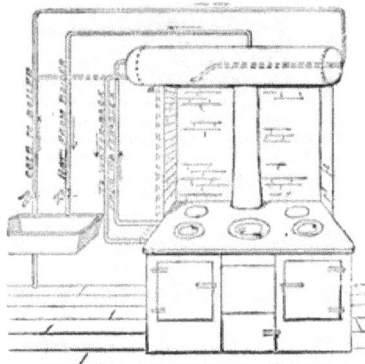

boiler an extra "spud" can be attached by tapping. Drill a three-quarter inch hole and expand or open it with a drift-pin until the three-quarter pipe tap will enter. This will thicken the edge of the hole sufficiently in the thin iron to get two or three good turns of the thread.]

DRAWING COLD WATER FROM HOT-WATER PIPE.

A PLUMBER, Liverpool, O., writes:
"I should feel obliged if you can explain the following occurrence. The accompanying sketch is a diagram of the water supply (hot) in a house. There are two lines of return circulation taken off the expansion pipe. The returning ends of these lines join together, and, as one pipe, are branched into the flow pipe to water-back (from cylinder).
"When the stop-cock is open one can only draw cool water from the cylinder, although the upper part of it may be quite hot; so that it appears that the water from the bottom of the cylinder is drawn at the taps instead of from the top of the cylinder. Upon closing the stop-cock hot water only is drawn at the same taps. What I wish to know is, by virtue

of what law does the flow of the water reverse itself when the stop-cock is opened and shut?

"The water level in tank will be about 35 feet above the top of water-back.

"The point A on expansion pipe will be about 3 feet 6 inches below water level in tank. Cylinder holds 50 English gallons."

[The fact that water will flow most readily in the direction of least resistance seems to be the solution of this question. The water flows from the tank downwards through the pipe *a*, and enters the boiler at the bottom. Should the stop-cock be closed, the water to leave the boiler (cylinder) and go to either faucet must go by the pipes *b* and *b*[1], and hence only hot water is drawn. Should the cock, however, be opened, the circuit by the pipe *a* through the bottom of the cylinder, and thence to the pipe *c*, *c*[1], and *c*[2], are presumably much the shorter than by the pipe *b* and *b*[1]; hence the flow of cold water at the faucets.

If your pipes throughout were of large diameter, excepting the return pipes, this probably would not occur, but they would have to be of sufficient size to supply the warm water by circulation in the proper direction. Several possible contributory causes may be given in addition to those caused by friction: (1) There may be partial obstruction caused by solder at a joint. (2) If the loss of head within the pipes when drawing freely at a faucet is sufficient to lower the water below the point A, then no hot water can be expected at the faucet *d*, though it might still run at the faucet *e*. If it does not run warm at the faucet *e* when drawing slowly, then our opinion is a partial stoppage will be found between the cylinder and the branch *b*[1] If the stoppage exists, remove it and then "choke down" the stop-cock until it is only sufficiently open to maintain circulation of the water at the temperature required. This will hold back on the flow of the water by the return pipe *c*, but if this

is not sufficient, to put a swinging or easily worked check valve in near the stop-valve.]

TROUBLE WITH A KITCHEN AND LAUNDRY BOILER SYSTEM.

DAVID S. COWAN, Bath-on-the-Hudson, N. Y., writes:

"About a year ago I fitted up a house in Albany with water throughout the house. There was a 40-gallon boiler placed in the kitchen connected to range. About a month ago I placed a boiler range in the laundry, which was on the floor below. The idea was to use the laundry range during the summer. Connections were made to the ends of supplies over laundry tubs for laundry boiler, and a stop-cock was

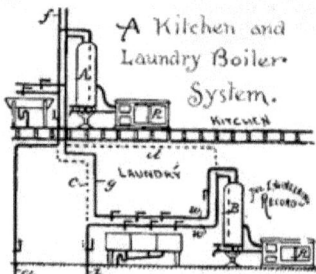

A Kitchen and Laundry Boiler System.

placed on the hot-water supply from laundry tubs, because when the laundry range was not in use and kitchen range was, cold water came through the laundry boiler and chilled the hot water over tubs unless this stop-cock was closed.

"The trouble now is that when laundry range is in use they draw hot water from the cold-water cock over the kitchen sink. I find that the hot water circulates through the hot-water pipe into kitchen boiler, and thence through the boiler tube into cold-water pipe and over sink. What I propose to do is to place a check valve on the cold-water pipe over the kitchen boiler to keep the hot water from entering the cold-water pipe. I would like to know if that would remedy the trouble, and also if this is the proper way to connect the laundry boiler. If not, please let me know what is."

[In the accompanying sketch *a* represents the cold supply to upper boiler A; *b*, the cold supply to tubs and lower boiler B; R, range for upper, and R[1] range for lower boiler; *f*, supply to bathroom; *g*, hot water supply to tubs in laundry; and *w w* are the points where connections were made to the ends of supply pipes over the laundry tubs for the laundry boiler.

If the apparatus was arranged with a circulating pipe, we should object to the use of the check valve at once, as it will undoubtedly offer resistance enough to spoil the circulation. The diagram shows no circulating pipe, and therefore we assume there is none, and think the check valve the readiest means of curing the trouble described in the letter. The cause of the trouble is undoubtedly that there is a better and stronger flow of cold water to the boiler in the laundry than there is to the boiler in the kitchen, and that when it is attempted to draw cold water at kitchen sink a large part of it is obtained by the way

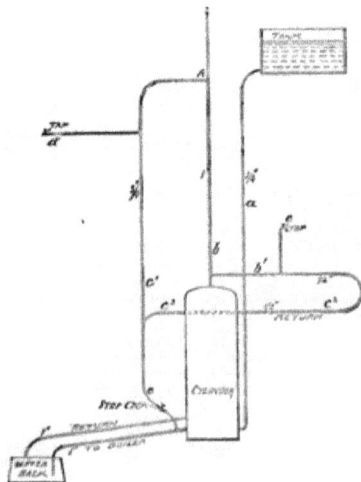

of the cold supply to the lower boiler; thence through hot-water pipe to upper boiler; thence through cold-water pipe to the sink. A partial stoppage in the cold-water supply to the upper sink and upper boiler will account for it, or a contracted or small supply pipe. There are, in our judgment, two methods of remedying the difficulty without the use of a check valve—namely, the removal of the supply pipe *a* and the joining of the supply pipe *b*, as shown by the arrow and dotted line *c*. The second method is to remove the supply pipe *b*, and the connection of the pipe *a* with the lower boiler, as shown by the dotted line *d*. We favor the second method proposed, however, as we are of the opinion the supply *b* is more ample and vigorous than the supply *a*, notwithstanding that they probably come from the same source.]

TROUBLE WITH A KITCHEN AND LAUNDRY BOILER SYSTEM.

J. A. ROSSMAN, of the firm of Rossman & Bracken, of New York, writes:

"After reading Mr. David S. Cowan's letter, and seeing his diagram for the kitchen and laundry boiler system, in the RECORD, I conclude that his whole trouble is caused by the position of the stopcock under the kitchen sink. He says the object is to use the laundry range during the summer, and appears to understand the necessity of shutting off the source of that supply when the range is not in use. He must certainly do the same with the other range and boiler when they are not in use. You will readily see by the position of the stop-cock, which he must use, that it would prevent the street supply from coming to his kitchen sink, as well as to his kitchen boiler. If he should place a check valve where he proposes, he would still have to close the stop-cock under the sink to prevent the cold water from passing through the kitchen boiler into hot-water system. He then would get no cold water to the kitchen sink or his bathroom.

"If he will place a stop-cock where he proposes to put the check valve—viz , on the cold-water branch to kitchen boiler, where it should have been put originally, he will be able to overcome his trouble. I think if he placed a check valve where he proposes, and in the autumn when he abandons the use of the laundry range and shuts that source of supply off he will very soon be asking you why his boiler or range pipes burst, or some other weaker point in his hot-water system gives way, because there would be no possible chance for expansion.

"I assume from the drawing, of course, that this is a direct-pressure system, and there is no tank in the house where a relief pipe could be taken. It is possible that my conclusions are wrong, from the fact that your drawing may be misleading, as it would hardly seem credible that any plumber would put two stop-cocks on supply *a* with no branches between, where in using either of them you would necessarily shut the entire system off the house, excepting the laundry tubs as it was originally fitted, and not have

one on cold supply to boiler. According to your drawing, in order to shut the water off the kitchen boiler, you are obliged to shut it from the kitchen sink as well as the bathroom and whatever fixtures there may be upstairs. Referring to your comments thereon, I can hardly see how a check valve placed as he proposes would interfere with the circulation of the hot-water system, if there was one in the building properly arranged. I mean by that, taken from a high point of the main hot-water riser and returned to the bottom of the boiler. The check would have no effect upon the circulation whatever. I also fail to see why you should favor your proposed 'second method' and abandon the supply pipe *b*, which you consider 'more ample and vigorous' than *a*, which you retain, except on the theory that two wrongs make one right.'

HOT-WATER CIRCULATION IN A GREENHOUSE.

A CORRESPONDENT at Washington, D. C., writes:

"I am figuring on a greenhouse job and want to run my pipes in accordance with the annexed sketch. Will the arrangement work? C is a 100-foot coil under a hot-bed, and it must be supplied from the main A on the other side of the path P without obstructing the passageway. This coil is the last at the end of pipe A.

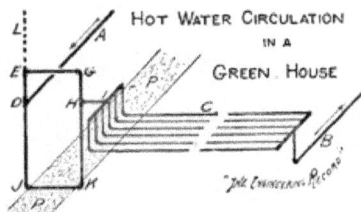

HOT WATER CIRCULATION IN A GREEN HOUSE

"I think it would circulate all right by going over the walk and having another bend around under it as shown. The main A is higher up than the branch to the coil. If this will not work, what would you suggest ?"

[The lower loops D, J, K are unnecessary. The arrangement will circulate through the upper loops D, E, G, H if the air be prevented from collecting there. This can be readily done by connecting a vent pipe E L to reach up above the expansion tank and open at the top. This can be the more easily done since in a greenhouse system the tank is rarely very much elevated.]

TANK-WATER SUPPLY FOR A KITCHEN BOILER.

READER, St. Louis, Mo., writes:

"Please give me your valuable opinion on the following arrangement of hot-water pipes. Is it safe and advisable in a small house in the country to have a tank in the attic, of a capacity of 200 gallons, and supplied through pump worked by hand, if in the kitchen there is a 40-gallon boiler connected with water-back in the usual way? There is a ¼-inch pipe from tank supplying boiler, and an expansion pipe is carried up from highest point on hot-water pipe over top of tank. Would any bad results follow temporary failure of supply, as such stoppages are very likely to occur?"

[If the work is done in the usual manner about the boiler, and the small hole for preventing syphonage is open in the cold-water pipe within the boiler near the top, all that can follow short interruptions to the supply of water will be the heating of the water which remains in the boiler to a point at which it may give off steam. This steam will find vent through the supply from tank if there is no check valve in it, or it will escape through the nearest hot-water faucet that is opened. If this state of affairs long continues the water in the boiler and water-back will evaporate and the back become burned or cracked.

There is a possible danger from explosion by over-pressure, but this is pretty remote. However, it is best not to take any risks, and therefore we advise an unlimited supply of water to the boiler at all times, and only give the above explanation that the question may be intelligently understood.]

A LOOP ABOVE AND A CIRCUIT BELOW A HOT-WATER BOILER.

WILLIAM McNAIR, Westbrook, Me., writes:

"I send a drawing of a hot-water job which appears to be unlike anything described in "Plumbing and House-Drainage Problems." In Fig. 1 the return after coming 50 feet from the end of the cellar,

has to rise above the second floor, and then drop to enter the boiler. The boiler supply and return have each an open pipe from the top of the boiler over the tank. The cellar circuit is carried along the timbers, with a fall of 6 inches to the return end, where a draw-off is put in. Another plan provides a connection of the return with the lower pipe to the boiler, shown by the dotted pipe A, with a shut-off to prevent drawing out the contents of the boiler when making repairs. Will either of these systems maintain a circulation in the cellar?"

[Our advice is not to attempt what you propose. It will not work and circulate in any form. The loop shown above the boiler will not help you. The up leg of the loop will lose as much heat, and theoretically a little more, than the down leg, and therefore will fully balance any gain due to the latter. It is a waste of time and material to construct circuits below the boiler. Run a circuit as shown in Fig. 2, and return below the floor, if you desire, and it will circulate if the pipe is free from air pockets.]

HOT WATER FROM THE RETURN PIPES.

SELIM, Piscataquis, N. H., writes:

"My opinion was asked as to a proposed plan for bringing hot water to washbowls and sinks. The engineer considered it a 'happy thought.' His plan was this: He has a low pressure or gravity system heating the building by steam, and was to make return pipes (one or more) supply the hot water to the bowls, etc. Would a man having any clear idea of the principle of steam heating attempt such things? I gave him my opinion in very plain English. I then asked how high above 'water line' his bowls and sinks were located; how much pressure he proposed to carry on his boiler; if he was to have a fireman in constant attendance, or control by automatic damper regulator; where his hot water was to come from? I asked if he had a feed-water heater and pump or injector, and if so, why he called it a gravity job; and

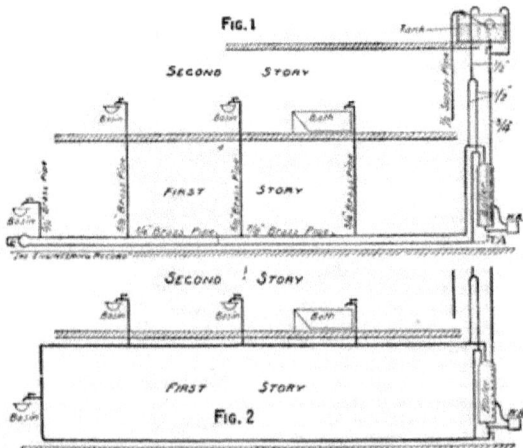

A LOOP ABOVE AND A CIRCUIT BELOW A HOT-WATER BOILER.

finally, why he did not put in a small hot-water boiler or tank, with brass coil connecting with his steam and return, and thus safely supply hot water to his bowls and sinks? He has a horizontal tubular boiler of 40 horse power. With the hot-water at several sinks running—left running thoughtlessly, as they are very likely to be—what would be the very probable result?"

[This proposed plan of hot-water supply is too ridiculous to be entertained, and but for the fact that just such men as would plan a job of this sort often, by their unskillfulness and ignorance, cause great inconvenience and injury to others, even placing human life in jeopardy, we would not feel justified in going into details in answering the query of our correspondent. No person properly trained as a heating engineer would lay out such a job, and employers should consult their own interests by not entrusting work to such impracticable and dangerous men. Assuming that the job was installed upon the plan indicated, only steam could be drawn upon the top floors, steam and water from the cocks near the water line, and water from those below the water line. Water drawn from such a system would not be fit for domestic use. It would be full of rust and at times would emit a disagreeable odor such as is often detected where air is drawn from gravity coils.

One of the first laws of steam heating which a fitter should learn is to allow no water to be taken from the returns. Experience has taught that this practice has caused the "burning" of more boilers than all other causes combined. Many heating contractors in recognition of this danger will not connect a "blow-off" directly to a sewer. This restriction we heartily indorse for small jobs or places where an engineer is not employed.

Your plan of a hot-water tank with brass heating pipes through which the steam and return pipes would connect, is very proper and is the best that can be done under some conditions. We would suggest in this case, using a hot-water circulating boiler of sufficient size and of the character used in the plumbing of dwellings. If there is sufficient pressure in the main service pipe, it will force the hot water from the boiler to the several points for use; or if not a tank should be placed sufficiently high and so connected that when in service it would act as a head, giving the desired pressure. The water in this boiler or tank may be heated by connecting flow and return pipes into the firebox of the steam boiler, on the same general plan as is used in connecting a kitchen range and tank. The pipes can be laid against the bridge wall. The hotter the place the better, if much hot water is required, but great care must be taken to have the connecting pipes properly run, otherwise there will be endless noises and repairs. Any good plumber should know how to arrange the job. You ask what would be the probable result of drawing hot-water service from the returns of a gravity system. It might be annoyance, stench, dirty water, scalding by steam, with chances favoring a burned or cracked boiler with a heavy boilermaker's bill, or an exploded boiler with attendant damage to property and peril to life, and the incidental inquiry—after the event—"How did it happen? Who is to blame?"]

HEATING A BOILER FROM TWO WATER-BACKS.

In a summer cottage at Far Rockaway, L. I., it was desired to have one boiler supply both kitchen and laundry and be heated at will from either range. K is the kitchen and L is the laundry range set back

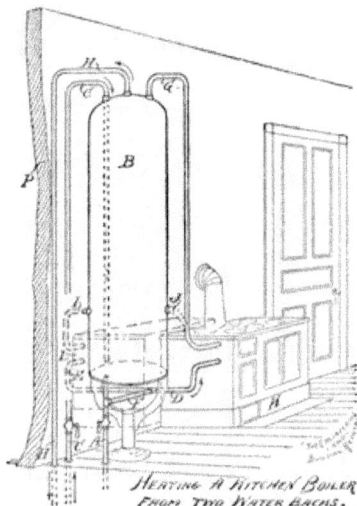

HEATING A KITCHEN BOILER FROM TWO WATER-BACKS.

HEATING A KITCHEN BOILER FROM TWO WATER-BACKS.

to back against a partition P. B is a 50-gallon boiler receiving cold water through a pipe C and supplying hot water through a pipe H. The sediment pipe A has branches D and E to both water-backs. When the kitchen range is used the circulation is through the pipes F, D, and G, but when the laundry range is used it is through the pipes F, E, and I. The pipe G was first connected to the boiler at K, as shown by the dotted line, but did not give good circulation. The present arrangement is satisfactory. The diagram is made from data furnished by John Renehan, New York City, who designed and executed the work.

WHAT CRACKED THE WATER-BACK AND CAUSED THE RUMBLING NOISE?

J. R. S. writes:

"I am a reader of your valuable paper and you will confer a favor by giving the whys and wherefores of the following difficulties a brother plumber experienced not long since with a range boiler and back. He asks me, Would leaving the sediment cock open burst the back when the supply was sufficient to keep the boiler full, regardless of waste through sediment cock. He gives me a sketch as follows:

"Boiler is in basement. A is a supply pipe and connected in the bottom of a tank on the fourth story of the building. B is a check valve to keep water in tank from washing back into the main. C is sediment cock. Our brother came to repair a burst in pipe D, and in order to do this he shut off the supply and emptied boiler through sediment cock C; wiped the joint, turned the supply on, and left the job as being all right, forgetting to close sediment cock. In a few hours he returned to see how the job was, and found a hot fire, but boiler and circulating pipes cold and a fearful water hammer in the boiler. On dumping the fire he found the water-back had cracked and then discovered the sediment cock open. On putting in a new back and closing the sediment cock the job proved to be all right.

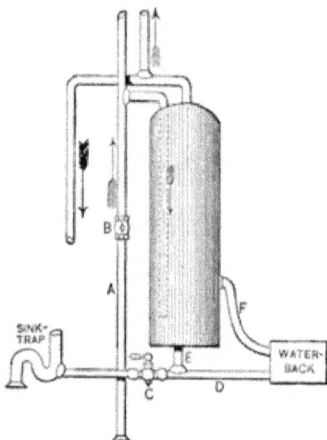

" I explained the cause of the trouble to my friend as follows: That the escape of the water through pipe E and sediment cock C created a vacuum in pipe D, making a downward flow of cold water in circulating pipe F, through back and pipe D, and out at sediment cock. The flow of cold water on the hot back cracked it, and also caused the water hammer in boiler. Am I right? If not, please favor us with a correct reason, and oblige."

[If there were a downward flow of water in the pipe F, thence through the water-back and pipe D, and out through the sediment cock, the chances are the back would not be injured, unless, indeed, this flow suddenly started when the back was red-hot.

If the boiler and circulating pipes F and D were cold, when the condition of things was first noticed, it is likely the water was passing as you say, otherwise they would be very hot.

The water hammer you mention certainly cannot be caused by steam formed in the back (as you say the pipes and boiler were cold), therefore the rumbling or water hammer must be caused by air in a partly filled boiler; or you may be misinformed about the boiler and pipes F and D being cold, and the water hammer may have been really caused by steam, as we assume it was by water running into the back through pipe D and then being driven violently from the back through both pipes in the form of steam,

If enough water remained in the boiler (when the sediment cock was open) to cover the upper end of the pipe F, the back would not be cracked or burned.]

TROUBLE WITH A HOT-WATER SUPPLY BOILER.

A. C., of Malden, Mass., writes:

" I inclose you a diagram of boiler connected to range as at present. The boiler continually snaps and breaks the pipes. The boiler holds about 120 gallons and is raised as high as it is possible to get it. Can you make any suggestions to obviate the trouble, and what is the best thing to do to get the best results?"

[Judging from your diagram, your sole and sufficient cause of trouble is that the circulation is so sluggish that steam forms in the water-back and accumulates there until it comes in contact with the colder water in the boiler, when it is suddenly condensed, causing a violent water hammer, and breaking the pipes as you describe. The remedy is to raise the flow pipe from the water-back to the boiler, so as to enter the boiler some distance above the water-back as shown by the dotted lines at A, and the higher the point at which the flow pipe enters the boiler the more rapid and efficient will be the circulation, though usually a difference of elevation of 2 feet or so will give sufficient circulation to heat the water rapidly and prevent it from remaining long enough in the water-back to be turned into steam.

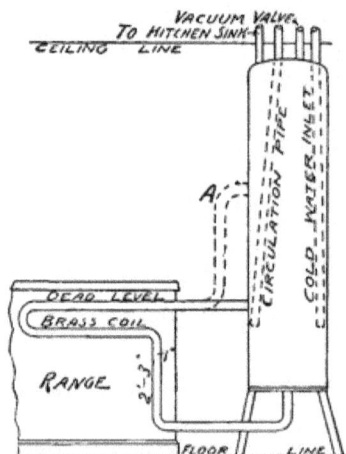

The only efficient cause of the circulation between a water-back and boiler lies in the fact that there is a column of water in the flow pipe and water-back which is hotter than that at a corresponding elevation in the boiler. The higher this column, and the greater the difference of temperature, the more rapid the circulation will be. It is the rise of the flow pipe, not its length, that does the work, and the direction taken by the lower pipe leading from the boiler to the water-back has no effect on the circulation whatever. It must be remembered, however, that while increasing the height of the flow pipe tends to accelerate the circulation, it cannot effectnally overcome the resistance offered by too many bends or fittings, or too small size of pipe, or, perhaps, some obstacle that has lodged in it, so that if after raising the flow pipe the circulation is still unsatisfactory, you may be sure that some obstruction such as above mentioned exists in the pipes, and if carefully sought for and removed there will be no further trouble. In this connection you will do well to read carefully the answer given in our issue of March 16, 1889, to the question " How to connect a water-back."]

STEAM AND RANGE WATER-BACK ON KITCHEN BOILER AT HAMPTON, VA.

In the governor's house of the Home for Volunteer Soldiers, at Hampton, Va., it was desired to provide heating capacity for the large kitchen boiler B, in addition to that furnished by the ordinary water-back D of the adjacent range A.

This was effected by connecting the boiler circulation pipes F and G with a steam water-back S, which received steam from the house-heating apparatus through pipe E and returned it through R. The steam water-back S consists simply of an iron

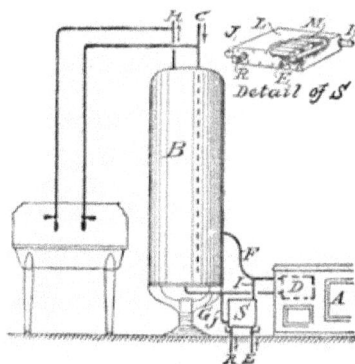

case L containing a steam coil M. This is surrounded by water admitted at the bottom through the branch J from the cold circulation pipe G. After being heated it flows out through branch I into the hot circulation pipe of kitchen boiler B.

This arrangement was made by Bartlett & Hayward, of Baltimore Md., and is said to operate satisfactorily, heating a large quantity of water rapidly

with both water-backs together, and working well with either alone.

DEFECTIVE HOT-WATER CIRCULATION.

H., Architect, Hartford, Conn., writes:

"A plumber has set a horizontal 40-gallon boiler over a kitchen range, as per sketch. It takes three hours, with the ample water-back and a hot fire, to provide tepid water at the top of boiler, and consequently at the fixtures, which have a circulating pipe. Hot water is absolutely unobtainable. The pipes are of brass, with $\frac{3}{4}$-inch bore. The pipe from

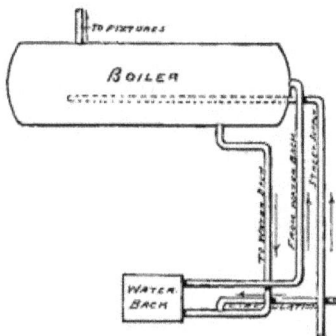

water-back becomes very hot, but that doesn't affect the water in the boiler.

"I claim that the apparatus will work all right if the boiler is turned upside down. Will it? If not, what would you recommend?

P. S.—Is there any difference in the hot-water supply from horizontal or from vertical boilers?"

[If the pipes are arranged as shown in your sketch there is nothing in the arrangement that should prevent your getting hot water and securing proper circulation. The trouble is probably due to some stoppage in the pipes, either dirt, shavings, or some other obstruction. To turn the boiler upside down would make no appreciable difference. At any rate, we should first look for the cause of the stoppage in the pipes leading to and from the boiler.

There should not be any difference, other things being the same.]

AN ERROR IN CONNECTING A DOUBLE BOILER.

PLUMBER writes:

"As a recent experience of mine in fitting up a double-boiler system may serve to give a useful hint to others or save them a like mistake, I send you an account of it.

"After renewing some plumbing in a private residence here, I turned the tank water on to the inside boiler A only, of the usual double boiler shown in the diagram. Very soon water began to run from cocks connected with the outside boiler B only. Careful search revealed no by-pass, and the inside boiler, though just tested, was declared to leak. When removing it to be retested I found that the sediment cocks C and D were of different patterns, so that when the handles were in the position shown, C was

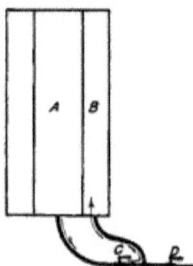

open and D was closed. Supposing both to be like D, I had turned them as shown, hence the trouble.

"This discovery of course solved the mystery, as the water passed through C and filled B as indicated by the arrows. C was immediately replaced by a cock like D, when everything worked properly, and no harm was done, except in losing considerable time to find the trouble."

RUMBLING IN A KITCHEN BOILER.

E. E. M., South Orange, N. J., writes:

"A friend of mine has been much troubled lately with a rumbling noise in the hot-water boiler connected with the plumbing in his house. In speaking of the matter to me I resolved to look into the connections and see if I could discover any reason for it.

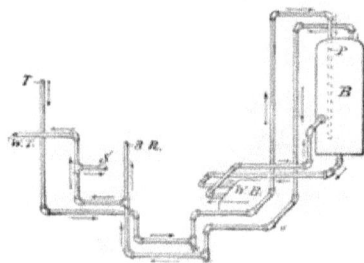

Accordingly, I sketched out the run of the pipes, which sketch is herewith transmitted. I can find no explanation of the difficulty, and have resolved to trespass upon your space and see if you can assist me. The system is entirely clear from dirt. The water used is rain or artesian-well water."

[In the sketch, which we reproduce, B represents the boiler; T and P the supply pipe from tank; W, B, water-back; W, T, pipe to washtubs; S, pipe to sink; and B, R, pipe to bathroom. Rumbling noises in

kitchen boilers, as complained of in this case, are not uncommon, and are due simply to the formation of steam brought about by a bright fire and a comparatively small demand for hot water from the boiler. The steam collects in the upper part of the boiler driving out the small quantity of water under the curved head through the small hole in the cold-water supply pipe P, a little below the point of entrance into the boiler. A very simple remedy for the noise would be to draw off some of the hot water from the boiler, allowing it to escape, say, into the kitchen sink, where it would, moreover, exercise a cleansing influence. The trouble, however, can scarcely be considered a serious one, indicating simply, as already stated, that there has been a falling off in the demand for hot water.]

HOT-WATER CIRCULATION QUESTION.

The following has been referred to us for reply:

St. John, N. B., April 15, 1887.

Sir: A range boiler, with cistern, bath, etc., fitted in same flat, is fitted with circulation pipe as per sketch. How can there be a circulation returning to

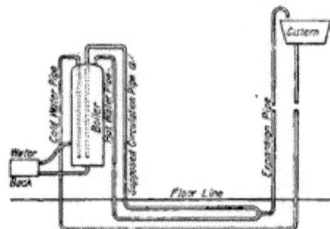

the boiler? As we understand it, the circulation can only be by gravitation. We hold that the tank can have no effect, even if placed on floor above. We can see no use in the circulation pipe fitted as it is —architects to the contrary. We cannot see that the cistern has any effect. There are rooms between kitchen and bathrooms through which pipes on walls might be objectionable.   C. E.

[An apparatus such as shown will not circulate, when by the word circulate we allude to the flow of warm water that goes on within properly arranged hot-water pipe from the head of a boiler to some distant faucet and returning again by a parallel pipe called a "circulation pipe." Therefore, it would be just as well, and would save something in first cost, to omit the pipe a in the apparatus shown in the sketch.

The increased height of a tank or cistern above the boiler has little or nothing to do in this case with the question of circulation so long as it remains above the level of the boiler, and the only effect it can have is to increase the pressures in the pipes generally. Anything, however, that will increase the pressure of the water in a properly arranged system will be the means of allowing the water to become relatively hotter before it forms steam, and as an increase of temperature may be assumed to be tributary to the increase of circulation, other things being favorable, a high tank may **then** be considered as more advantageous than a low one, but this will not apply in this case, as the apparatus will not "circulate."]

WHY WATER IS MILKY WHEN FIRST DRAWN.

BOSTON, May 20, 1886.

SIR: As a subscriber of your valuable paper, I would like to inquire through the columns of the same the cause of the water at my house (in Sharon, Mass.) coming from the faucet in a milky form or appearance. It comes from the regular town water-works, and has recently been put in, and I know of no other on the street that has this trouble. I wish to know if it would be considered objectionable for drinking or domestic purposes? Any explanation of the above would be thankfully received through the columns of your paper. Very respectfully yours,
C. T. DEERY.

[The milky appearance is probably due to confined air in the water under the pressure in the pipes which escapes when the water runs from the faucet. It doubtless becomes clear after it has stood in a vessel a moment. Under these circumstances it is not objectionable for drinking or domestic purposes.]

WATER WASTED THROUGH IMPROPER ARRANGEMENT OF BOILER VENT AND TANK OVERFLOW.

IN looking **over the** plumbing of a New York building in 1891 a condition of affairs was observed which may, in more **than one** case, account for a large water bill, and may in others account for want of warm water in the boiler **at** times when the fires are good and when plenty of **hot** water is expected.

In the case in question we found the air and vent pipe *a* brought in a goose-neck fashion *b* over **the** edge of the tank, and down into the top of the standing overflow pipe *c* as shown. We asked why the pipe was brought close down to the overflow, and were informed "that it was to insure the keeping of water in the trap under the tank," and were told "**that** when the water got very warm in the hot-water pipe and circulation, that it would run over **and add water to** the trap." This may be true, and **if the** goose-neck **is not** carried sufficiently high **above the water in** the tank there is no doubt of it, **but it did not seem to** strike the man who designed **it that** should the density of water in the hot-water **pipe from** the boiler decrease, by being warmed, **until it reached the top** of the goose-neck and run over, **then the water, as it** run down the short leg of **the goose-neck, would balance an** equal height of

water in the rising pipe, and that this syphon might then run continually, taking water from the tank and actually discharging it at a higher level into the overflow pipe, just as long as heat was applied to the boiler, or until the water in the tank fell so low as not to be able to balance a column of warm water equal to the height of the open end of the goose-neck. To make this plain, assume the height of water from the water-back to the top of the water in the tank to be 60 feet, and its temperature to be, say, 40° Fahr. This will exert a pressure of 26.11 pounds per square inch in the water-back. Consider, again,

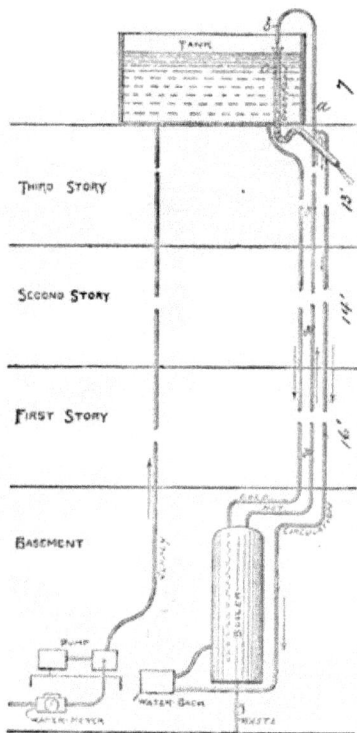

the height of a column of water in **the** hot-water pipe at a mean temperature, say 200° Fahr., that will have the pressure necessary to balance a 60-foot column at 40 degrees. Then as the expansion of water from 40 to 200 is equal to an increase of its bulk of 0.0386 (without increasing its weight), we have $60' \times 0.0386 + 60 = 62.31$ feet as the height of the warmed column.

From this it is evident that unless a difference of fully 2 feet 4 inches exists between the level of the water in the **tank** and the lower end of the goose-

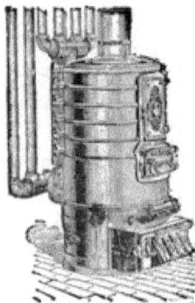

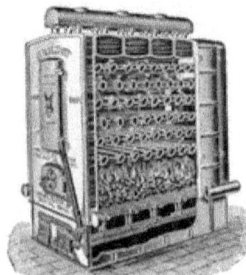

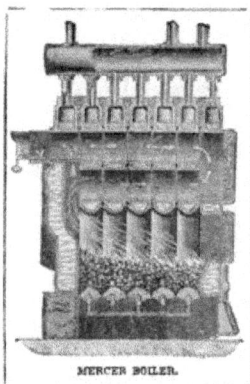

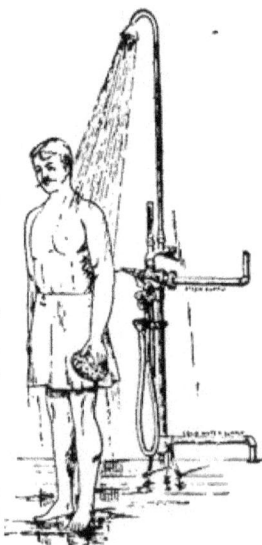

NEW YORK FILTER

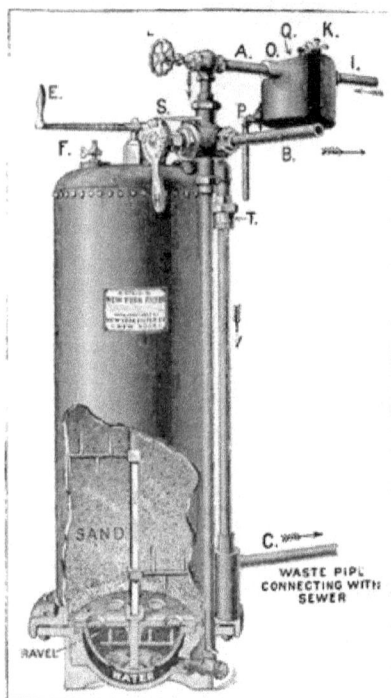

WASTE PIPE CONNECTING WITH SEWER

SAND

RAVEL

WATER

This is our type of HOUSE FILTER, for ¾ to 1-inch connections. It is connected to Service Pipe in basement and all the water entering the house passes through the Filter, is thoroughly purified, and drawn bright and sparkling from every spigot, free from all BACTERIA AND VEGETABLE IMPURITIES. IT DOES NOT REDUCE THE PRESSURE.

Only one valve (H) to handle in operating the Filter.

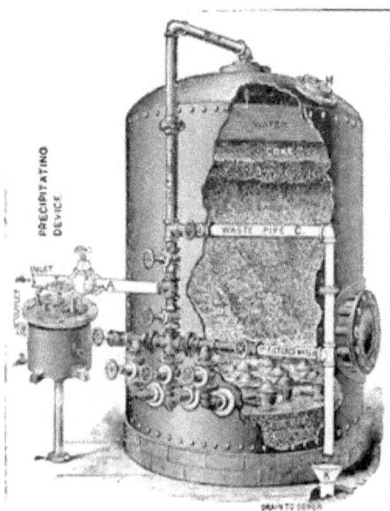

This cut shows our improved SECTIONAL WASHING FILTER, VERTICAL TYPE, 40 inches up to 10 feet diameter, as used by the principal Hotels, Asylums, Mills, Laundries, etc., etc., rendering the most impure water clear, bright and sparkling at all times.

VENTILATION AND HEATING.

By JOHN S. BILLINGS, A. M., M. D.,

LL. D. Edinb. and Harvard. D. C. L. Oxon. Member of the
National Academy of Sciences. Surgeon, U. S. Army, etc.

FROM THE PREFACE.

IN preparing this volume my object has been to produce a book which will not only be useful to students of architecture and engineering, and be convenient for reference by those engaged in the practice of these professions, but which can also be understood by non-professional men who may be interested in the important subjects of which it treats; and hence technical expressions have been avoided as much as possible, and only the simplest formulæ have been employed. It includes all that is practically important of my book on the Principles of Ventilation and Heating, the last edition of which appeared in 1889; but it is substantially a new work, with numerous illustrations of recent practice. For many of these I am indebted to THE ENGINEERING RECORD, in which the descriptions first appeared.

I am also indebted to Dr. A. C. Abbott for much valuable assistance in its preparation, and to the architects and heating engineers who have furnished me with plans and information, and whose names are mentioned in connection with the descriptions of the several buildings, etc., referred to in the text.

WASHINGTON, D. C.,   JOHN S. BILLINGS.
December, 1892.

TABLE OF CONTENTS.

ADDRESS, BOOK DEPARTMENT,

THE ENGINEERING RECORD,

100 WILLIAM ST., NEW YORK.

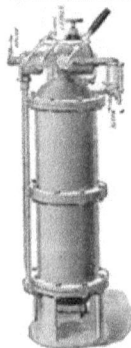

A FACT

Admitted now by all, that Porcelain Enameled Iron Baths are superior to all others.

THE BRIGHTON.

BEAUTIFUL, SANITARY, AND DURABLE.

We make all styles but only one quality---THE BEST. They are handled by all first-class supply houses. Architects who require the best specify them.

DAWES & MYLER,

Manufacturers,

NEW BRIGHTON, PA.

PLUMBING PROBLEMS;

OR,

Questions, Answers and Descriptions,

FROM

THE ENGINEERING RECORD,

ESTABLISHED 1877.

(Prior to 1887, THE SANITARY ENGINEER.)

With 142 Illustrations.

"A feature of THE ENGINEERING RECORD (prior to 1887, *The Sanitary Engineer*), is its replies to questions on topics that come within its scope, included in which are Water-Supply, Sewage Disposal, Ventilation, Heating, Lighting, House-Drainage and Plumbing. Repeated inquiries concerning matters often explained in its columns, suggested the desirability of putting in a convenient form for reference a selection from its pages of questions and comments on various problems met with in house-drainage and plumbing, improper work being illustrated and explained as well as correct methods It is, therefore, hoped that this book will be useful to those interested in this branch of Sanitary Engineering."

TABLE OF CONTENTS:

PLUMBING PROBLEMS.

Large 8vo. cloth. $2.00.

Address, Book Department.

THE ENGINEERING RECORD,

P. O. Box 3037.

100 William St., New York.

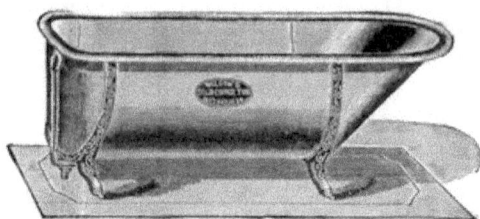

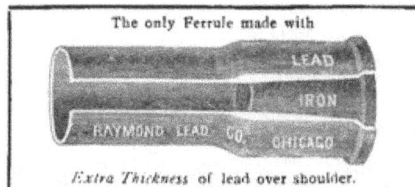

AMERICAN
Steam and Hot-Water Heating
PRACTICE.

From THE ENGINEERING RECORD.
(Prior to 188 *The Sanitary Engineer.*)

A SELECTED REPRINT OF DESCRIPTIVE ARTICLES, QUESTIONS, AND ANSWERS.

WITH FIVE HUNDRED AND EIGHTY-FIVE ILLUSTRATIONS.

PREFACE.

THE ENGINEERING RECORD (prior to 1887 THE SANITARY ENGINEER) has for sixteen years made its department of Steam and Hot-Water Heating and Ventilation a prominent feature. Besides the weekly illustrated descriptions of notable and interesting current work, a great variety of questions in this field have been answered. In 1888 Steam-Heating Problems was published. This was a selection of questions, answers, and descriptions that had been published during the preceding nine years, and dealt mainly with steam heating. The present book is intended to supplement this former publication, and includes a selection of the descriptions of hot-water, steam-heating, and ventilating installations in the different classes of buildings in the United States, prepared by the staff of THE ENGINEERING RECORD, besides a collection of questions and answers on problems arising in this department of building engineering, covering the period since 1888, in which the heating of dwellings by hot water has become popular in the United States. The favor with which Steam-Heating Problems has been received encourages the hope that American Steam and Hot-Water Practice may likewise prove useful to those who design, construct, and have charge of ventilating and heating apparatus.

TABLE OF CONTENTS.

AMERICAN STEAM AND HOT-WATER HEATING PRACTICE.

AMERICAN STEAM AND HOT-WATER HEATING PRACTICE.

SENT POSTPAID ON RECEIPT OF $4.00.

Address Book Department, THE ENGINEERING RECORD, 100 William St., New York.

THE TRENTON POTTERIES CO.

TRENTON, N. J.

CRESCENT
POTTERY.

DELAWARE
POTTERY.

EMPIRE
POTTERY.

ENTERPRISE
POTTERY.

EQUITABLE
POTTERY.

IDEAL
POTTERY.

THE Largest Manufacturers of Sanitary Earthenware in the World, and make the Leading Specialties in America. Our

VITREOUS CHINA,

Sanitary, is recognized by the Architects and Engineers of the Country as the very best Ware ever produced.

Our Stamp on Ware is a guarantee of quality. We beg to caution our patrons not to confound our Ware with that produced by other Trenton potteries.

OUR PRODUCT

GUARANTEED

SUPERIOR

TO ANY

IMPORTED

WARE

AS TO STYLE,

QUALITY,

AND

DURABILITY.

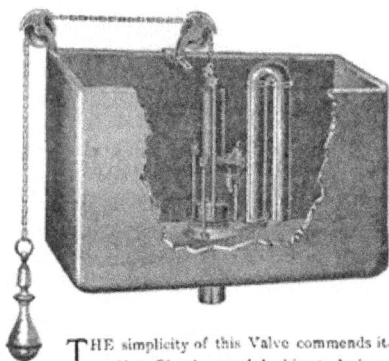
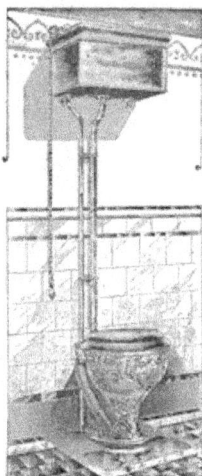

Sent Post-paid on Receipt of $3.00.

STEAM-HEATING PROBLEMS;

OR,

Questions, Answers and Descriptions Relating to Steam-Heating and Steam-Fitting,

FROM

THE ENGINEERING RECORD,

ESTABLISHED 1877.

(Prior to 1887, THE SANITARY ENGINEER.)

With 109 Illustrations.

PREFACE.

THE ENGINEERING RECORD, while devoted to Engineering, Architecture, Construction, and Sanitation, has always made a special feature of its departments of Steam and Hot-Water Heating, in which a great variety of questions have been answered and descriptions of the work in various buildings have been given. The favor with which a recent publication from this office, entitled "Plumbing and House-Drainage Problems," has been received suggested the publication of "STEAM-HEATING PROBLEMS," which, though dealing with another branch of industry, is similar in character. It consists of a selection from the pages of THE ENGINEERING RECORD of questions and answers, besides comments on various problems met with in the designing and construction of steam-heating apparatus, and descriptions of steam-heating work in notable buildings.

It is hoped that this book will prove useful to those who design, construct, and have the charge of steam-heating apparatus.

CONTENTS:

STEAM-HEATING PROBLEMS.

Address, BOOK DEPARTMENT,   *Sent Post-paid on Receipt of 3.00.*

THE ENGINEERING RECORD,

P. O. BOX 3037.   100 WILLIAM ST., NEW YORK.

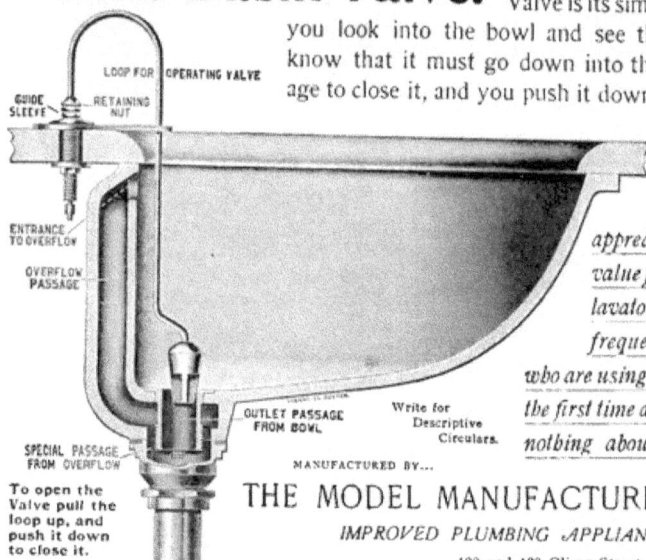

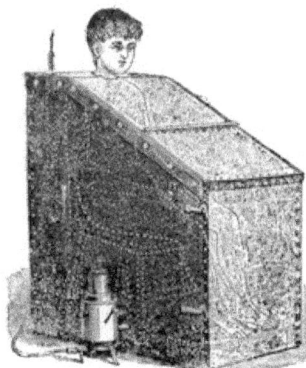

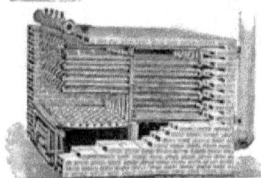

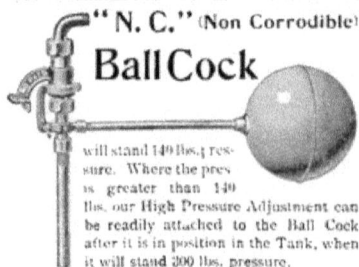

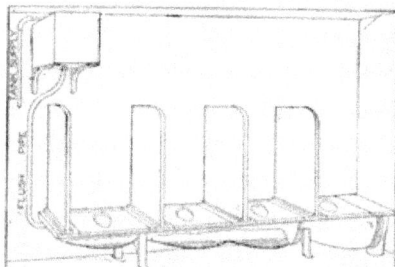

HOT-WATER HEATING AND FITTING;

OR,

WARMING BUILDINGS BY HOT-WATER.

A DESCRIPTION OF

Modern Hot-Water Heating Apparatus—The Methods of their Construction and the Principles Involved.

WITH OVER TWO HUNDRED ILLUSTRATIONS, DIAGRAMS, AND TABLES.

BY WILLIAM J. BALDWIN, *M. Am. Soc. C. E.*,
Member American Society Mechanical Engineers,
AUTHOR OF "STEAM-HEATING FOR BUILDINGS," ETC., ETC.

Graphical methods are used to illustrate many of the important principles that are to be remembered by the Hot-Water Engineer.

The volume is 8vo., of 385 pages, besides the index; handsomely bound in cloth, and will be sent postpaid on receipt of $4.00

Among the questions treated are the following:

Laws of Hot-Water Circulation.

Flow of Water in the Pipes of an Apparatus.

Graphical Illustration of the Expansion of Water.

Graphical Illustration of the Theoretical Velocity of Water in Flow-Pipes.

Efflux of Water Through Apertures.

Passage of Water Through Short Parallel Pipes.

Passage of Water Through Long Pipes.

Friction of Water in Long Pipes.

Quantity of Water that will Pass Through Pipes under Different Pressures.

Diminution of the Flow of Water by Friction in Long Pipes.

Loss of Pressure by Friction of Elbows and Fittings.

How the Friction of Elbows and Fittings may be Reduced to a Minimum.

Flow of Water Through the Mains of an Apparatus, Considered under its Various Practical Conditions.

How to Find the Total Head Required when the Quantity of Water to be Passed and the Size and Length of the Pipes are Known.

How to Find the Quantity of Water in U. S. Gallons that will Pass Through a Pipe when the Total Head and Length and the Diameter of the Pipe is Known.

To Find the Diameter of the Pipes for a Given Passage of Water.

How to Find the Direct Radiating Surface Required for Buildings.

How Heat is Lost from the Rooms of a Building.
Simple Formula for Finding the Radiating Surfaces for Buildings.
Experiments by Different Authorities on Radiating Surfaces.
To Find the Amount of Water that should Pass Through a Radiator for a Certain Duty.
How to Determine the Size of Inlet and Outlet Pipes for Hot-Water Radiators.
Diagrams Giving Graphical Methods for Finding the Diameters and Lengths of Flow and Return Pipes for Hot-Water Apparatus.
Proportioning Coils and Radiators of an Apparatus for Direct Radiation.
Description of Different Systems of Piping in Use.
Proportioning an Apparatus for Indirect Heating.
Illustrations of Boilers.
Hot-Water Heating in the State, War, and Navy Department Building.
Hot-Water Heating in Private Residences.
Boilers Used for Hot-Water Heating.
Direct Radiators Used for Hot-Water Heating.
Indirect Radiators Used for Hot-Water Heating.
The Effect of Air-Traps in Hot-Water Pipes.
Expansion Tanks—and How they should be Prepared.
Danger of Closed Expansion Tanks.
The Various Valves Used for Hot-Water Heating.
Air-Vents Used for Hot-Water Radiators.
Automatic Regulators Used in Hot-Water Heating.
Special Fittings for Hot-Water Heating.
How to Conduct Tests of Hot-Water Radiators.
Method of Connecting Thermometers with Hot-Water Pipes and Radiators.
Tables of Contents of the Pipes of an Apparatus.
Table of Co-efficients of the Expansion of Water from Various Sources, with an Ample Table of Contents from which the above Items were Selected; also an Alphabetically Arranged Index, the Whole Containing a Large Amount of Useful Information of Great Value to the Engineer, Architect, Mechanic, and Householder. No Architect, Engineer, Steam-Fitter, or Plumber throughout the United States should be without a copy of this book. It is written in the simple style of Mr. Baldwin's former book, "Steam-Heating for Buildings," and is within the ready comprehension of all.

Address, BOOK DEPARTMENT,

THE ENGINEERING RECORD,

P. O. Box 3037.   100 WILLIAM ST., NEW YORK.

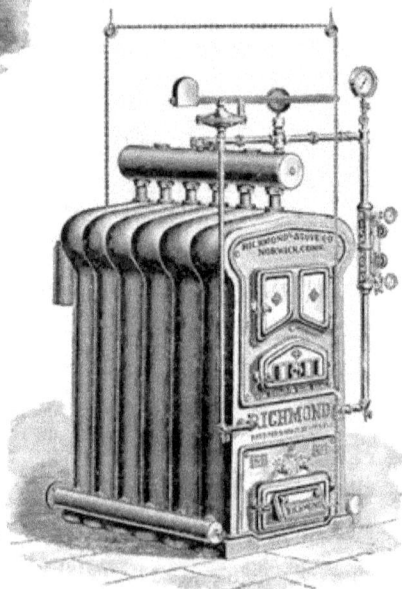

www.ingramcontent.com/pod-product-compliance
Lightning Source LLC
Chambersburg PA
CBHW021512210326
41599CB00012B/1228